MULTIMODAL SAFETY MANAGEMENT AND HUMAN FACTORS

To the Youth, who can teach all of us to discern meaning and to my co-pilgrims in the journey: Luisa, Athena, Sophia, Aletheia, Federico and Inigo

Multimodal Safety Management and Human Factors

Crossing the Borders of Medical, Aviation, Road and Rail Industries

Edited by

JOSÉ M. ANCA JR
Swinburne University, Australia

ASHGATE

Published by
Ashgate Publishing Limited
Gower House
Croft Road
Aldershot
Hampshire GU11 3HR
England

Ashgate Publishing Company
Suite 420
101 Cherry Street
Burlington, VT 05401-4405
USA

Ashgate website: http://www.ashgate.com

British Library Cataloguing in Publication Data
Multimodeal safety management and human factors : crossing
the borders of medical, aviation, road and rail industries
1. Transportation - Safety measures 2. Transportation
accidents 3. Risk-taking (Psychology) 4. Accident
investigation
I. Anca, José M.
363.1'207

Library of Congress Cataloging-in-Publication Data
Anca, José M.
Multimodal safety management and human factors : crossing the borders of medical, aviation, road and rail industries / by José M. Anca, Jr.
p. cm.
Includes index.
ISBN 978-0-7546-7021-6
1. Traffic safety--Case studies. 2. Traffic engineering--Case studies. 3. Human engineering--Case studies. I. Title.

HE5614.A59 2007
363.12--dc22

2006034813

ISBN-13: 978 0 7546 7021 6

Printed and bound in Great Britain by TJ International Ltd, Padstow, Cornwall.

Contents

List of Figures

List of Tables

Foreword

With this collection, Joey Anca has made a great contribution to safety and human factors. He has brought together contributions from a distinguished group of safety experts in four safety-critical domains. He provides the most comprehensive compilation of information on safety management systems.

I have known Joey for more than two decades in a variety of cultures ranging from the Philippines to Singapore to Australia and domains from aviation to university to the rail industry. His breadth of experience makes him uniquely able to integrate the critical issues regarding human factors, safety, and safety management systems.

The multi-modal format provides understanding and contrasts far beyond what focus on a single domain could offer. Those with experience in one will gain insights into the breadth of human factors and safety concepts through exposure to the dominant issues in the others

This volume should be required reading for all who work or plan to work in human factors and safety – students, educators, managers, and especially those who toil at the sharp end of these professions.

Robert L. Helmreich, PhD, FRAeS
Professor of Psychology
Director University of Texas Human Factors Research Project
The University of Texas at Austin
September 2006

Preface

There is meaning in this book. It is a technical book about safety management and human factors. And yet one would have missed its entire message if it were to be just *technical*. The book has to be wrung and squeezed to discern the meaning from all its authors. It is to drill deep, as in mining for a concatenation of old and new clues that outlines the practical applications of safety management and human factors this time, in rail, medical, aviation and road modalities.

Assembling rail, aviation, medical and road experts is a formidable task. Experts, researchers and practitioners have come together from the different perspectives and offer a balance of tools and theory from their areas of expertise. Beyond the literal messages of the authors, the exploration for meaning leads the reader into four core agenda about safety management and human factors:

- safety management and human factors foundations across modes are universal;
- There are emerging tools such as Monitoring of Normal Operations, that are interoperable in all the modes;
- The pre-eminence of data to drive conclusive concepts in safety management is of utmost importance; and,
- Protecting lives, the quest to minimize risk, and the audacity to learn from the 'other mode' are essential to progressing safety management and human factors into an enduring science.

The book is classified into four parts:

Part 1: Multimodal Characteristics of Safety Management Systems (SMS)

The goal of Part 1 is to illustrate in a definitive manner that SMS foundations such as investigation, safety change transformation, SMS elements and unsafe act classifications share the same characteristics. These include the rigor to link the undesired outcomes of tasks, with system flaws that reside deep in complex operations. It will also tackle the influence of governance and regulation in mapping the boundaries of risk and safety.

Part 2: Safety Management Metrics, Analysis and Reporting Tools

There are two approaches in dealing with data collection. The first is ubiquitously apparent. It is that which the veracity of one piece is influenced by another, likened to seeking linkage across a lateral *data plain*, if you will. The second is perhaps more important. And that is about surveying the *data plain* and at some points, drilling into the subplain to find multiple connections, with the curiosity to discover newer contexts and meanings.

Prefacing the part, novel approaches such as using insurance data, theory of constraints, the use of techniques such as ICAM and minimizing road user errors provide an interesting flavour to safety data management and meaning.

Part 3: Normal Operations Monitoring and Surveillance Tools

A colleague of mine from academia once said that the squirrel is an iconic representation of SMS. A squirrel is in contant motion, anxiously scanning the environment for threats and opportunities: threats to mean predators, and opportunities to mean food. The essence to the mad anxiety is an incessant monitoring of what goes as expected (normal operations) and what doesn't (deviations).

Part 3 explores the air traffic control system, the use of Line Operations Safety Audits (LOSA) from aviation into the rail context, error management and training applications.

Part 4: The Modality of Human Factors: Exploring the Management of Human Error

There is a perception from industry that human factors is a veiled activity of the nebulous. I suppose that there is truth in the perception to the extent that human error remains to be the colony of the quasi-predictable and the qualified professional. This part will reinforce these points and propose that whilst human factors tools are important, the solutions to human error are a bond between the pure science (the statistic) and the application (the context).

Driver distraction, situational awareness, fatigue management, medical team resource management and rail/aviation human factors programs are some of the topics in Part 4.

As the four parts give a broad classification of SMS and human factors, each chapter delivers a modal facet of SMS or human factors information with a wide spectrum of applications – from those seeking practical tools to scientists who are developing emerging research agenda in rail, road, aviation and the medical fields.

Stepping back and looking at all the chapters with *gestalt* lenses, it is like a shopping mall with a full collection of specialty shops. The shopper could well be a student, seeking a leading edge thesis topic; a train driver, looking for a connection between fatigue and error; a safety specialist embarking on safety culture change; or a doctor wanting to gain more knowledge about patient safety. The reader is urged to draw meaning from each chapter in this 'multimodal shopping mall' and avoid the path personified in a 1960s hit song by The Seekers: '*hey there, Georgy Girl... always window shopping but never stopping to buy.*'

José M. Anca Jr
Sydney, Australia
September 2006

Acknowledgements

The editor wishes to acknowledge the contribution of the chapter authors who have laboured both in the development of solutions to SMS and human factors in their fields and the crafting of their respective chapters. Specific and profuse gratitude to Bob Helmreich of the University of Texas at Austin, a long time friend and mentor of 20 years that started in Leningrad. Bob guided the creation of the book. To Kathy Abbott of the FAA, one who has contributed tremendously to flight safety and remains as one of the most astute Human Factors professionals I've met; thank you for the keynote that heralds the future of safety and human factors. To Jean-Jacques Speyer of Airbus, a colleague whom one can always rely on; thank you for being there, relentlessly representing the Airbus brand.

Lastly, but most importantly, the book would not have been born if not for the support of Swinburne University of Technology in Melbourne, Australia. Swinburne, as in all learning institutions, is the wellspring of inspiration from both mentors and students.

List of Abbreviations

ADAS	Advanced Driver Assistance Systems
AME	Aircraft Mechanic Engineer
AOC	Air Operators Certificate
ASFA	Aviation Safety Foundation Australasia
ASK	Available Seat Kilometres
ATO	Approved Testing Officer
ATSB	Australian Transport Safety Bureau
CAA	Civil Aviation Authority
CASA	Civil Aviation Safety Authority, Australia
CDU	Control Display Unit
CFIT	Controlled Flight Into Terrain
CORS	Confidential Observations of Rail Safety
CRM	Crew Resource Management
ETOPS	Extended Twin Engine Operations
FDW	Following Distance Warning
FMS	Flight Management System
GA	General Aviation
GPWS	Ground Proximity Warning System
HF	Human Factors
HFACS	Human Factors Analysis Classification System
ICAM	Incident Cause Analysis Method
ICAO	International Civil Aviation Organization
ISA	Intelligent Speed Adaptation
ITS	Intelligent Transport Systems
LAME	Licensed Aircraft Mechanic Engineer
LOSA	Line Operational Safety Audits
MUARC	Monash University Accident Research Centre
NOTAM	Notice to Airmen
NTSB	US National Transport Safety Board
PSF	Patient Safety Framework
SA	Situational Awareness
SAQ	Safety Attitudes Questionnaire

SCP	Safety Change Process
SHERPA	Systematic Human Error Reduction and Prediction Approach
SMS	Safety Management System
SSA	Situational Safety Awareness
TAFE	Technical and Further Education
TAAATS	The Australian Advanced Air Traffic System
TSB	Transportation Safety Board of Canada
TSI	Transport Safety Investigation Act

PART 1
Multimodal Characteristics of Safety Management Systems

Part 1 contains insights on the systemic nature of safety management and how different industry modes (medical, rail, road and aviation) have introduced SMS in varying degrees of maturation. It will submit that the foundations of SMS are the same in these modes, however, its tools and applications vary because of the disparate contexts of these industries.

The Introduction will also highlight anecdotal findings from these different industries of the need for more tools rather than theories on SMS. Safety management systems will be defined in this chapter from the standpoint of equipping organizations with risk-resistant strategies.

Some examples of high-reliable organizations undergoing massive change will be discussed and both successes and pitfalls will be characterized in the process.

Chapter 1

Can Simply Correcting the Deficiencies Found through Incident Investigation Reduce Error?

Gerry Gibb
Safety Wise Solutions Pty Ltd

Safety wise solutions

ICAM (Incident Cause Analysis Method) is a holistic incident investigation method which aims to identify local factors and failures within an entire organization system (*e.g.* communication, training, procedures, incompatible goals, equipment, *etc.*) which contribute to an incident.

ICAM as a tool complements existing processes of error prevention, error containment and error mitigation. Applying ICAM allows the precursors to errors to be identified and corrected. ICAM provides the ability to identify what really went wrong and to make recommendations on what needs to be done to prevent recurrence. It is directed towards building error-tolerant defences against future incidents.

Can simply correcting the deficiencies found through incident investigation reduce error? This chapter will apply ICAM to review an aviation accident that occurred to determine whether the precursors to the accident could have been identified.

The chapter will also discuss whether the application of ICAM before the accident could have identified the precursors to the incident and thereby have prevented it.

Introduction

> Our investigation and correction activities tend to be based on the amount of damage and injury – which is random. We don't really have prevention programs; we have accidents correction programs (Wood, 1997).

Modern safety theory would suggest that relying on correcting deficiencies found though incident investigation as a means to reduce error is somewhat restrictive (Klietz, 1994; Wood, 1997; Reason, 2000; Wiegmann and Shappell, 2003). Significant factors leading to the accident and the subsequent actions necessary to prevent recurrence do not always emerge from the physical evidence of the case alone. Many accidents occur, not because they cannot be prevented, but because the organization did not appreciate the gaps in their safety systems, nor learn from – or perhaps did not retain the lessons from – past accidents within or outside their organization. The 'best fit' tool, identifying and resolving precursors to error and learning from past accidents, can assist the organization to gain an appreciation of the strengths and limitations of their safety systems and reduce errors that lead to incidents and accidents. It is necessary for organizations to select the 'best fit' investigative tool, to identify and resolve the precursors to error as well as focus on improving corporate memory to avoid potential accidents.

The selection of the investigative tool

The purpose of selecting a suitable investigative tool is to be able to manage exposure and risks across the organization – at both operational and strategic levels. Selecting the most appropriate tool will improve the likelihood for determining the precursors to a specific incident. An investigative tool provides a framework for data collection and organization and allows the investigation to follow a logical path. Each organization has a unique set of goals, constraints (*e.g.* budgets) and existing conditions (*e.g.* business structure), so will select a tool (or investigator) that suits them best. A comprehensive safety investigation not only looks at how an accident occurred, but also looks at why it occurred. Most importantly, the investigation recommends corrective action that can be taken across the organization to prevent such occurrences happening again.

There is no 'one size fits all' solution for designing and implementing an organization-wide safety management program. The principle objective of incident investigation is to prevent recurrence, reduce risk and advance health, safety and environmental performance. It is not for the purpose of apportioning blame or liability.

Organizational safety systems

Many models for accident causation have been used to propose ideas to minimize loss. The 'systems' approach was developed to manage safety. This approach promotes a balance between the assessment of people, behaviour and the infrastructure to support the operation. However, given the adoption of safety systems in many organizations, there are questions surrounding organizational incidents that remain unanswered by many of these systems.

- How can absent or failed defences be identified?
- What are the successful approaches to eliminating or altering the barriers to error management?
- What are the precursor tasks or environmental conditions that lead to incidents?
- What organizational factors predict success in prevention programs?
- What are the barriers to the acceptance of new error control strategies?
- What are the best techniques to encourage the implementation of recommendations?
- How can the effectiveness of implemented recommendations be evaluated?

How can the retention of corporate learning be achieved?

Whilst this chapter will not answer all of these questions, it proposes that investigative tools can be used both pre-incident to answer the 'why' and post-accident to answer the 'how'.

Standard investigative tools look at the physical evidence of a particular incident and where it is specifically relevant to the particular investigation; they also examine the pertinent organizational contributing factors. Whilst this should be effective at determining the causes of the incident and hopefully be effective at preventing the same incident from occurring again, these techniques are not focused on, and therefore not effective at making, safer organizations. To create safer organizations, a more holistic tool, which examines the full range of organizational elements in depth, is required. Importantly, if one wishes to reduce the pre-cursors to error within an organization, such a tool should be used proactively and not be limited just to post-incident application. A number of investigative tools are available to an assessor. This chapter will explore the Incident Cause Analysis Method (ICAM) tool for the purposes of discussion.

Incident Cause Analysis Method (ICAM)

Professor James Reason and his colleagues from the University of Manchester in the United Kingdom developed a conceptual and theoretical approach to the safety of large, complex, socio-technical systems. Drawing on the work of Reason, ICAM identifies the workplace factors that contribute to an incident and the organizational deficiencies within the system that act as its precursors.

ICAM allows the investigator to gather data, organize it and carry out an effective analysis by organizing causal factors into four manageable elements. These are:

- absent or failed defences;
- individual or team actions;
- task/environmental conditions; and
- organizational factors.

ICAM is designed to ensure that the investigation is not restricted to the errors and violations of operational personnel. It identifies the local factors that contributed to the incident and the latent hazards within the system and the organization (Gibb and DeLandre, 2002).

Before starting, the level of investigation required is determined, then the investigation team is assembled and, with the assistance of the client organization, the investigation commences.

Specific objectives of ICAM

The specific objectives of investigations using ICAM are:

- to establish all the relevant and material facts surrounding the event;

- to ensure the investigation is not restricted to the errors and violations of operational personnel;
- to identify underlying or latent causes of the event;
- to review the adequacy of existing controls and procedures;
- to recommend corrective actions which, when applied, can reduce risk, prevent recurrence and, by default, improve operational efficiency;
- to detect developing trends that can be analysed to identify specific or recurring problems;
- to ensure that it is not the purpose of the investigation to apportion blame or liability – where a criminal act or an act of wilful negligence is discovered, the information will be passed to the appropriate authority; and
- to meet relevant statutory requirements for incident investigation and reporting.

Once the data has been gathered, it is organized into a sequence of events, validated, and then analysed. From the analysis of findings, the facts can be classified and charted in the ICAM model for inclusion in the investigation report and for briefing management on the investigation findings.

The investigators work with clients to design and implement customized solutions based on their company strategies, structure and culture to enhance performance and optimize costs. A fundamental concept of ICAM is the acceptance of the inevitability of human error. As stated by Reason (2000), an organization cannot change the human condition, but they can change the conditions under which humans work, thereby making the system more error tolerant.

The investigation identifies recommendations for corrective actions to prevent recurrence, reduce risk and advance safety. The investigation report is presented to assist management to understand the factors contributing to the incident. The investigation team works with management to review the recommendations and develop corrective actions for system deficiencies and ineffective organizational processes. These should integrate with existing systems and time lines to achieve the expected outcomes of the organization's business case. Actions and target dates need to be realistic and achievable to ensure follow-through and completion.

Applying ICAM as a predictive tool allows the organization to identify and retain information on causal factors, so the organization remembers that the accident occurred, what caused it, and why procedures were changed. R. Wood (1997) states: 'Beyond question, the most difficult and troublesome aspect of aircraft accident investigation is the determination of cause.'

It is difficult to reduce the causes of an accident to a single sentence. To improve the methodology for cause determination, an appropriate investigative tool and reporting style will encourage the organization to look at precursors and causes the same way investigators do.

Case study: Learjet accident, Aberdeen, USA, 1999

Had an ICAM analysis occurred before the incident, could the precursors to error and the deficiencies in the safety system been identified, and thus prevented the accident? The precursors to error need to be identified and resolved, thereby breaking the chain of latent conditions and its effects. To conduct a complete ICAM analysis of a particular incident, a full investigation would need to be conducted by the investigation team as soon as possible after the incident had occurred.

It should be noted, however, that this case study analysis has been applied after the event and based on existing documentation and reports. Applying ICAM to the Payne Stewart Learjet accident in 1999 (Airsafe, 2005), it is possible, for illustrative purposes, to identify organizational factors and task/environmental factors contributing to the incident. This incident has been selected as it represents some of the difficulties in applying standard post-incident investigative techniques. The following is an excerpt from the National Transportation Safety Board (US NTSB) summary report on the Learjet incident.

Case study: summary

A Learjet crashed near Aberdeen and Mina, South Dakota on 25 October 1999. The aircraft, a Learjet model 35, registration number N47BA was operated by Sun Jet Aviation of Sanford, Florida. The aircraft flew from Sanford to Orlando, Florida on the morning of the accident, where it picked up its passengers. The flight departed Orlando with two pilots and four passengers, including professional golfer Payne Stewart, about 09.19 destined for 'Love Field' in Dallas, Texas. The planned flight time was two hours. The airplane had about four hours and forty-five minutes of fuel aboard.

Air traffic control lost radio contact with the flight at 09.44 Eastern Daylight Time, when the airplane was climbing through and located north-west of Gainesville, Florida. The flight was cleared to 39 000 feet. The aircraft proceeded on a north-west heading. The aircraft was intercepted at about 45 000 feet by military aircraft, which followed the plane until it crashed near Aberdeen. Preliminary reports from the Federal Aviation Administration place the crash time at 13.26 Eastern Daylight Time.

NTSB provides the following notes to confirm their difficulty with the investigation.[1] This investigation took thirteen months to complete and was hampered by several factors:

- Because the aircraft impacted at nearly supersonic speed and at an extremely steep angle, none of its components remained intact. Therefore, our

1 The website http://www.ntsb.gov/events/aberdeen/ppt_presentations.htm includes presentations to the NTSB about the accident. There may be individual and team actions and failed defences that contributed to the incident, however these may never truly be known. As the speed of impact was near supersonic, it destroyed much of the physical evidence that could have provided a more definitive analysis of technical and human causal factors.

investigators had to painstakingly examine the fragmented valves, connectors, and portions of other aircraft parts before they could draw any conclusions about the accident's cause.

- The airplane was not equipped with a flight data recorder, an invaluable tool in most major investigations, and it had only a 30-minute cockpit voice recorder, which was of limited use during this investigation.
- All of the investigators involved in this investigation were also investigating other accidents. The Investigator-in-Charge was working on four other investigations in addition to this one. (US NTSB, 2001)

Applying ICAM

Absent or failed defences

Absent or failed defences are those that failed to detect and protect the system against technical and human factors. These are the control measures that did not prevent the incident or limit its consequences. A strong focus on the investigation was the supply of oxygen to the cabin and flight crew. The NTSB aircraft accident brief DCA00MA005 (2005) states: 'The accident could not have occurred without both the loss of cabin pressure and the failure of flight crew to receive supplemental oxygen.'

The absent or failed defences identified in the case study include:

- cabin pressurization was unable to be maintained;
- lack of bleed air supply to the cabin;
- closed flow control valve (supplying warm air to windshield);
- timeliness of warning for donning oxygen masks;
- oxygen quality/quantity;
- incomplete standardized manual and procedures;
- ambiguous maintenance procedures, some verbal, some written, some not signed off;
- poorly written reporting relationship between aircrew and maintenance staff;
- non-adherence to company policy for maintenance reporting/signoff;
- inadequate procedures for emergency oxygen supply;
- unclear whether guidance was provided for procedures anomaly – switching to auto pilot;
- procedures for reporting maintenance issues inadequate;
- written records for maintenance reports incomplete;
- previous inconsistencies in application of SOPs;
- crew pairing of inexperienced captain and first officer; and
- limited flying time on particular type of plane by the Captain.

Individual/team actions

Individual/Team Actions are the errors or violations that led directly to the incident. They are typically associated with personnel having direct contact with the equipment, such as operators or maintenance personnel. They are always committed actively (someone did or did not do something) and have a direct relation with the incident. The Individual/Team Actions identified in the case study include:

- lack of timeliness in donning oxygen masks;
- lack of verbal response from cabin crew to ATC radio contact; and
- not adhering to maintenance procedures – some verbal, some written incompletely, some not signed off.

Task/environmental conditions

These are the conditions in existence immediately before or at the time of the incident that directly influence human and equipment performance in the workplace. These are the circumstances under which the errors and violations took place and can be embedded in task demands, the work environment, individual capabilities and human factors. The Task/Environmental Conditions identified in the case study include:

- potential difference in corporate culture for new staff members;
- limited flying time on type by the Captain;
- windows frosted over, well below freezing;
- no flight data recorder;
- low pressure evident in cabin (via alarm);
- ambiguity surrounding effectiveness of alarm system; and
- recycling of tape voice recorder – every half hour.

Organizational factors

These are the underlying organizational factors that produce the conditions that affect performance in the workplace. They may lie dormant or undetected for a long time within an organization and only become apparent when they combine with other contributing factors and lead to an incident. The organizational factors identified in the case study include:

- deficient monitoring and auditing of maintenance item completion;
- equipment not fit for purpose – suspected valve problems with closure of flow control valve;
- process of (managing) introduction of new aircraft;
- incident reporting system deficiencies;
- inadequate procedures for checking quality and quantity of on board emergency oxygen bottle;
- deficiencies in maintenance control – use of MEL, etc.;

- no evidence of risk appreciation process used;
- ambiguous monitoring by management of resources, climate and processes of a safe working environment;
- incomplete corporate commitment to safety;
- failure to revise maintenance strategy;
- failure to appreciate risk exposure or vulnerability within the organization; and
- no follow-up from previously failed defences – identification, tracking and resolving maintenance items and adverse trends.

Discussion

Could the application of ICAM before the accident have identified the precursors to the incident? We cannot say conclusively. The lack of physical evidence at the crash site, the lack of complete documentation, data and records leaves sufficient doubt about the exact cause of the accident.

ICAM applied, as a preventive tool, would have identified the absent and failed defences noted in the NTSB report. These include the poor follow-through on work procedures regarding maintenance reporting, identification, tracking and resolving of maintenance items and adverse trends. Furthermore, applying ICAM would have examined the failed and absent systems defences with more rigour and examined a broader range of organizational factors such as individual/team actions and the human factors elements of task and environmental conditions.

For example, a preventive ICAM would have identified change management issues with the purchase of a new aircraft and advised the operators to be alert for any follow-on impact. By implementing absent defences and improving failed defences within the safety systems, the risk of future error could have been reduced.

ICAM could have been applied before the purchase of the aircraft to identify task and environmental conditions regarding the condition of equipment and highlight the ventilation issues previously identified and repaired as 'for consideration/follow up'. If used by an investigator with strong technical knowledge of the aircraft systems, applying ICAM may have brought up questions about the absence of a flight data recorder and the half hour loop on the tape voice recorder, amongst other factors. These recommendations would be aimed at reducing the potential for error and preventing the recurrence of failure.

From a human factors perspective, the findings may have questioned the experience/knowledge or skills of the Captain, or the anomaly of practices regarding the timing of switching to auto pilot.

Above all, when making recommendations arising from applying ICAM, an investigation team is conscious that not all contributing factors can be completely eliminated. The team works with the organization to develop recommendations using the acronym 'SMARTER' – specific, measurable, accountable, reasonable, timely, effective and reviewed. The team is conscious that residual risk may remain. This needs to be recognized and managed. The impact of and potential benefit of the corrective actions are discussed with the organization and interim recommendations for immediate actions are given priority.

ICAM is a holistic method which aims to identify local factors and failures within the entire organization system. ICAM can be used both as an investigative tool and applied before any incident as a means of creating a safer organization and reduce the potential for error.

Where to from here?

The identification of causes serves no purpose unless there some defensive action is taken. If we apply preventive efforts only to events with the 'highest' likelihood of disaster, we miss the opportunity to prevent a regularly recurring event of 'low' probability which may have latent effects – for example, in the case study, the maintenance reporting and follow through procedures.

While recommendations may arise from a specific occurrence it is important for all operators to recognize the benefit of learning from other's mistakes and realizing that the implementation of such learning could benefit their operations and prevent an occurrence (Gibb and DeLandre, 2003). Proactive use of ICAM provides safety learning to the organization without the costs associated with an occurrence. Applying ICAM as a preventive tool allows an organization to shift its focus from accident investigation findings to preventative safety. By designing error tolerant workplaces that reduce error, organizations have the potential to mitigate error consequences and therefore proactively prevent incidents.

Additional benefit is gained by applying the model to a number of similar low-consequence events, to assess what would otherwise be a set of unrelated safety concerns and to develop strategic recommendations for safety improvements (Gibb and DeLandre, 2003).

Conclusion

Would the application of ICAM before the incident have prevented the accident from happening? We cannot say conclusively. The lack of physical evidence at the crash site, the lack of complete documentation, data and records leaves sufficient doubt about the exact cause of the accident. However, there is sufficient evidence to say that the application of ICAM before the incident would have made it a safer organization and reduced the risk of this accident or a range of other potential incidents from occurring.

Standard investigative tools look at the physical evidence of a particular incident and whilst they may be effective at preventing the same incident from occurring again, they are not focused on, and therefore not effective at making safer organizations.

Correcting the deficiencies found using standard incident investigation methods should reduce the errors that caused a particular incident; but, to reduce the precursors to error and make an organization a safer organization, a more holistic tool, such as ICAM, which looks at the full range of organizational systems, is required.

References

Airsafe.com (2005), Payne Stewart Accident Information, <http://www.airsafe.com/stewart.htm>, accessed 15 November 2005.

Gibb, G. and De Landre, J. (2002), Lessons from the Mining Industry, paper prepared for the Human Error, Safety and Systems Development (HESSD) Conference, Newcastle, NSW, 17 June 2002.

Gibb, G. and De Landre, J. (2003), Designing an Error Tolerant Workplace: Using ICAM Proactively, paper prepared for NSW Mining OHS Conference, August 2003.

Klietz, T. (1994), *Lessons from Disaster*, Published by the Institution of Chemical Engineers (UK) – ISBN 0 85295 307 0.

National Transportation Safety Board (2001), Aircraft Accident Brief, DCA00MA005, February 2001.

National Transportation Safety Board (2005), Presentation of Findings, <http://www.ntsb.gov/events/aberdeen/ppt_presentations.htm>, accessed 15 November 2005.

Reason, J. (2000), Human Error: Models and Management, *British Medical Journal*, vol. 320, 768–770 [PubMed 10720363] [DOI: 10.1136/bmj.320.7237.768].

Wiegmann, D.A. and Shappell, S.A. (2003), *A Human Error Approach to Aviation Accident Analysis*, Aldershot, UK: Ashgate.

Wood, R. (1997), *Aviation Safety Programs: A Management Handbook*, second ed., Colorado, USA: Jeppesen.

Chapter 2

Moving Up the SMS Down Escalator

Bruce Tesmer
Continental Airlines

If you are not enjoying what you do, you should not be doing it! I have been working in new safety data collection programs for the last 10 years and have enjoyed it all: the inch-at-a-time progress, savouring the victories, learning from the defeats, hanging in here to find another way. It is a journey I have not made alone. I have learned a lot from the top down and even more from the bottom up. I have been with those that have said 'this will never work' to those saying 'don't stop here, push the envelope further'. They were both correct.

I have had input from many to write this chapter, but special enlightenment from those I have worked closest with: Dr Robert Helmreich and exceptional talent of those members of the UTHF Research Project.

All the concepts, ideas, processes and items in this chapter can be improved. I hope you find something useful that you can improve. The title of this chapter reflects the fun of the challenge and the satisfaction of accomplishment. The first time I saw an escalator, I could not resist the challenge of running up the down escalator just to get to the top. I found the same difficulty and challenge with safety data programs and SMS. I hope you pursue a similar path.

What defines how safe we are? Is it accidents, incidents, and negative events? Lets talk about accidents. The commercial aviation system has the lowest accident rate of any transportation system in terms of miles travelled. That's the good news. However, because there are so few accidents, they cannot be used for statistical analysis. There are more incidents than accidents, which make incidents a better source of data, and there are more negative events than incidents. Close calls, near misses, and complaints can all be used as a data source if they are reported and recorded. However, we still lack knowledge of what really occurs in normal operations.

How do we know about safety issues? Does the boss tell us? Do we find out from the insider network (Offices of the CFO, CLO, or COO)? Does the grapevine tell us ('I had the big guy on my jumpseat and he said …')? How about the regulator, who just accomplished 10 line checks and made 'the following conclusions'? Or, worst of all, is it by seeing an event, with your logo, broadcast on worldwide CNN? Perhaps it is the Safety Department, who is tasked to keep us safe. We will learn much more with data collection programs designed to feed into a Safety Management System (SMS).

I define a safety management system as the process of removing what is out of date (obsolete) and installing what is up to date (new). It is change in terms of safety philosophies, policies, procedures, and practices. Examples from aviation include the following concerning automation. The current philosophy is to use automation at the most appropriate level required by the situation. That is a change from the past philosophy, which was to increase automation as much as possible. As a policy, after top of descent, only one pilot at a time can be head down working the FMS CDU. As a procedure, during the approach, the flying pilot will display the descent page on the FMC and the monitoring pilot will display the legs page. We continually monitor crews to see if the normal practices are in accordance with the philosophies, policies, and procedures. The SMS updates these areas concerning philosophies, policies, and procedures in a continuous cycle to manage safety change before the trauma. You can be aggressive and make changes rapidly (on the leading/bleeding edge), or you can be conservative and make changes slowly (and risk becoming antiquated). As soon as changes stabilize (one day, one week, one month, one year) the process or evaluating how the changes worked can begin.

How is your SMS doing?

Question: do you have an SMS? Answer: yes is the only answer. Having no SMS is still having an SMS, because change happens, either through your people or through trauma. One way or the other change will happen. Can you define your SMS? Answer: probably not. Can you describe it? Answer: maybe. You cannot be expected to know what your SMS is unless it is defined and advertised. It is difficult for anyone to define because it encompasses all levels of the organization from the CEO to the line employee. To be effective, all employees need to be informed about how it works, with emphasis upon their individual level and one layer above and below them. They have to know how they can impact safety. If you submit a proposed safety change, you need and deserve a response on how your suggestion was handled. If you do not receive a response, chances are you may not make another safety suggestion ever, or you will just complain to the informal organization (grapevine). An employee not answered is a safety resource lost.

The first three SMS steps

1. Collect valid data that tells you what is happening in your normal operations.
2. Define your safety change process that converts the data into corrective action changes. Then, track those changes and re-measure after they are stable.
3. Ensure there is focus and a process in every division, so safety information gets to the CEO and back to the line employees.

Step 1: build and define your SMS by knowing your normal operations

Intuition and experience can no longer be the only driving source for safety change. There are safety issues in normal operations that do not meet your standards. They are known by the line employees. But they will not be discovered until you measure. Until you measure you are only guessing at the first step. Information about operations must be given freely, without shooting the messenger, even when there are negative consequences and errors involved. This allows for learning a better way of doing the task next time. That's safety progress. However, in today's business environment blame and punishment are a routine management gift to the reporter. Because of this, soon there will be no information coming in at all. To gain the most accurate safety data requires a non jeopardy environment.

How can I measure? You can interview. Ask what is happening in normal operations. What's good? What's bad? What's ugly? You can survey, to discover safety attitudes, behaviours, culture, and information flow. You can use safety normal operations direct observations (SNODO/LOSA). You can also use employee self recording systems (APSAP or close call). If you interview, use their office not yours. If you survey, survey the whole group. Direct observation by trained observers needs to include observers of the same job field, which will produce the biggest picture with the most detail. Employee self reporting systems provide prospective of the operation and safety where the job is being done. Remember it is their view only. If a team is involved you must get everyone's view – they will be different.

The key to normal operations data is to collect it in a non jeopardy, de-identified manner. Non jeopardy is not immunity! I say again – non jeopardy is not immunity! The data collected are all in terms of the threat and air management structure. Threats are external to the individual. Errors are internal, made by the individual. Individuals do not see or report most of the threats that they deal with because they deal with them daily. Example: weather; if it is 120 degrees on the ramp or if it is minus 20 with ice and snow, the bags still need to be loaded, the engine still needs to be fixed and the crew still needs to do a walk around inspection. Most do not see the weather as a threat; they see it as just another lousy day. Mistakes, slips, lapses, commissions, omissions, and other unintentional errors, on the other hand most likely go unreported due to fear of punishment or simply embarrassment.

Threat and error management data show the individual that threats are everywhere and unintentional errors are a fact of life if you are human. Blame is not the focus. Finding counter-measures that will prevent the same thing from happening again is the focus. It is more important to collect threat and error data to understand what is happening in the system than to blame and punish the individual. However, if the errors are intentional, deal with illegal/criminal operations, drugs/alcohol or falsification of records/documents/statements, then there will be jeopardy.

Step 2: turn data into safety change: the safety change process (SCP)

The safety change process begins with data collection. The data is then analysed and sent to the assignment authority. Next the possible solutions are investigated to precipitate a prioritized list of correct actions. The list is then sent to senior management for approval and funding. Installation occurs after funding the change is allowed to stabilize. Confirmation is then completed by re-measuring.

Step 3: ensure all level/divisions/departments have an SCP

The CEO is chair of the Corporate Safety Review Board. The board is made up of all Senior Vice Presidents from all divisions. All divisions and departments participate through their safety action teams. Note: until the CEO is part of the process, you only have safety change functions. Until you have all divisions included, you only have a partial program. SMS requires everyone to complete all three steps.

The SMS advantage

Safety changes also produce a more effective and efficient operation. Cost savings and reduced flight incidents, ground damage, improved fuel use and other expense reductions. A reduced number of unstable approaches (non-conforming to standard), will have everyone sleeping better. Reduced checklist errors and reduced number of Undesired Aircraft States where the aircraft is headed for a negative event and the crew is unaware, greatly reduces the exposure to approach and landing accidents.

The SMS level

Are we absolutely safe? Answer: NO! But we can be safer. Your exposure to negative safety events, negative press, and negative speculation about your continued existence can be reduced as you move into more effective and higher levels of SMS. You just have to keep charging up the down SMS escalator. If it were easy we would already have done it!

Chapter 3

Unsafe or Safety-Significant Acts?

David McNair
Transportation Safety Board of Canada

Reason's Model serves the accident investigator well in determining the causes of occurrences. The relation of human activity to the operational environment or pre-existing conditions can be analysed, leading to the identification of safety defences that may not be adequate. However, in some cases, the terminology used to categorize human behaviour can be an impediment to an acceptable or complete analysis of an occurrence. For example, the term 'unsafe act' is used because the result of the act being described is known. Some have questioned the adjective 'unsafe' for behaviour that could be considered to be normal or expected. Yes, there are unsafe acts, but there are also positive safety acts that may limit the consequences of an occurrence. Whether unsafe or positive, the human act under study is a safety significant act. This chapter addresses the investigation of the influence of human behaviour during an occurrence by considering these safety significant acts.

Introduction

Accident investigators are tasked with identifying safety deficiencies by conducting investigations into occurrences. Occurrences, either accidents or incidents, have many elements of human activity that can contribute to, or mitigate the effects of latent or active failures. Various methods have been developed to illustrate or categorize human behaviour in a variety of systems such as transportation, medical care, plant operations and other specialized areas. When things go wrong, leading to accident and incidents, safety investigators are called upon to determine the reason(s) for the occurrence and suggest actions to prevent a recurrence.

The Transportation Safety Board of Canada (TSB) is a government agency which is independent of other government departments. It is the object of the TSB to advance transportation safety for marine, rail and aviation modes. The TSB has developed tools to allow its investigators to effectively investigate all facets of occurrences, including the actions of humans. We know that human performance is integral to all occurrences, and thus to the investigations of occurrences. This chapter will describe the investigation process from an aviation perspective because flight safety is the specialty of the author.

This chapter will outline the general methodology for investigating occurrences from a human factors perspective, and will introduce the terms and definitions used by the TSB. These human factors investigation methods and definitions have been derived from well-known theories and practices such as the SHEL Model and the works of James Reason and others. Much of the previously mentioned human factors knowledge has been included in manuals used by the aviation safety investigators. A prime example of such a document is the *Manual of Human Factors Investigation* published by the International Civil Aviation Organization (ICAO). The TSB, like other comparable safety investigation agencies, also has manuals that provide human factors information for investigators whose specialties are operations (pilots), technical (aircraft maintenance engineers or other types of engineers), or air traffic control.

For large or 'major' occurrences, human factors (human performance) can be investigated either by human factors groups or by subject groups (operations, maintenance, *etc.*) with human factors specialists included in the subject groups. In smaller investigations, the investigation 'team' may consist of only one or two investigators, and these investigators are expected to have at least some basic human factors knowledge.

The TSB has developed data base tools which also include modules to track the influence of human factors and to guide the investigators in determining when a safety deficiency exists and for the preparation of safety action to mitigate the safety problems. The methodologies developed for investigators facilitate logical thinking and specific influences on an occurrence chain of events. In fact, identifying all the events of an occurrence is one of the first steps in establishing what needs to be investigated in more detail. 'What precipitated an event?' is one of the first questions asked by investigators. The reason, or the possible reasons, for the event normally lead to further in-depth investigation. Often, the in-depth investigation researches the actions of humans to determine why the person did what he or she did.

This chapter introduces the idea of replacing the well-known and often-used term 'unsafe act' by the words 'safety significant act'. The starting point for the discussion is a quick and very abbreviated discussion of the investigation methodology used by the TSB. Next, examples will be used to see what effect the substitution of 'safety-significant' for 'unsafe' will have upon the understanding or investigation of human performance. The discussion is prepared to present 'food for thought', hopefully without giving indigestion to human factors purists. The opinions expressed in this chapter are those of the author and not the TSB.

Investigation methodology

The human factors information available to the aviation safety investigator uses terms that are well-known to anyone who has seen the graphical depictions of Reason's Model. The versions presented in Figures 3.1 and 3.2 of this chapter are contained in the ICAO *Manual of Human Factors Investigation*. The representations are powerful depictions of how accidents happen. An unsafe act is depicted as an active failure.

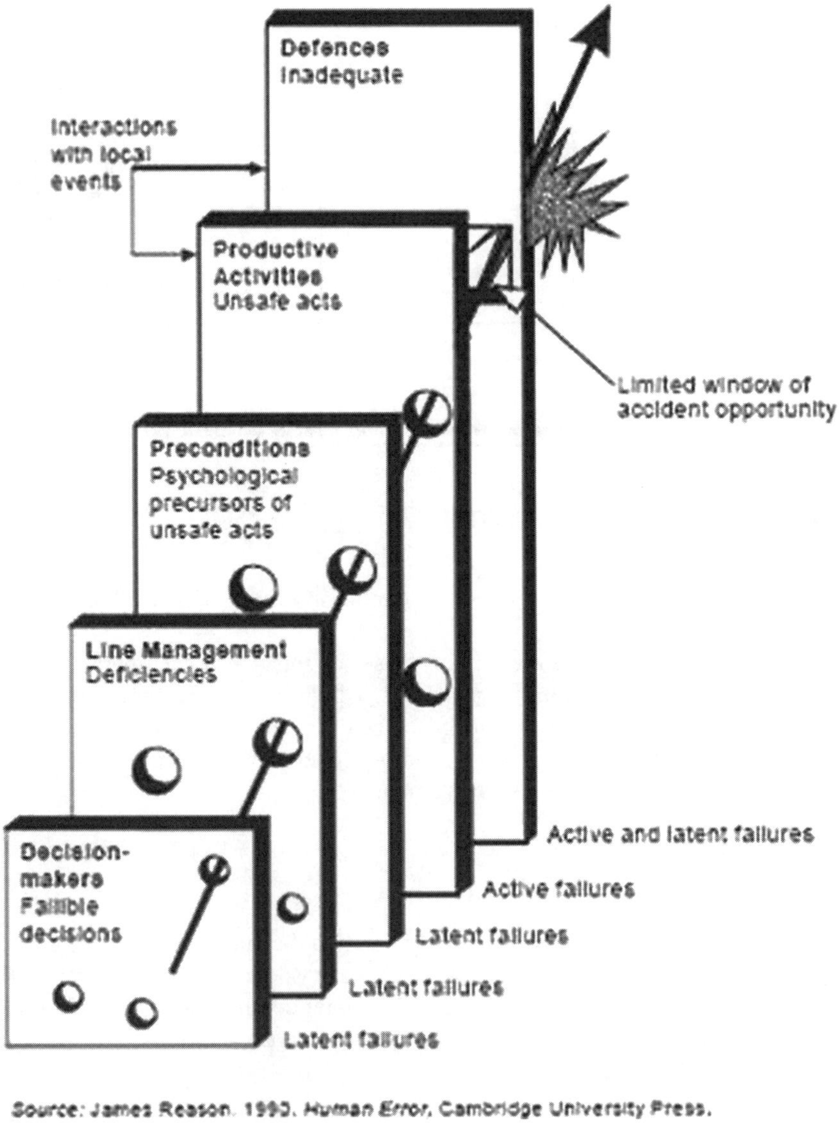

Figure 3.1 The Reason Model (A)

Before describing the investigation methodology, some definitions used by the TSB are presented below.

- Unsafe condition (UC) – a situation or condition that has the potential to initiate, exacerbate, or otherwise facilitate an undesirable event, including an unsafe act;

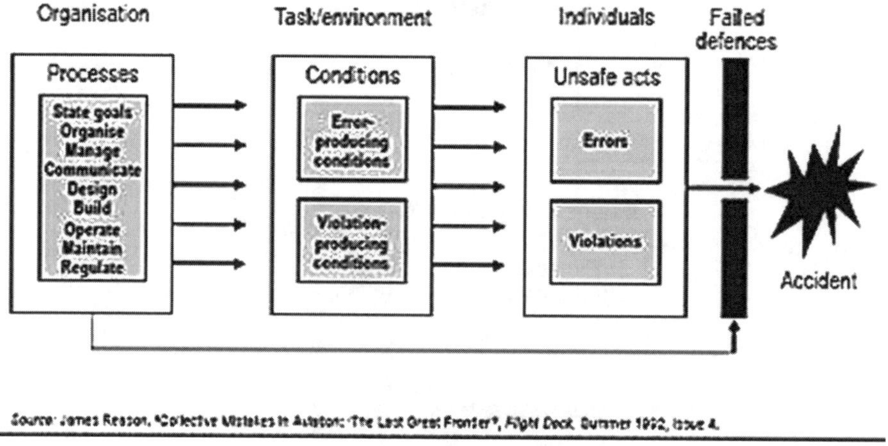

Figure 3.2 The Reason Model (B)

- Underlying factor (UF) – an unsafe condition for which no further unsafe acts or unsafe conditions apply;
- Unsafe act (SA) – an error (slip, lapse, or mistake) or deliberate deviation from prescribed operating procedures, which, in the presence of a potential unsafe condition, leads to an occurrence or creates occurrence potential;
- Safety deficiency – an unsafe condition or underlying factor with risks for which the defences are less than adequate.

Identifying safety deficiencies

The goal of the investigation is to identify safety deficiencies. Safety action is then developed to eliminate, or at least mitigate, any safety deficiencies. The safety actions can be recommendations, safety advisories, or other communiqués to Civil Aviation Authorities, manufacturers, or others who can make changes for safety.

Sequence of events

One of the first steps in the investigation process is the development of a sequence of events for the occurrence. This sequence of events can be produced by the use of sophisticated computer tools, or by very basic methods such as Post-It Notes® on a wall or on a large sheet of paper. The making of a sequence of events is done in brainstorming sessions. The sequence of events diagram can be created by one investigator, along with his or her supervisor, for small investigations, or by input from many groups for major investigations. The events themselves are identified and some possible reasons for the event are listed. Certain events are seen to be of high importance or interest and these are labelled as 'safety significant events'.

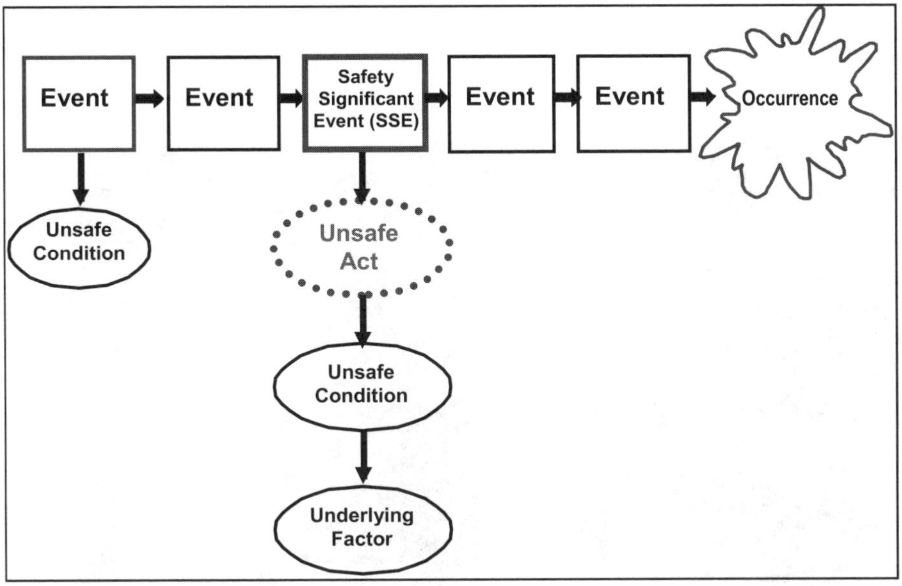

Figure 3.3 Sequence of events diagram

When a safety significant event is identified, an attempt is made to determine why the event happened. Was it the result of an unsafe condition (latent failure) or an unsafe act (active failure)? The active failure(s) may stem from an unsafe condition(s). The investigators drill down until no other items remain (the last underlying factor). An example of a sequence of events diagram is given in Figure 3.3.

Often the first identified item that leads to the safety-significant event is an action by a human. In the present scheme of things, such an action receives the label 'unsafe act' if it had an apparent negative influence on the occurrence. Events that had positive safety influence can also be considered to be safety-significant. These safety-positive events also often result from human activity (actions that 'save the day'). To analyse these safety-positive acts, should they be labelled as 'safe acts'? Sometimes things that have a positive influence on one occurrence can be inherently unsafe in other situations, so the label 'safe act' does not seem appropriate. One of the problems with labelling the actions as 'unsafe' or 'safe' is the instant inference of the nature of the act, commencing at a very early stage of the investigation. It seems to this author that the mere fact that an action is being studied means that the act has safety significance, negatively or positively, and the label 'safety-significant act' is more appropriate.

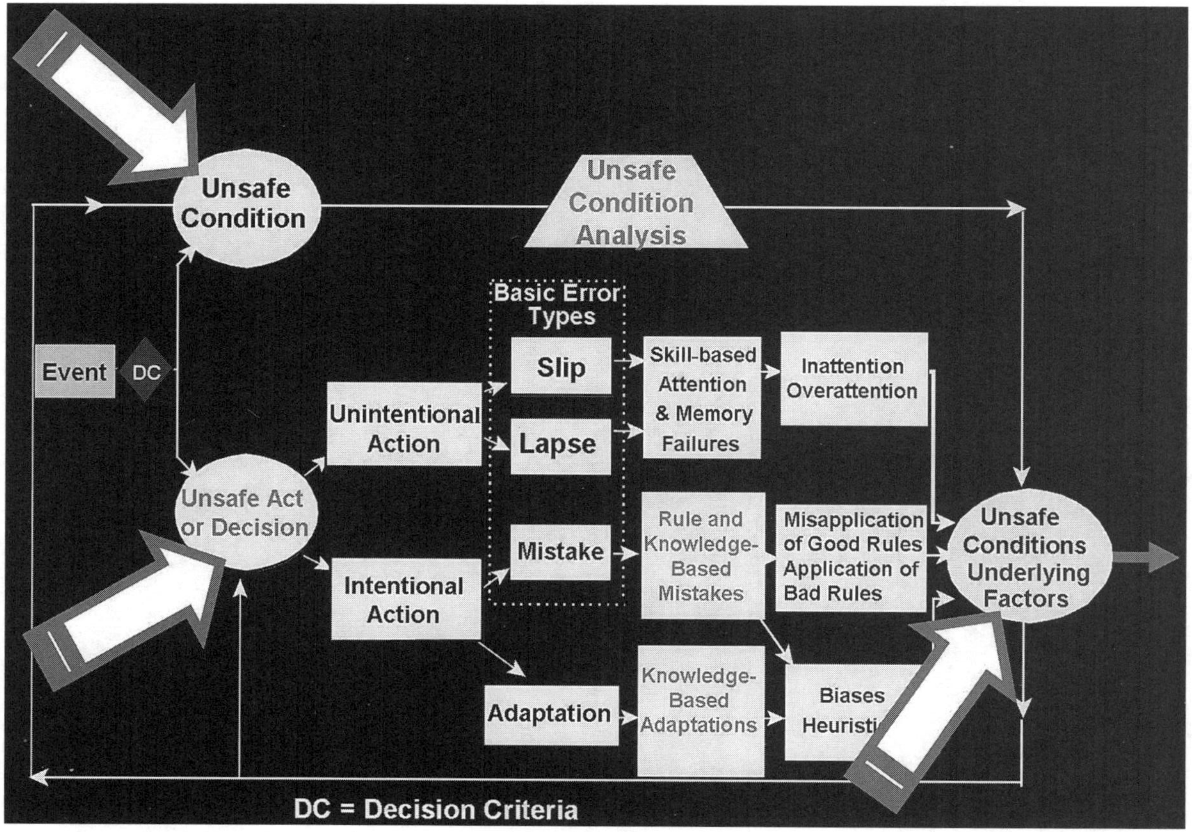

Figure 3.4 Investigation process from TSB manuals

Error analysis

Any 'unsafe act' that has been identified is analysed to determine the reason for the act or decision. These analyses lead to a determination of the type of error. When reviewing Figure 3.4, which is contained in TSB investigation manuals, the process of evaluating human error, starting with an 'unsafe act or decision', is shown, along with the interaction of unsafe conditions. Would this error and unsafe condition analysis be degraded if the human act being analysed was labelled as 'safety-significant' instead of 'unsafe'? It seems to this author that it would not.

Next, a study of some occurrences will be used to assess the labelling of the human acts during investigations. Perhaps the best way to illustrate the problems and thoughts regarding the analysis of human acts, be they 'unsafe' or 'safety-significant', is by reviewing a few examples. These examples will include one experience from long ago in the author's career as a pilot, as well as some selected well-known aircraft accidents. Most of the examples are based on aviation occurrences, but there is also a medical example given.

Personal example – grand finale

Most pilots, when looking back at their flying experiences, can think of moments that were not particularly safe. This author is no exception. The following narrative is written in the first person for reading ease.

The first example to be discussed involves an air show. Not a large, well-organized air show, but a low-key event held at a small airport in western Canada. With a total of about 600 hours of flying time in my logbook, I was asked by my Squadron Commander to do some fly pasts in a T-33 jet trainer for the air show. After being told not to do anything dangerous, and that I would be flying the only jet in the show, I departed home base for the small airfield. In order to get below the 2000-foot cloud deck at the airport, I had to fly an instrument approach through a thin layer of stratus that was about 1000 feet thick. For the next couple of minutes I did a few fly pasts at low speed, medium speed and then performed a reasonably tight turn around the airport. However, I had convinced myself that the fly pasts were probably boring for the crowd, so I would create a bit of a grand finale for my portion of the air show. Knowing that the stratus cloud was not thick, I decided to perform a vertical roll through the thin layer of stratus as I departed back to home base. I got the T-33 going as fast as it would go in the space available, about 410 knots, and pulled the required vertical g and got the manoeuvre going with no problem. As expected, I entered the thin layer of stratus with lots of airspeed. I realized, after some considerable period of time, that my aircraft was still in cloud and that the airspeed was decaying rapidly. I started an unusual attitude recovery and got the aircraft back into the normal flight regime just as I popped out of the cloud. Looking back, I could see that my vertical roll was carried out in the only area where a cloud towered well above the stratus layer.

For the scenario just given, it is not difficult to use 'unsafe act' to describe what led to the need for a low-energy unusual-attitude recovery. In fact, after the air show experience I said to myself 'what was I thinking'. Fortunately, the recovery was effective or else there would have truly been a grand finale that day. Even though the decision to perform the impromptu aerobatic manoeuvre fits the label 'unsafe', the recovery action does not. Yet both are safety-significant acts.

DC-10 accident – Chicago

On 25 May 1979, the pilots operating AA Flight 192, a DC-10, had just rotated the aircraft for take-off when the left engine and pylon separated from the wing. The aircraft climbed to a maximum height of about 325 feet above the ground and began to roll to the left. The left roll continued past the vertical, and the aircraft descended into a field and was destroyed. All 272 persons on board the aircraft perished, and two people on the ground also lost their lives.

The engine separation resulted from maintenance action, which led to the failure of pylon structure. Engines were being changed using forklifts and this method did not produce the precision required. The result was an overload fracture, as well as fatigue cracking, of the pylon aft bulkhead upper flange. When the engine left the aircraft, it led to the retraction of the left slats and this led to asymmetric stall at the speed flown by crew. Figure 3.5 shows a small segment of a sequence of events for the accident.

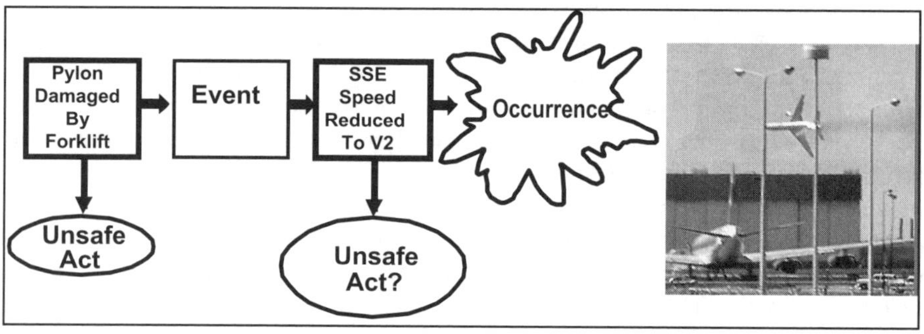

Figure 3.5 DC-10 Chicago sequence of events example

The crew attempted to fly the aircraft in accordance with procedures they had practised many times in the DC-10 simulator. Unfortunately, these procedures called for a speed that was too slow, and the aircraft stalled and crashed. Finding 10 from the NTSB final report of their investigation into the Chicago accident is given below:

> The flight crew flew the aircraft in accordance with the prescribed emergency procedure which called for the climb out to be flown at V2 speed. V2 speed was 6 KIAS below the stall speed for the left wing. The deceleration to V2 speed caused the aircraft to stall. The start of the left roll was the only warning the pilot had of the onset of the stall.

By studying the acts and decision of the pilots using error analysis techniques, it could be argued that they misapplied a good rule because of a lack of knowledge regarding the aircraft's condition. If we can use error analysis to conclude the presence of error when the accident crew did exactly what other well-trained pilots would have done, why do we need to label their acts and decisions as 'unsafe'? Why not just call their actions safety-significant acts?

Swissair 111 – Peggy's Cove

On 2 September 1998, a Swissair MD11 was *en route* from New York's JFK Airport to Geneva, Switzerland. When the flight was passing the east coast of Canada, the pilots detected an unusual odour in the cockpit. A diversion to Boston was initiated, but before turning back, the crew decided to land at Halifax, Nova Scotia. The Swissair crew quickly began preparations for a landing at Halifax. The aircraft was overweight for landing and the cabin meal service was in progress. The accident sequence for the accident lasted about 21 minutes. The flight never made it to a runway at Halifax and it crashed into water about five nautical miles south of Peggy's Cove, Nova Scotia. The 229 people on board the flight were killed.

Many events were analysed in detail, and an extensive sequence of events was created during the investigation. Analysis of events that were deemed to be safety-significant was carried out. The methods and the terminology used were consistent with current human-factors investigation practices. However, at times, the term 'unsafe act' posed a problem. For some investigators, any act or decision that resulted in an apparent delay in getting the aircraft on the ground was considered to be 'unsafe'. For some other investigators who were not familiar with the term 'unsafe act', such labelling appeared to assign unwarranted blame. Figure 3.6 provides a short representation of a few events from the accident sequence.

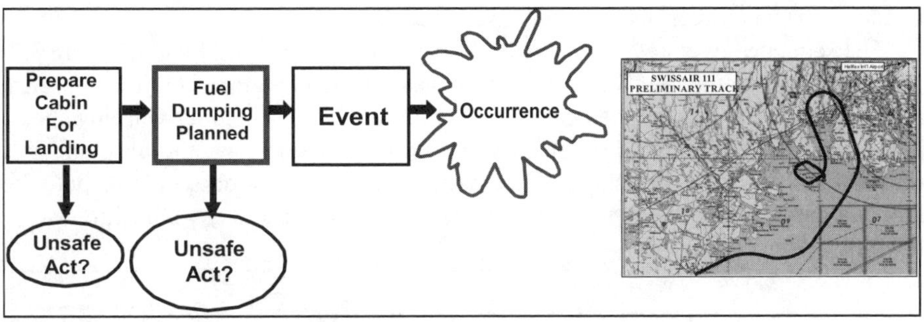

Figure 3.6 Swissair 111 sequence of events example

Would it have lessened the effectiveness of the investigation if the actions shown in the diagram in Figure 3.6 had been labelled as 'safety-significant acts'? Ultimately the investigation concluded, in Other Finding 5, that:

> From any point along the Swissair Flight 111 flight path after the initial odour in the cockpit, the time required to complete an approach and landing to the Halifax International

Airport would have exceeded the time available before the fire-related conditions in the aircraft cockpit would have precluded a safe landing.

Medical example – drug levels following renal transplant

The medical example involves a renal-transplant patient who received a living-related-donor kidney on 3 December 1998. During the admission process, the patient was asked whether he was on a special diet. He replied that he was on renal diet, but indicated that he expected that this would change after the transplant, which was scheduled for the next day. The transplant surgery was successful, but there were problems during the recovery in hospital. The blood levels of cyclosporine were unstable and this appears to have led to elevated blood pressures. The drug doses were varied greatly based on the twice-daily blood work results, and no one could understand why the drug instability was occurring.

On the fifth day following the transplant, one day before the scheduled discharge of the patient, the hospital dietician briefed the patient on his new diet. The new diet was a pleasant change from the renal diet, and the patient was very content with his new dietary freedom. Just as the briefing was ending, the patient asked the dietician if there was something that should not be consumed. The dietician then informed the patient that grapefruit juice should be avoided because it interferes with cyclosporine levels. It turns out that the patient had been having a glass of grapefruit juice with every breakfast since the transplant. This, despite the fact that the patient did not even like grapefruit juice, he had ordered every juice but grapefruit, but had consumed it because he assumed that if this had been given to him by the hospital, it must be good for him. When grapefruit juice was withdrawn from his diet, the cyclosporine levels stabilized and remained stable at desired levels. This medical incident was never reported and did not cause lasting harm.

The sequence of events for this simple medical example is shown in Figure 3.7. The patient did not know that cyclosporine was affected by grapefruit juice because he was merely handed the medication by the nursing staff. The drug-warning sheets packaged with the drug were not provided until the patient left the hospital. Is it reasonable to classify the actions of the patient as unsafe? No one could dispute the fact that the actions of the patient had safety significance for his own health.

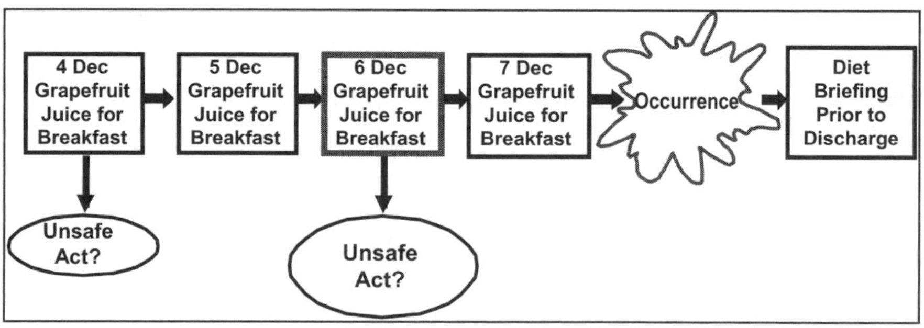

Figure 3.7 Medical incident sequence of events

Air Canada B767 at Gimli – fuel exhaustion

On 23 July 1983, an Air Canada Boeing 767-200 was on a domestic flight from Ottawa, Ontario to Edmonton, Alberta. Approximately halfway along the route, one engine failed, followed quickly by the second engine. The aircraft had run out of fuel. The captain completed a dead-stick landing to a former air force base. This airport, located near Gimli, Manitoba, had been converted into a race track, but the captain was able to find enough runway to land the aircraft. There was limited damage to the aircraft and only minor injuries to a few passengers.

The report of the Commission of Inquiry (1983) into this occurrence details the human and mechanical factors that led up to the fuel exhaustion in flight. The aircraft had fuel-measuring system problems. Through the actions of the maintenance and flight crews, some of which involved the use of metric measurements (the other models in the Air Canada measured fuel in pounds and the Boeing 767 was new the airline's fleet), the aircraft departed Ottawa with too little fuel. Figure 3.8 shows a very small portion of the occurrence sequence of events.

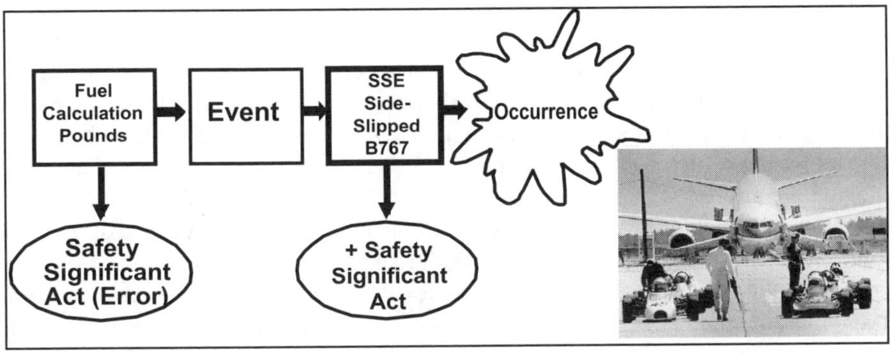

Figure 3.8 B767 Gimli fuel exhaustion sequence of events example

Some of the actions and decisions made by the maintenance personnel and the flight crew contained errors, and could be categorized as unsafe acts without too much argument, but is there any reason why they could not be called safety-significant acts? Also, how does the investigator label acts and decisions that helped save the day? The captain was able to land 800 feet down the runway by employing techniques of side-slip and cross-controlling. The captain used skills that he had learned as a glider pilot, to get the 'Gimli glider' to the threshold of the runway. Using such techniques to control the aircraft's descent, which are definitely not encouraged by the B767's flight manual, prevented a much more serious and potentially fatal accident. The recovery acts of the captain certainly had safety significance.

Air Transat A330 in Azores – fuel exhaustion

Eighteen years after the fuel-exhaustion occurrence at Gimli, another Canadian-operated jet transport aircraft landed with no fuel in the tanks. On 24 August 2001, an Air Transat Airbus 330-200 was *en route* from Toronto, Ontario, Canada, to Lisbon, Portugal. As the flight was crossing the Atlantic, the crew noted a fuel imbalance. About 15 minutes later, the crew decided to divert to Lajes, Azores. Less than 30 minutes later, the right engine flamed out. The left engine then quit because of fuel exhaustion when the aircraft was still about 65 nautical miles from Lajes and descending through 34 500 feet. With no engines operating, the crew, with the assistance of air traffic control, was able to complete the night landing on runway 33 at Lajes. The aircraft received some damage and there were injuries during the evacuation following the landing, but the outcome was much better than the very real possibility of a ditching in the ocean with 306 people on board.

The investigation of this accident was conducted by Portugal, with the assistance of Canada and other countries in accordance with the protocols of Annex 13 of the International Civil Aviation Organization (ICAO). The investigation determined that a recent engine change resulted in a mismatched installation of fuel and hydraulic lines, which led to the chaffing of the fuel line in the right engine. A large fuel leak commenced from the right engine fuel line. This also led to unusual and distracting oil readings for the right engine. Through a series of actions and decisions by the crew, combined with checklists and their use, all the fuel in the aircraft left via the right-engine fuel-line leak. There were a series of safety-significant acts that allowed the aircraft to run out of fuel. Some of these acts belonged to maintenance personnel who installed the mismatched fuel and hydraulic lines. Actions of the crew contributed to the loss of all the fuel on the aircraft. Through a series of positive safety-significant acts, starting with the quick decision to divert, the crew was able to salvage a landing in very difficult conditions. Safety significant acts work for both safety-negative and safety-positive human behaviour.

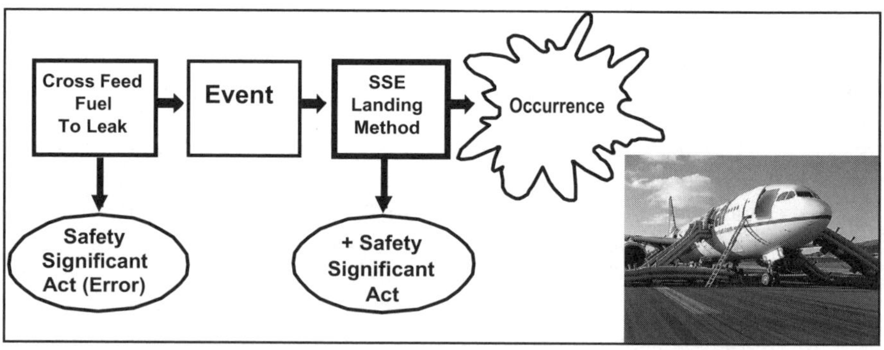

Figure 3.9 Air Transat A330 Azores fuel exhaustion sequence of events example

Conclusion

There are some who are probably thinking that the suggestion to use 'safety significant' in lieu of 'unsafe' is just another example of political correctness gone mad. Certainly, that is not the reason for the suggestion. The examples cited in this chapter are intended to show that the term 'unsafe act' can be replaced with 'safety-significant act' without degrading the capability of analysing human performance. Given that the investigation of human acts will often lead to error analysis, it would appear that there is no real necessity to label acts as 'unsafe'. Calling acts 'safety significant' does not degrade the ability of investigators to understand human performance.

The use of 'safety significant' *versus* 'unsafe' avoids the perception of by those involved in an occurrence that they are somehow unsafe by nature. The labelling of the acts by participants in occurrences as 'safety significant' *versus* 'unsafe' can remove impediments in discussions regarding human performance. We also saw that the term 'unsafe act' may prejudge its real influence upon the occurrence sequence. Using the label 'safety-significant act' can facilitate the study of actions that are safety-positive.

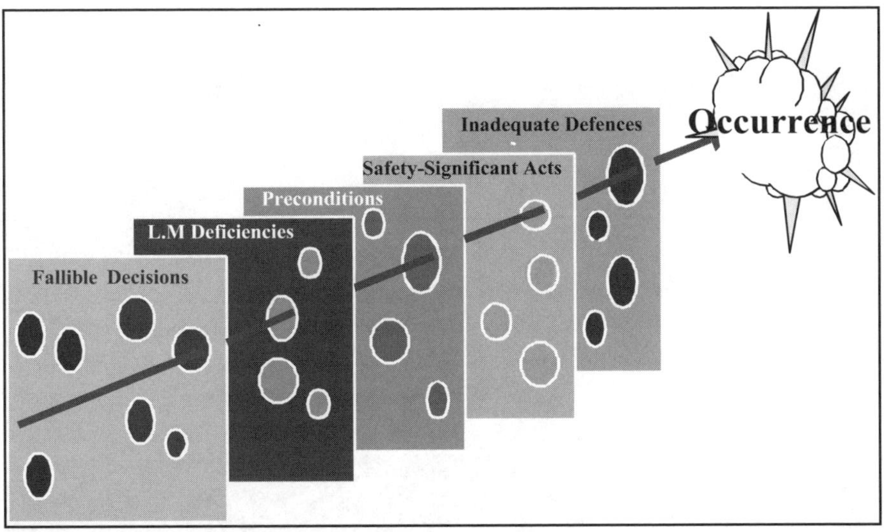

Figure 3.10 A slight revision of the Reason Model

The author has taken the liberty of slightly modifying the traditional depiction of latent and active failures. The changing of 'unsafe acts' to 'safety-significant acts' does not appear to degrade the understanding of this often-used and valued diagram.

References

Government of Portugal, Accident Investigation Final Report. 22/ ACCID / GPIAA / 2001, Lisbon, Portugal.

International Civil Aviation Organization (ICAO), Doc 9683-AN/950: *Human Factors Training Manual*, Montréal, Canada.

National Transportation Safety Board (NTSB), *Aircraft Accident Report*, Washington: United States of America, AAR-79-AAR-17.

Reason, J. (1997), *Human Error*, Cambridge: Cambridge University Press.

The Honourable Mr Justice George H. Lockwood, (1983), *Final Report of the Commission of Inquiry Investigating the Circumstances of an Accident Involving the Air Canada Boeing 767 Aircraft C-GAUN that Effected an Emergency Landing at Gimli, Manitoba on the 23rd Day of July, 1983*, Ottawa: Transport Canada.

Transportation Safety Board of Canada, *Manuals of Investigation*, Gatineau, Canada.

Transportation Safety Board of Canada, *Report Number A98H0003*, Gatineau, Canada.

Transportation Safety Board of Canada, *The Guide to Investigating Human Factors*, Gatineau, Canada.

Chapter 4

The Calgary Health Region: Transforming the Management of Safety

Jack Davis, *Calgary Health Region*
Jan M. Davies, *University of Calgary*
Ward Flemons, *Calgary Health Region*

In Calgary in 2004 two patients died unexpectedly from excess potassium chloride in pharmacy-prepared dialysate. Internal and external safety reviews produced 121 individual recommendations. The deaths and reviews provided a safety 'wake-up call' and an opportunity to re-examine the Region's safety practices and procedures. Since then, many changes have been undertaken, starting with top management's recognition that safety was not identified as a unique priority. The Region has created a patient safety framework (like a safety management system), started to recognize the tensions between various aspects of care (such as access and safety), and developed and implemented new policies (for reporting, disclosing, informing, and a just and trusting culture). Rather than specifying a 'no-blame' culture, the Region has chosen to emphasize responsibility and accountability. Taken together, all these changes are helping to move the Region along the trajectory of a safety culture, in the hope of providing safer patient care.

Introduction

Situated in the south western part of the Province of Alberta, the Calgary Health Region (the 'Region') is one of the largest integrated health regions in Canada providing health services to a population of 1.14 million residents, as well as providing tertiary care services to other residents of southern Alberta. The Region's principle catchment comprises a geographic area of 39,260 km^2 and includes 14 major urban and rural communities. The health services provided by the Region are delivered from more than 100 locations and include 12 acute care sites, 40 care centres, and a variety of community and continuing care centres, with a total of 7,836 acute care beds, continuing care space, and general and psychiatric rehabilitation and recovery beds. More than 24,000 individuals are employed by the Region in a variety of health professional practices and support service areas. The Region grants privileges to practice medicine to 2,150 doctors. The majority of doctors are not employed by the Region but practice in a fee-for-service capacity under the terms of provincial legislation and the federal Canada Health Act. More than 4,000 volunteers are registered with the Region and provide over 250,000 hours of voluntary service.

The 1,143,368 residents served by the Region represent 36 per cent of the total population of the province, in an urban to rural ratio of about five to one. The majority of patients (88.5 per cent) live within the Region, with 7.1 per cent residing outside the Region but within the province, and approximately 5 per cent come from other western Canadian provinces (namely southern Saskatchewan to the east and south-eastern British Columbia to the west).

In 2004 to 2005, more than 354,000 visits were made to the Region's emergency departments. During the same period there were more than 112,000 hospital admissions. Patients underwent more than 62,000 surgical operations, with an average length of stay in hospital of 6.55 days. There were 14,473 births (in 2003) and the life expectancy (also 2003) was 83 for females and 78 for males.

Alberta is the sixth largest province by geographic area, totalling 661,848 square kilometres, of which less than 3 per cent is covered by water. The provincial population of 3,223,400 is the fourth largest in Canada. With a strong economy combined with the natural beauty of the Rocky Mountains, rolling foothills, and prairie grasslands, the provincial population is rapidly growing. The province continues to attract large numbers of newcomers to Alberta and especially to the Calgary and surrounding areas. The Region's population growth is outlined in Table 4.1:

Table 4.1 Increase in population in the Calgary Health Region, 2001–2005

Year	Population	Increase
2001	1,067,058	24,992
2002	1,098,149	31.091
2003	1,122,521	24.372
2004	1,143,368	20,847
2005	1,168,426	25,058

The increase in the population has been accompanied by an increase in the health care budget for Region health expenditures, of approximately 9 to 10 per cent annually, although some specific items, such as drugs and medical/surgical supplies, have increased by 8 to 10 per cent each year. The budget for drugs was about $75 million for 2005, and in 2004–2005 the total regional operating cost was $2,082,817,000. The fastest growing expenditure is attributed to healthcare provider salaries and benefits, which are driven by inflation, influenced by shortages in the supply of available workers, and relate back to the Region's need to further attract and retain healthcare providers to meet the requirements of its rapidly growing population (Report to the Community, 2005). This profile is provided to help describe how the Region has planned, managed, and delivered integrated health services to a large and geographically dispersed population.

Background

These details are also given to describe the context of the journey the Region has taken since early 2004. The unexpected deaths of two patients who were receiving specialized renal dialysis treatment in one of the Region's hospitals and the subsequent events were the tipping point (Gladwell, 2000) for the Calgary Health Region. Many major system changes and redesigns in fundamental areas of quality and safety have since been undertaken within the Region.

Death is always a risk in healthcare treatment and sometimes a reality during treatment as a result of the natural progression of certain illnesses and the ageing process. However, although both patients were seriously ill and receiving care in two of the Region's Intensive Care Units, they were not expected to die. Investigations of these deaths determined that while undergoing continuous renal replacement therapy, the patients were dialysed with an incorrect dialysate solution containing potassium chloride rather than sodium chloride. Further investigation revealed that a substitution error had been made in the preparation of the special dialysate fluid that both patients received. The dialysate fluid was not available commercially and had been prepared in the Region's central pharmacy.

After appropriate investigations were completed, information about the findings was disclosed to both families. Thereafter, permission was sought and received from the families to inform others, including the public and other healthcare organizations. The public's reaction, as judged by the response of some representatives of the media, was not immediately sympathetic to the Region's initiatives to improve safety. Furthermore, although media attention was directed at some of those involved in the care of the patients who died, as well as to management, many other healthcare providers in the Region felt that they were also being blamed for these deaths. The Region needed to deal with the public's reaction to a 'shame and blame' attitude, in the same way that an individual would. This helped to provide some of the impetus for further development of safety initiatives.

These initiatives included both internal and external reviews, with the generation of 121 unique recommendations, which extended well beyond the safety measures for the handling of KCl. These recommendations included the need for a re-examination of technical issues, as well as safety practices and procedures.

The response to these recommendations from the Region included recognition that these deaths were not the 'fault' of the 'workers', but rather symptoms of problems within the system. In addition, there was also recognition that a substantial shift in organizational culture was required. The question was raised as to whether safety was considered to be a priority within the Region or if, in fact, it was buried and/or assumed within other priorities and professional obligations. Upon further examination, safety was not identified as a unique priority within the Region. In every effort to meet the health demands of a rapidly growing population, the Region's priorities had been focused on improving access to services and meeting these needs within a defined and strict budgetary process. While quality was considered to be an important element and assured through various activities, including a national accreditation process for healthcare facilities (CCHSA, 2005), safety was thought to be intrinsic to the provision of healthcare. 'We don't come to work to be unsafe' was a common response. Once again the Region had never formally listed safety as a priority (although it was included as a dimension of quality).

Table 4.2 outlines the six dimensions of quality of care. Given the demands placed on the Region by a growing population demanding more healthcare services in a timely fashion, the focus of the Region had been on providing accessible, affordable, and efficient care. The dimensions of effectiveness, appropriateness and safety were lower on the list of priorities. After these deaths and the ensuing events of 2004, there was recognition of the need to re-examine and re-order these dimensions of quality. There was also an acceptance of the innate tensions between specific pairings of these dimensions. Safety may come at a cost to access. Care that is appropriate may not be affordable. Care that is effective may not be efficient. For example, the tension between safety and access could shift toward access, through the provision of elective surgery 24 hours/day, seven days/week. Conversely, there could be a shift toward safety, with night-time operations restricted to emergencies. The tension between 'appropriate' and 'affordable' is reflected in discussions such as 'just because we can, should we?' Sending patients home the same day after an operation may be efficient but may not be effective if the patient deteriorates without specialized care.

Table 4.2 Dimensions of quality of care

Accessible
Affordable
Efficient
Safe
Appropriate
Effective

Source: Derived from Eagle and Davies, 1993; adapted from Macintosh AM, McCutcheon DJ, 1992

Although many individuals in the Region had been trying to ensure that the health care provided reflected all these dimensions of care, demonstration of a shift in top management's understanding of the problems was necessary to effect a change of this magnitude. Without this understanding, without a change in the knowledge and attitudes of top level and executive management, the Region would have continued to provide care as it did in early 2004. Leadership became the critical success factor.

The Patient Safety Framework

However, this was not the only change. A Patient Safety Framework (PSF) was developed to serve as the safety blueprint (Figure 4.1) and a guide for safety planning within the Region (Flemons, Eagle and Davis, 2005). As shown in Table 4.3, this PSF is not unlike a safety management system, with principles, policies, procedures and practices (Degani and Wiener, Transport Canada, 1994).

Figure 4.1 The Patient Safety Framework

At the core of the Region's PSF is a safety management cycle that starts with the first phase of hazard identification. Once hazards are identified, then hazard analysis and management can be undertaken, through better understanding of the contributing factors, prioritization of which hazards should be addressed first, and development of recommendations for system improvements. The second phase concerns system improvements, their strategies and design, testing and implementation. The third phase involves continually checking the performance of the system through:

- a series of safety performance process or outcome measures;
- a formal evaluation of recommended system improvements, the effect of their implementation and whether they have contributed to other unanticipated hazards; and
- research of new methods of delivering safer healthcare.

The Patient Safety Framework has four fundamental cornerstones that provide the foundation for implementation and longer-term success. These elements are:

- Leadership/Accountability
- Organizational Structure
- Culture and
- Resources.

Table 4.3 The '4Ps' of safety management

Philosophy • There will always be threats to safety • Safety is everyone's responsibility • The organization sets the standard
Policy • There are clear statements about responsibility, authority, and accountability • Organizational processes/structures incorporate safety goals into every aspect of the operation • Everyone develops and maintains the necessary knowledge, skills and attitudes
Procedures • There are clear directions for all • There are means for planning, organizing and controlling • There are means for monitoring, assessing and evaluating safety status
Practices • Everyone follows well designed, effective procedures • Everyone avoids the shortcuts that can detract from safety • Everyone takes appropriate action when a safety concern is identified

Source: Derived from Transport Canada; Degani and Wiener, 1994

Leadership/accountability

The Calgary Health Region has a single Board that provides governance for all of its operations. The Board and the management teams all have important roles to play in ensuring appropriate safety management. Initially, the Region's Board established a Safety Task Force to oversee the Region's safety strategy. Region management reviewed its Balanced Scorecard and made changes to further emphasize its priority on quality and safety. Leadership walkrounds (Frankel *et al.*, 2003) were initiated. Patient safety events are now actively promoted and supported by the Executive Management Team members, who routinely participate in patient safety symposia and conferences. With respect to accountability, a more formalized process has been instituted to address safety issues. For example, reports that track the progress of the implementation of safety recommendations are now produced for management and are shared with the Board.

Organizational structure

Of the many changes in this area, one that should be especially noted involves the establishment of a Regional Clinical Safety Committee, linking and integrating the Region's five clinical portfolios and all the key organizational support areas. Another important change is that department-based quality assurance committees, with their traditional doctor-only membership, are being transformed to become interdisciplinary service clinical safety committees. The Regional Clinical Safety Committee now ties all this work together and reports to the CEO's Executive Vice President and Chief Clinical Officer.

Culture

To help transform the Region's safety culture, Reason's approach and focus on reporting, learning, a just (and trusting) culture, and flexibility has been adopted (Reason, 1997). A key goal in the Region is to ensure that improvements are made that address system safety deficiencies, including those identified by healthcare providers. To foster a culture where people feel safe in reporting hazards, the Region developed a safety-related policy specifically addressing the ways in which employees and doctors will be supported in situations where hazards have been identified or where harm has occurred. The policy outlines in a transparent manner the Region's commitment to just and fair processes for reviewing and making system improvements, from both a system and individual perspective (see below).

Resources

The Region has invested several million dollars to build the infrastructure, training, communication and equipment required to support this new safety strategy. Contingency funding has also been made available for ongoing system improvements. These funds allow each of the operational portfolios and their services to act quickly on and invest in safer systems, rather than wait for the annual budgeting approval process.

In addition to these four cornerstones, the Region's Safety Framework also emphasizes two additional characteristics of a safer organization. These are the need for appropriate safety policies and procedures and the fostering and development of safety communication and safety related education.

Policies and procedures

Since 2004, the Region has undertaken much work in the area of policy development and preparation of guidelines and procedures for implementation. This has included a new governance policy on safety and the handling of potassium chloride and identification of other high hazard medications, as well as others (see below).

Communication

To address the ethical issues of maintaining communication between the Region and its patients, its providers, and its key partners and stakeholders, several policies were also developed to promote safety through communication and transparency in actions. To avoid confusion over terminology, three types of communication have been defined. Reporting relates to communication between healthcare providers and the Region and concerns hazards and harm. Disclosure relates to communication between the Region (including its healthcare providers) and patients about circumstances when patients have been harmed or nearly harmed by the care that they have received. Informing relates to communication between the Region and its key partners and stakeholders.

These three safety-related policies, as well as the 'just and trusting' culture policy, have come to be seen as a social and ethical contract. This contract lies between:

- the Region, providers and patients (Disclosure of harm policy);
- Providers and the Region (Reporting of harm and hazards policy);
- the Region and its principal healthcare partners/stakeholders (Informing policy); and
- the Region and its providers (Just and trusting culture policy).

In effect, development of these new polices and the focus on the social and ethical contract represents the start of the Region's journey on this safety culture road. As clearly delineated by Reason (1997), an organization with a safety culture equates to one with an 'informed culture', that is, having an effective safety information system. A safety culture is also one that supports reporting, is just, demonstrates flexibility, and encourages learning.

With respect to the latter policy, the term 'trust' was specifically included, to place the emphasis on this particular attribute. The Region hopes that as an organization it will be characterized by an atmosphere of trust. The counter to this is that the Region's healthcare providers are encouraged, and even rewarded, for providing essential safety related information. However, the workers are also clear about where the line must be drawn between acceptable and unacceptable behaviour.

In developing these policies, the Region is not aiming at building a 'no-blame' culture. In fact, the Region has chosen not to use this term in its safety policies and guidelines. Rather, the attributes of responsibility and accountability are emphasized. Every healthcare employee and doctor has a responsibility to provide care and services that:

- are safe;
- are at or above professional and/or Region standards;
- are respectful of the needs of patients/families; and
- engage patients/families as partners.

The Region in turn has the responsibility to provide structures and processes that support the delivery of safe care and services. At the same time, the Region, its employees and doctors should all be accountable for their actions.

Communication is vital and intrinsic to every part of the organization. As a Region, and as individual healthcare providers, we are striving to become more transparent in what we do and say, with disclosure to patients/families in situations where patients have been harmed or nearly harmed (Davies, 2005) becoming more frequent. We have also detailed how we will inform the public, while being mindful of our legal requirements for patient privacy. We are using the power of our own stories, to describe who we are, what we do, and where we hope to go. And we are learning to value and encourage 'intelligent wariness' (Reason, 2000) but not to be afraid, either as leaders or as workers.

Conclusion

The tragic deaths and ensuing events of 2004 and 2005 were not the effectors of a single point change. Rather, they have acted as the tipping point (Gladwell, 2000) for an ongoing and dynamic process of continuous improvement of the quality of care in the Region. This process, however, continues to require leadership – at all levels, from top management to bedside teams. Communication is vital and intrinsic to every part of the organization, in the form of both structural elements and actions. As a Region, and as individual care providers, each must become more transparent in what is done and said, including disclosure to patients of what happened, and sometimes what nearly happened (Davies, 2005), becoming more frequent. The Region now has detailed how the public will be informed, while being mindful of the legal requirements for patient privacy, including the Health Information Act (HIA, 2000) and the Public Health Act (PHA, 2000). The Region is now using the power of its own stories, to describe who the Region represents, what the Region does, and where the Region hopes to go. Finally, all are learning to value and encourage 'intelligent wariness' (Reason, 2000) but not to be afraid, either as leaders or as workers.

Acknowledgements

The authors thank Avril Derbyshire, Glenn McRae and Sharon Nettleton for their assistance in the preparation of this chapter.

References

Calgary Health Region (2005), Leading the Way. Report to the Community. http://www.calgaryhealthregion.ca, accessed 6 January 2006.

CCHSA, Canadian Council for Health Services Accreditation (2005), <http://www.cchsa.ca/>, accessed 6 January 2006.

Davies, J.M. (2005), *Disclosure, Acta Scandinavica Anesthesiologica*, 2005, No. 49, 725–727. [DOI: 10.1111/j.1399-6576.2005.00805.x]

Degani, A. and Wiener, E.L. (2004), *Philosophy, Policies, Procedures and Practices: The Four "P's" of Flight Deck Operations*, in *Aviation Psychology in Practice* eds Johnston, N., McDonald, N. and Fuller, R., Aldershot, UK: Avebury Technical. Ashgate Publishing Ltd.

Eagle, C.J. and Davies, J.M. (1993), Current Models of "Quality": an Introduction for Anaesthetists, *Canadian Journal of Anesthesia*, 1993, 40, 851.

Flemons, W.W., Eagle, C.J. and Davis, J.C. (2005), Developing a Comprehensive Patient Safety Strategy for an Integrated Canadian Healthcare Region, *Healthcare*

Quarterly, 2005, No. 8 Special Issue, 122–127.

Frankel, A., Graydon-Baker, E., Neppl, C., Simmonds, T., Gustafson, M. and Gandhi, T.K. (2003), Patient Safety Leadership WalkRounds, *Joint Commission Journal on Quality and Safety*, 2003, No. 29, 16–26.

Gladwell, M. (2000), *The Tipping Point*, Boston: Little, Brown and Company.

HIA., Health Information Act (2000), Chapter/Regulation: H-5 RSA.

Macintosh, A.M. and McCutcheon, D.J. (1992), 'Stretching' to Continuous Quality Improvement from Quality Assurance: a Framework for Quality Management, *Canadian Journal of Quality in Health Care*, 9, 19–22.

PHA., Public Health Act, P-37 RSA (2000). <http://www.qp.gov.ab.ca/catalogue/>, accessed 6 January 2006.

Reason, J. (1997), *Managing the Risks of Organizational Accidents*, Aldershot, UK: Ashgate.

Reason, J. (2000), Human Error: Models and Management, *British Medical Journal*, 2000, No. 320, 768–770. [DOI: 10.1136/bmj.320.7237.768]

Transport Canada Safety Management System, <http://www.tc.gc.ca/CivilAviation/systemSafety/pubs/tp13739/SMS/what.htm>, accessed 6 January 2006.

Chapter 5

Governance and Safety Management

Greg Marshall
National Air Support

In recent years, the term 'Corporate Governance' has received much attention in the media, and for all the wrong reasons. The failure of companies such as HIH and OneTel at the regional level, and of WorldCom and Enron at a global level, has raised the spectre of accountability of both directors and officers of companies, particularly those of public corporations.

Although the term is used freely, 'Corporate Governance' is not necessarily understood in terms of its definition, or its intent. Perhaps alarmingly, many from the ranks of both directors and management are not fully aware of roles and responsibilities in terms of risk management within the context of corporate governance.

Introduction

The vulnerability of an organization to disaster and financial crisis is directly associated with its degree of resilience to risk influences, both internal and external. The key for any organization is to move from a position of risk to a position of safety; from a position of vulnerability to a position of resilience.

Risk management is a key element of the corporate governance mix, and the approach that organizations take toward risk management programs can determine their resilience to undesirable influences, both internal and external. The subject of corporate governance is very broad and involves a complex array of topics, each of which could easily occupy entire semesters of study at university level. This presentation is not a dissertation about company law or corporate reform. It is designed to provide a very broad overview of the correlation between corporate governance and some of the risks that exist within organizations. More importantly, it considers its relationship in terms of the responsibility of directors, officers of companies, shareholders and owners.

Corporate governance – what is it?

Corporate governance, from the broader perspective, can be defined as the sets of systems and associated processes and rules that govern an organization. These processes, systems and rules are designed to support the business of the organization in the interest of, firstly, its shareholders and, secondly, other stakeholders.

The responsibility for corporate governance lies with the Board of Directors and, although they can delegate the day-to-day management of the organization to managers and other personnel, they cannot abdicate their overall responsibility.

Under the Corporations Act, responsibility for the actions of the company is vested upon both the directors and the officers of the company. Officers may be defined as those managers appointed by the Board to manage the day-to-day activities of the company. The law binds '…a person, by whatever name called and whether or not he (or she) is a director of the board who is concerned, or takes part, in the body's management' (AICD, 2003). Further, employees may also be defined to be officers of the corporation where their actions, in respect of occupational health, safety and welfare legislation, have a direct legal liability towards each other (AICD, 2003). This demonstrates that responsibility can extend beyond the company's directors to other members of the organization as well.

Essentially, corporate governance focuses on three key objectives:

- to protect and reinforce the rights and interests of stakeholders, especially in areas concerning conflicts of interest;
- to ensure the board properly fulfils its primary responsibility to direct strategy and monitor the performance of the organization, particularly with regard to assessing the performance of senior managers; and,
- to ensure that management controls and reporting procedures are satisfactory and reliable.

A vital component of corporate governance is risk management, including both business and safety risk. Management control, including the development of an effective response to risk, is interwoven into these three elements of corporate governance (AICD, 2002).

Risk management

Business risk management, from a broad perspective, includes an understanding of the strategic fit of the organization within its market and the assessment of any emerging business opportunities with respect to their potential or probable impact on the short- and long-term viability of the organization. Prudent business risk management will seek to mitigate risk associated with strategies such as acquisitions, divestments, organic growth, and the development of new business opportunities.

Safety risk management typically includes the practices aimed at ensuring the safety and welfare of employees and contract personnel, under occupational health, safety and welfare legislation embodied within state and federal statute law. Safety risk management is also established to mitigate the indirect costs associated with occurrences that may have an impact on the business of the organization. These indirect costs may include, for example, a loss of goodwill from existing or potential clients. Importantly, safety risk management is also imposed to protect the interests of the organizations directors and officers by ensuring that appropriate management strategies are established to minimize, not only the organizations exposure to potential litigation, but also the exposure of its directors and officers to litigation where they may be held liable under the provisions of the Corporations Act.

Whilst directors cannot be responsible for the day-to-day risk management of a company, they should determine the company's risk management policies and their associated overall strategies. Their role is to actively monitor risk and ensure that the appropriate risk mitigation strategies are in place. They should also be responsible for clearly communicating the policy and strategy to management and informing all key stakeholders, including employees, of the company's risk management philosophy.

The correlation between corporate governance and risk is in understanding the varying level of risks that exist within a business, or set of businesses, in relation to a given return on investment. At some point, a line is drawn in the sand, showing the acceptable level of risk *versus* a level of return on investment. Above that line, the business proposition is considered undesirable; the risks are considered to be unacceptable. Below that line, the business proposition is considered desirable; the risks are considered to be acceptable.

Sometimes, that line may not be all that clear, requiring risk mitigation strategies to pursue a particular business activity. Or, more alarmingly, organizations may be tempted to move that line to fit in with a new 'rationalized' risk model to suit the pursuit of a particular business endeavour. This is particularly true of organizations that exist in highly competitive and dynamic marketplaces where revenue is marginal.

Risk and safety management

A safety management system is a process adopted by management to mitigate the safety risks associated with the conduct of the organization's activities. These systems have evolved over the course of time, generally with their origins borne of compliance regulations imposed upon industry. The effectiveness of safety management systems can only be measured in terms of the organization's safety culture. Safety culture is inextricably linked to the organizational culture of the firm. For example, if commercial imperatives are continually placed before safety, the organization's safety culture will diminish.

The safety culture of an organization infers that all personnel accept ownership for matters relating to safety and that this process is actively supported by all, including executive management personnel. Traditional safety management systems were typically reactive and placed a heavy emphasis on compliance. These systems were established to meet a minimum requirement simply because certain organizations had to have them. They were typically viewed as cost centres affording little in the way of value-add to the organization.

These systems evolved to those encompassing quality management principles and focused mainly on plant, equipment, and management systems. The tendency was for management to focus on specifics, or adopt a 'tunnel vision' type of approach to safety. Safety imperatives were assessed individualistically, rather than by viewing the overall interaction of people, processes and systems as a whole.

These days, safety management systems and associated infrastructures must be viewed, not solely in terms of cost, but in terms of savings to the organization in preventing occurrences that lead to both direct and, importantly, the more significant indirect costs, many of which are intangible. These leading types of systems develop a holistic approach to safety, assessing the organization and how it performs as a whole. Ownership for safety is accepted by everyone within the organization, and not just by management and the Board.

Risk rationalization

To demonstrate a correlation between hazard and risk within a business proposition, let's look at a simple analogy. Consider, if you will, a mouse.

The basic business proposition of a mouse is simple: it is to find a supply of food to satisfy its demand expressed by hunger. Ideally, this food source should be convenient and relatively easily attainable. For the mouse, searching and acquiring food carries with it certain risks, such as food sources and availability; competition with other mice from other colonies; and predators and others who wish to eradicate them.

Some sources of food may be particularly convenient but may have very high associated risks. Other sources may not be as convenient but may have relatively lower levels of associated risk. A particularly convenient, although very high-risk, source of food can be found with mousetraps. Now, through a process of observing others, an astute mouse may have concluded that the risk of serious harm from these devices is very high and, without a mitigating strategy, the risk is considered to be unacceptable. Assuming there is an abundant source of food elsewhere, and with relatively little associated risk, obtaining food from the mousetrap would be disregarded.

But since food sources may suddenly become scarce, through increased competition and dwindling supply, hunger may force the mouse to re-evaluate his risk assessment and re-consider the mousetrap as a viable source of food. In other words, the mouse has been forced to rationalize his former assessment of the risk.

Let's consider that the mouse in our above example represents an organization and that obtaining food represents the economic business of the organization. The organization originally established a risk assessment and management strategy aimed at preventing accidents and incidents that might bring harm to its employees and its business. The organization conducts its business with prudence. But what if there is a sudden downturn in the economy, and/or what if the business loses some of its key contracts? Is it likely to be forced into a position where it reconsiders its previous business propositions, originally rejected because the risks associated with them were, at the time, considered too high and therefore unacceptable? Could this organization be forced to rationalize its approach to risk simply to survive?

What does this have to say about the organization's culture, its approach to risk, and the integrity of its oversight in terms of corporate governance?

'Can-do' and workarounds

A number of articles have been published recently on the subject of 'can-do' cultures. These cultures are typically focused on the delivery of a product or service on time and/or within budget to meet internal or external client expectations. Whilst there is nothing insidious with this philosophy in itself, the practicalities of can-do cultures can often lead to errors and mistakes occurring at the implementation levels of the organization. Whilst many of these may be borne of systemic deficiencies, such as poor project planning or inadequate staffing levels, many more are related to the attitudes, behaviours and perceptions that arise within the workplace at both the employee and supervisory levels, often through the poor communication of intent.

Whilst management may espouse that the company employs a 'can-do' approach to its business, it is often not stated that this means safely and in a compliant manner. Consequently, those at the workplace level may interpret 'can-do' as 'must-do,' even if this means employing workarounds by taking shortcuts in procedures and processes.

Often, these workarounds are achieved by using some lateral thought by employing some novel techniques to solve a problem or to achieve a task in a timely manner. Whilst these workarounds are employed with good intent, they pose significant risks to the organization that may not be apparent to those engaged in the tasks at the time. Typically, these workarounds are employed without the knowledge of those at the executive levels of the organization. However, sometimes, executive management may be complicit in such practices.

The following is a de-identified news item used to provide an example of a workaround that was carried out through the non-availability of appropriate equipment. It concerns an accident that occurred when an employee attached an improper towbar between an aircraft and a tug. The towbar broke and the aircraft rolled forward, crushing the employee.

> The National Transportation Safety Board concluded that an employee made a 'decision to use improper equipment' to push back an airplane from the gate, causing the accident that crushed the employee against the nose of the plane. The report stated that the employee used a towbar that was too short for the tug that was being used. But the union contends the employee was using the only equipment the airline provided to do the job. A union official stated that the employee 'had no choice to make'.

The federal report conflicts with another agency's findings. An Occupational Safety and Health authority fined the airline for what it called a 'serious' violation of workplace safety laws. The airline is contesting those findings.

The union argued that, because the proper piece of equipment was not available, employees had learnt to make do with what was available. A number of questions arise. Did the employees voice their concerns at the inadequacy of the equipment, and what did management do in response? Did the employees learn to manage with inadequate equipment because of a management culture that dictated that they just had to make do with what they had? Or, did management rationalize the risk based on the premise that this 'workaround' had been used many times before without incident, and therefore, it deemed this practice acceptable, even 'safe'.

Situational violations, or straying from accepted practices by bending a rule or procedure, can normally result from one or more of the following:

Time pressures exist when inadequate maintenance scheduling processes do not permit all required tasks to be carried out in accordance with maintenance manuals or other documentation. Or aircraft scheduling may mean that a deviation from checklists or procedures is required to minimize delays in getting the aircraft out of maintenance, or to ensure 'on-time' departures.

A lack of adequate supervision may result in complex tasks being carried out by inexperienced staff, without the benefit of assistance and guidance from senior supervisory personnel. Or, improper processes or procedures are being carried out which are not detected at the supervisory level; or supervisors turn a 'blind-eye' for the sake of expediency.

The non-availability of equipment, tools and parts may result in the fabrication of improvised tooling, or by using other equipment not designed for the given task. Incorrect parts may have been ordered and delays in re-ordering correct parts may result in financial penalties to the organization.

Insufficient staff to carry out a task may result in conditions where personnel are directed to carry out tasks for which they are not licensed, or not trained. These tasks may also be carried out willingly, in the interest of meeting the customer's need, without due regard for the associated risks. Perhaps these personnel may improvise 'short-cuts' to expedite the completion of a task, or set of tasks.

These aberrant behaviours and actions are insidious and, if not detected and trapped through a robust safety and quality audit regimen, become unofficially sanctioned procedures. Diane Vaughan (1996), in her definitive work on the loss of the space shuttle Challenger referred to this as 'the normalisation of deviance'.

Learning organization

> Learning difficulties are tragic in children, but they are fatal in organizations. Because of them, few corporations live even half as long as the person – most die before they reach the age of forty (Senge, 1990).

Peter Senge's quote is not only applicable to the general business sphere; it is particularly apt for both the aviation organization and the aviation system as a whole. A learning organization can be defined as 'one which has developed the continuous capacity to adapt and change'. The organization must not only learn from its own mistakes, it must also learn from the mistakes of others. To learn from these mistakes infers that change is necessary to correct, or improve upon, the various systems employed by the organization. The idealistic learning organization not only realizes that it must adopt change; it embraces change. But what if the organization fails to accept responsibility for deficiencies within its own systems and procedures and chooses the blaming route?

In an article published in the *British Medical Journal* in March 2000, Professor James Reason noted that blaming individuals is emotionally more satisfying than targeting institutions. It also serves to divert the focus on error away from the organization's systems, thereby absolving managers of any responsibility.

Uncoupling a person's unsafe acts from any institutional responsibility is clearly in the interests of managers. However, this approach serves to isolate unsafe acts from their system context. If organizations fail to recognize error in the system context, the organization fails to learn and history is doomed to repeat itself.

Culture and change management

Culture is variously defined, but is generally accepted as being the collective sets of values, beliefs and ideals that exist within an organization. In other words, 'It's the way we do things around here'. Essentially, culture is established by what the leaders of an organization pay systematic attention to. A pro-active safety culture is typified by management's stated commitment, and its observed actions, in support of positive safety initiatives and outcomes. However, if, for example, leaders place a heavy focus on cost containment, or if there is a loss of some key contracts, then cost will become the dominant consideration and may well be placed ahead of any other considerations, perhaps even its approach to safety.

Organizations, particularly large ones, with endemic poor safety cultures and poor quality management practices, cannot be rectified by change initiatives overnight. An organization that has developed a culture of discontent, borne of a philosophy of 'can-do' at any cost, cannot be influenced by wholesale change. Change must be measured and it must be introduced incrementally. Radical and discontinuous change can, in itself, be highly damaging to an organization.

Change can be a very difficult concept for some organizations to adjust to, or even accept, and may even be challenged and openly resisted. This is particularly true for large organizations with well-entrenched corporate philosophies or excessive bureaucratic processes. An organization's culture is a powerful force that persists, despite reorganization and the departure of key personnel. A close observer of NASA commented that 'cultural norms tend to be fairly resilient, the norms bounce back into shape after being stretched or bent. Beliefs held in common throughout the organization resist alteration' (CAIB, 2003).

But what if organizations fail to learn from their mistakes, or resist attempts at change aimed at mitigating overall risk?

And what part do commercial pressures play in resisting change efforts?

Let's look briefly at two examples of the rationalization of risk involving the loss of the space shuttles Challenger and Columbia. In both cases, the risk and safety management systems failed.

In the case of Challenger, the initial design did not predict the blow-by of hot gases past the 'O' rings that were used to ensure an integral seal between the individual segments that comprised the solid rocket boosters. However, experience derived from early flights showed evidence of 'blow-by' of these hot gases through the partial erosion of these 'O' rings. The design of the joint was subject to engineering re-evaluation and it was assessed that the design could tolerate the damage. As flight experience increased, and successively more damage was noted, and without a 'failure,' the increasing rates of damage were continually re-assessed as tolerable.

In the case of Columbia, the initial design did not predict that pieces of foam would break free from the external tank with the potential to strike and damage elements of the shuttle. Once again, experience showed otherwise and that, in fact, foam was breaking free, and strikes to the orbiter and other shuttle elements was occurring. However, once again, an engineering evaluation determined that the damage was relatively minimal and assessed that the design could tolerate the damage.

The CAIB noted that, in both cases, the initial decision to accept the engineering re-evaluation was a turning point. It established a precedent for accepting, rather than eliminating, these technical deviations. As a consequence, subsequent evidence of 'O' ring erosion and foam strikes were not defined as signals of danger, but as evidence that the design was acting as predicted. Engineers incorporated worsening anomalies into the engineering experience base, holding larger deviations from the intended design. In both cases, this 'rationalisation of risk' had fatal consequences. Commercial imperatives were significant factors in both cases.

Summary

All organizations exist for the purpose of obtaining an economic or social benefit, and effective cost management is a vital tool. Commercial organizations face a number of competitive pressures that must be balanced against good corporate governance principles.

Any re-evaluation of a business model in line with market forces, which results in a downsizing of facilities or human resources, brings about a degree of risk, particularly when the scope of that organization's activities, and demands in terms of customer service, remains unchanged. This should raise red warning flags.

Whilst it may be tempting to re-evaluate previously discarded business propositions, a re-evaluation, or rationalization, of previous risk models can be a dangerous precept. Any initial risk evaluation, or risk re-evaluation, must be supported by the use of an accepted risk management process. In Australia, the AS/NZS 4360:2004 for risk management is an example and a good starting point.

An organization's safety culture can serve as an effective safeguard against systemic risks by involving all personnel, management, supervisory, and at the shop floor levels, in the hazard identification process. But this can only be achieved by supporting the reporting of hazards and occurrences and investigating them from the systemic perspective. An organization with a culture of blame exists with a false sense of security. It will not only fail to learn from systemic lessons, it will decimate the reporting of occurrences and hazards from within.

The processes utilized by the safety and quality management systems should be viewed as an opportunity to trap, eliminate, or mitigate risk. Effective safety and quality processes identify and eliminate any undesired practices or procedures at the embryonic stage before they are allowed to become 'normalized'. An organization that is prepared to learn from its own mistakes, and the mistakes of others, will take action to mitigate risks when these become apparent. But it must be prepared to accept and adapt to change. A robust safety management system serves as a defence to the organization to provide an adequate warning of impending risks that may be associated with organizational change. But it can only do this with the utmost support of management and the board in their stated and demonstrated commitment to safety. Further, the reporting of matters pertaining to safety and risk must be facilitated on a regular basis to those officers and directors ultimately responsible for the efficacy of the organization's risk management system.

Not only is a robust safety management system a vital corporate governance tool, it is also a critical investment in an organization's future.

References

CAIB (Columbia Accident Investigation Board) (2003), Report, Washington, DC: Government Printing Office.

Pennell, R.C. and Adrian, S.L. (2002), *Directors Rights and Responsibilities*, Company Director Manual, Australian Institute of Company Directors, Thomson, Kew: Legal and Regulatory.

Reason, J. (2000), Human Error: Models and Management, *British Medical Journal*, March, 2000, 768–770. [DOI: 10.1136/bmj.320.7237.768]

Vaughan, D. (1996), *The Challenger Launch Decision: Risky Technology, Culture, and Deviance at NASA*, Chicago: University of Chicago Press.

Chapter 6

Overcoming the Short-Medium-Term Problems of Fleet Transition: The Expectation and Achievement of Economic Gains

Boyd Falconer and Christopher Reid
University of New South Wales and Jetstar Airways

Fleet transition is a required and ongoing process for every airline. While fleet transition has a large number of operational and economic benefits, it also causes some short-medium-term problems. In many instances, operational benefits translate into almost immediate economic gain, however fleet transition requires considerable planning and investment relating to engineering, airport preparation, training and operational requirements. This chapter discusses the theory and practice of normal operations monitoring, with a particular emphasis on Australia's successful low cost carrier Jetstar.

Introduction

There is a great expectation about the economic gains to be made relating to fleet transition. This is perpetuated both anecdotally – amongst aviation managers and corporate sales people representing aircraft manufacturers – and in the academic literature (Bazargan, 2004). Indeed, in the preparation of this chapter a review of the corporate sales pitch presented on the Airbus and Boeing websites was conducted to gain an appreciation for the magnitude of expected gains from transitioning to more modern aircraft. Perhaps not surprisingly, the benefits described by each of the manufacturers were significant in both economic and operational terms.

Clearly, fleet transition is a required and almost continuous process for every airline. Yet fleet transition, with a large number of operational and economic benefits, also causes some short-medium-term problems. Indeed, operational benefits translate into almost immediate economic gain; however, fleet transition requires considerable planning and investment relating to engineering, airport preparation, training and operational requirements.

This chapter discusses the theory and practice of normal operations monitoring, particularly fleet transition, through the looking glass of Australia's immensely successful low cost carrier, Jetstar. The chapter firstly provides an overview of fleet transition by examining the benefits – both operational and economic – of 'upgrading' to new-technology aircraft via a well-orchestrated fleet transition. Jetstar's transition from Boeing 717 aircraft to Airbus A320 aircraft provides a valuable case study in this context. Secondly, the chapter examines in some detail the typical challenges to normal operations monitoring in a period of fleet transition. Thirdly, the chapter provides a dedicated summary of Jetstar's experience with fleet transition, such that industry personnel of both large and small aviation organizations can learn the key issues that they may be able to implement in their own organizations.

Fleet transition – an overview

While initially a large up-front operational and economic expense, over the long term transitioning from an old fleet to a new fleet of aircraft provides an airline with opportunities for new destinations, increased payload/capacity and efficiency gains or improvements from fuel burn per seat and maintenance costs. Many airlines have recently undertaken massive fleet renewal programs such as easyJet in the UK. EasyJet is replacing its entire fleet of B737 aircraft with new A319 aircraft. More recently Qantas has announced a fleet renewal program involving an order for 65 Boeing 787 Dreamliner aircraft to replace the B767-300 fleet.

Fleet transition is an integral and ongoing role of every airlines operation. The replacement of older, inefficient aircraft with new and more efficient aircraft provides for myriad operational and economic benefits.

Benefits of fleet transition

Many operational benefits of a new fleet equate into economic benefits when the aircraft enters service. Using Jetstar as an example, the airline from July 2004 commenced a transition from 14 Boeing 717-200s to 23 Airbus A320 aircraft.

The operational benefits of the Airbus A320 over the Boeing 717 include:

- Increase in payload through containerized freight and baggage;
- Increase in capacity from 125 passengers on the Boeing 717 to 177 on the Airbus A320;
- Increased endurance or range allowing for new sectors previously out of reach of the 717 such as Melbourne-Cairns and trans-Tasman sectors to Christchurch using Jetstar's ETOPS certification; and
- Marketing and product enhancements such as leather seats, wider cabins and a standardized image.

It should perhaps be highlighted at this point that the operational (and economic) benefits of new sectors, described briefly in point three above, are not insignificant.

Since Jetstar started, aircraft capacity has grown more than at any other time in the Australian airline industry (see Table 6.1). Interestingly, the market is estimated to be able to absorb four to five per cent.

Table 6.1 Aircraft capacity growth

Airline	Estimated growth (ASK)
Jetstar	+ 32%
Qantas	+ 10%
Virgin	+ 8–10%
Overall	+ 12%

Source: JetStar estimates

Through the introduction of Jetstar into regional leisure markets and the airline's aggressive transition from Boeing 717 to Airbus A320, Jetstar has created an unprecedented increased capacity in the Australian domestic market for the Qantas Group (May 2004 *vs.* May 2005). These capacity increases are listed in Table 6.2, and are dominated by three-figure percentage increases.

Table 6.2 Growth in capacity to JetStar destinations

Region	Capacity increase
Whitsunday Coast	+ 250%
Cairns	+ 233%
Hamilton Island	+ 164%
Sunshine Coast	+ 136%
Newcastle (Hunter Region)	+ 100%
Melbourne (Avalon Airport)	+ 100%
Gold Coast	+ 28%
Tasmania (Hobart/Launceston)	+ 20%
Mackay	+ 19%

Source: JetStar estimates

Economic benefits

The significance or magnitude of the fleet upgrade will ultimately determine the economic benefit of a fleet transition. For example, the on-going economic benefits of upgrading from an old fleet of DC-9s or Boeing 737-200/300/400 series would be far greater than transitioning from a fleet of A320s to Boeing 737-800s.

Older aircraft like the before DC-9s and classic 737s have higher maintenance costs, increased fuel burn and limited flexibility in today's de-regulated and volatile airline industry. Transitioning to new aircraft would have an immediate impact on fuel and maintenance costs and provide airlines such as Jetstar the flexibility to change its network as market demands change.

One thing to consider is that not all the economic benefits come from savings in items such as fuel and maintenance (Bazargan, 2004). Newer aircraft of similar capacity to its older relatives can also provide operators with the ability to open up new markets and increase revenue that was otherwise unobtainable. Aircraft such as the new A340-500 and Boeing 777-200LR have provided airlines with opportunities to fly sectors and open new markets beyond anything that could have been achieved with their existing fleet of aircraft. Singapore Airlines previously operated to New York via another port and now with the A340-500 it operates non-stop on a 19-hour sector. The extended range aircraft with an all premium product onboard allows Singapore Airlines to now not only offer a new service, but also earn premium revenue. Qantas also recently investigated the idea of using ultra-long range aircraft for non-stop flights from Sydney to London, although at this stage decided the time and technology was not at a level to make the service viable.

For Jetstar, the economic benefits of switching to the A320 from the B717 are not as significant as upgrading from an over 20-year-old aircraft to the A320. Although the increase in capacity, payload and operational flexibility of an all A320 fleet, along with a lower cost per ASK, now at a level of 8.0 cents, provides the airline with an unprecedented level of operational and economic advantage over current and future competition.

Challenges

While fleet transition has a large number of operational and economic benefits, it also causes some short- to medium-term problems. In many instances, operational benefits translate into almost immediate economic gain; however, fleet transition requires considerable planning and investment relating to engineering, airport preparation, training and operational requirements.

Planning and investment challenges

The planning for and investment into new aircraft is an ongoing and expensive operation. Every aspect of the airline's operation has to be looked at to ensure the aircraft fits the airline's operation and business model. For low-cost, price-sensitive airlines such as Jetstar, making the right decision on its fleet is the foundation of the airline's success.

In planning for the introduction of a new aircraft type, every department within the airline has to be consulted and work as a team to ensure a smooth transition. There are so many aspects to new aircraft introduction it would be impossible to cover them all in this chapter, although the three biggest operational areas would include:

- engineering
- crewing
- airports.

Engineering

Beside the actual purchase of the aircraft itself, the second biggest economic outlay for an airline would be its engineering equipment, infrastructure and spares to support the requirements of the new aircraft. In the case of Jetstar, the airline now has two separate maintenance plans and spares programs in place for two aircraft manufacturers, one to maintain the existing fleet of Boeing 717s and the other to manage the ever-increasing fleet of A320s. Both aircraft are physically maintained differently, spares are ordered differently and manufacturers are consulted differently.

The biggest challenge for Jetstar Engineering is ensuring that the right level of spares and support is maintained for the B717s as they *exit*, and increasing the level of A320 spares and support as that fleet increases. The most undesirable thing an airline could do is have unnecessary capital caught up in expensive spare parts that exceed the current fleet requirement. Therefore, disposing of spares as the 717s are removed is just as challenging as ensuring the correct quantity of spares is purchased every time a new A320 enters the fleet as recommended by the respective aircraft manufacturers.

One other significant challenge for engineering is ensuring the engineers, *i.e.*, LAMEs and AMEs, are trained and endorsed in time for the aircraft introduction. At the time Jetstar was about to introduce the A320, Australia and South Asia has not seen a locally operated A320 in such numbers as Jetstar planned since Ansett Australia was in operation. A majority of training was completed with the assistance of Airbus and, given the history of Jetstar as Impulse, originally operating Beechcraft 1900s in regional New South Wales, a significant change in organizational culture was also required (Helmreich and Merritt, 1998; Falconer, 2005).

Crewing

From a crewing perspective, ensuring the right number of pilots and cabin crew to operate the proposed schedule is critical. For Jetstar, the biggest crewing challenge is planning the conversion training for B717 pilots over to the A320. Jetstar is building its operation around a number of bases where both aircraft 'overnight' and crew reside. These bases include the major capital cities in the states Jetstar operates, along with regional centres such as Newcastle and Cairns. As an A320 enters the Jetstar fleet, it replaces a B717 at the base, and this causes planning issues for the airline's operational resources.

To operate one aircraft, Jetstar requires five Captains and five First Officers. To convert a group of B717 pilots to the A320, they must be removed from Boeing 717 flying and complete two months of ground-based training in Brisbane followed by approximately 100 hours of line training in the flight deck with a training captain. The challenge that arises is that the training has to commence approximately three to four months before the new A320s' introduction at that base. With these pilots removed from B717 flying, pilots from other bases have to be positioned through that base to operate the remaining Boeing 717 schedule until the new A320 enters service. When multiple aircraft are delivered in a single month, this causes even more challenges for the airline crew planners.

Airports

A major challenge for Jetstar Airports when transitioning from the Boeing 717 to the A320 is ensuring the airport is A320 capable in time for the pending aircraft arrival. The Boeing 717 is a bulk-load aircraft, which means passenger baggage and freight is loosely loaded by hand in the baggage-hold of the aircraft. A minimum amount of airline-owned ground equipment is required for the general loading of the Boeing 717, basically a tug, baggage trolleys and forward stairs.

The A320, on the other hand, is a containerized aircraft that requires extensive capital expenditure in new equipment at the airport. The most expensive of these items is a 'TLD' Loading Device, a unit that loads/unloads the baggage containers (also known as an AKH). The TLD devices are made in France and typically cost up to $275,000 each and have a manufacture and delivery lead-time of some four months. The A320 also requires a set of baggage containers costing almost $1,000 each, up to 14 baggage container dollies to transport the containers at $7,000 each and two sets of A320 stairs for forward/aft dual-door boarding.

Ensuring the required equipment is available and delivered, a ground handling agent is in place, trained to Jetstar requirements, and the airport terminal and apron is set up to accept the A320 are all integral tasks airports must manage before a new aircraft introduction.

One final aspect regarding Airports that has to be considered is the physical readiness of the Airport operator/owner. Most of Australia's regional airports are operated by Local Government Councils, with little or no private equity involved in their management. A majority of these airports also rarely see an aircraft bigger than a Dash-8 or small propeller aircraft. For Jetstar to introduce a Boeing 717 and then a larger A320 causes a number of operational issues such as obtaining pavement concessions if required, and apron markings and terminal constraints being addressed.

The challenge of operational requirements

The final and single most important aspect of any new aircraft introduction for an Australian airline is the regulatory requirements from the Civil Aviation Safety Authority (CASA). The success of airlines meeting the safety and regulatory requirements imposed by CASA ultimately determine whether the new aircraft introduced departs with paying passengers.

The ultimate aim of ensuring all regulatory requirements are met is to make certain the new aircraft is added to the carrier's Air Operators Certificate (AOC). For a new aircraft type never before operated by the airline, this will require a CASA review of all aspects of the airline's operation, from airport set-up and training, crew training, certification and engineering processes and planning.

Future Jetstar transition

Ensuring all the airline planning, training and regulatory requirements are timed to meet the impending aircraft arrival is the basic aim of any airline when transitioning to a new aircraft type. For Jetstar, when the fleet transition is a process of some 18 months with continual new crew training and port upgrades for the A320, fleet transition is a full time job managed by every department every day until the last aircraft is delivered.

For Jetstar, the domestic fleet transition will finish in May 2006, and the next fleet chapter begins: the introduction of the Airbus A330 for long-haul operations and the impending transition to the new Boeing 787s from 2008. Most importantly, the planning for the A330 introduction has already begun.

Conclusion

This chapter has very clearly shown that fleet transition is a continuous process for every airline – and very necessary. Yet while successful fleet transition holds a large number of operational and economic benefits, it also causes some short-term and medium-term problems. Indeed, fleet transition requires considerable planning and investment relating to engineering, airport preparation, training and operational requirements.

This chapter discussed the theory and practice of normal operations monitoring, particularly fleet transition, through the looking glass of Australia's immensely successful low cost carrier, Jetstar. The chapter firstly provided an overview of fleet transition by examining the benefits – both operational and economic – of 'upgrading' to new-technology aircraft via a well-orchestrated fleet transition. Jetstar's transition from Boeing 717 aircraft to Airbus A320 aircraft provided a valuable case study in this context. Secondly, the chapter examined the typical challenges to normal operations monitoring in a period of fleet transition. Thirdly, the chapter provided a dedicated summary of Jetstar's experience with fleet transition, such that industry personnel of both large and small aviation organizations can learn the key issues that they may be able to implement in their own organizations.

Acknowledgements

The authors would like to thank numerous Jetstar personnel for their support, particularly Iain Ritchie, Fleet and Schedules Planning Manager; Peter Shepherd, Head of Engineering and Maintenance; and Matthew Bell, Manager Operations Delivery. We also gratefully acknowledge the enthusiastic support of this paper received from Simon Westaway, Manager Corporate Relations; Trevor Jensen, Jetstar's Head of Technical Operations; and Alan Joyce, Jetstar's Chief Executive Officer.

References

Bazargan, M. (2004), *Airline Operations and Scheduling*, Aldershot, UK: Ashgate.

Falconer, B.T. (2005), Cultural Challenges in Australian Military Aviation: Soft Issues at the Sharp End, *Human Factors and Aerospace Safety*, 5, No. 1, 61–79.

Helmreich, R.L. and Merritt, A.C. (1998), *Culture at Work in Aviation and Medicine: National, Organizational and Professional Influences*, Aldershot, UK: Ashgate.

PART 2
Safety Management Metrics, Analysis and Reporting Tools

Part 2 examines the perceived need from the different industry modes, for more practical tools on SMS. Indeed, SMS is a practical discipline with the objective of minimizing operational errors and risks. Whilst SMS theory is crucial, the chapter will demonstrate the current thrusts of some 'mature' SMS disciplines (such as aviation and rail) towards proposing legislation to implement SMS.

Part 2 will also focus also on determining 'meaning' from collected data – indicating that the goal of SMS is not to collect data per se, but rather to derive meaning from the data itself. Key potentials revolving around cross-border (or multi-modality) of these tools shall be discussed. The reader is invited to discover strategies in maintaining and protecting safety information data towards building a reporting culture.

Chapter 7

A New Reporting System: Was the Patient Harmed or Nearly Harmed?

Jan M. Davies
University of Calgary

Carmella Duchscherer and Glenn McRae
Calgary Health Region

After two unexpected deaths, system-wide changes to safety management were made in the Calgary Health Region. These included development of a new reporting system to replace incident reports. The new 'safety learning reports' focus on what happened to the patient: was the patient harmed or nearly harmed? The latter, representing close calls, are valuable sources of information about hazards and hazardous situations, for both patients and healthcare providers. Immediate safety reviews can be initiated by the reporting of both close calls and untoward events. Reporters are encouraged to describe 'what saved the day', as well as make recommendations for system improvement.

Unexpected deaths

In February 2004, two patients were undergoing continuous renal replacement therapy in one of the Intensive Therapy Units of the Calgary Health Region (the 'Region'). Although seriously ill, the patients were not expected to die, nor were they expected to develop severe hyperkalemia.[1] The unexpected deaths, coupled with a clinician's astute review of the cases, triggered an examination of the bags of dialysis fluid. Like the findings in the patients' blood, the concentration of potassium was found to be much greater than normal. Because bags of the special dialysis fluid were not available commercially, the Central Pharmacy of the Calgary Health Region had prepared a batch of 36 bags. Fearing that excess potassium would be found in the other bags and that more patients might die, the clinician had the remaining bags removed from use. Other acute care sites in Canada that prepared this kind of dialysate were also informed. Details of the events were disclosed to the families. With their permission, the public was informed (Johnston *et al.*, 2004).

---

1 An abnormally high concentration of potassium ions in the blood.

Investigations

A comprehensive internal investigation was undertaken and then an external review was commissioned. Membership of the latter review team included a clinician, a consulting pharmacist and a human factors expert (External Report, 2004). The Health Quality Council of Alberta (HQCA) also produced a review of the safe handling of potassium chloride (HQCA, 2004). As a result of these investigations, 121 unique recommendations (not all related to potassium chloride) were generated. The internal investigation, under the auspices of the Critical Incident Review Committee, revealed that a substitution error had been made during the preparation of the special dialysis fluid – of potassium chloride (KCl) for sodium chloride (NaCl). A number of factors related to the environment and equipment were identified. The external review identified other factors, relating to the patient, the personnel, the pharmacy, the organization, and the regulatory agencies. All three investigations identified 'labelling' as a major contributory factor. This included labelling of the boxes in which the bottles of NaCl and KCl were transported to the Region, and in which they were stored, and also the labelling of the bottles of NaCl and KCl themselves (see Figures 7.1 and 7.2).

The External Review recommended that the Region implement a comprehensive and systematic method for collecting and analysing incidents. This recommendation was coupled with one for the establishment of a method for the monitoring of the implementation of safety recommendations.

The Region's response

The response of the Region was to invoke a complete change in its organizational approach, including the development and implementation of a new system safety framework (Flemons, Eagle and Davis, 2005). This framework highlighted the areas that the Region needed to address. More importantly, however, the framework has served as a template for a safety management system for the Region. The framework starts with a safety management cycle of identifying hazards, effecting system improvements, and then ensuring performance. Integral to the effectiveness of the cycle are the Region's policies and procedures, as well as communication and education. Safety management does not exist, however, in isolation, but is embedded in a foundation with four components. These are shown in Figure 7.3 and are:

- leadership and accountability;
- organizational structure;
- resources; and
- culture.

Figure 7.1 Manufacturer's boxes of sodium chloride (NaCl) and potassium chloride (KCl), in which the bottles of NaCl and KCl were transported and stored

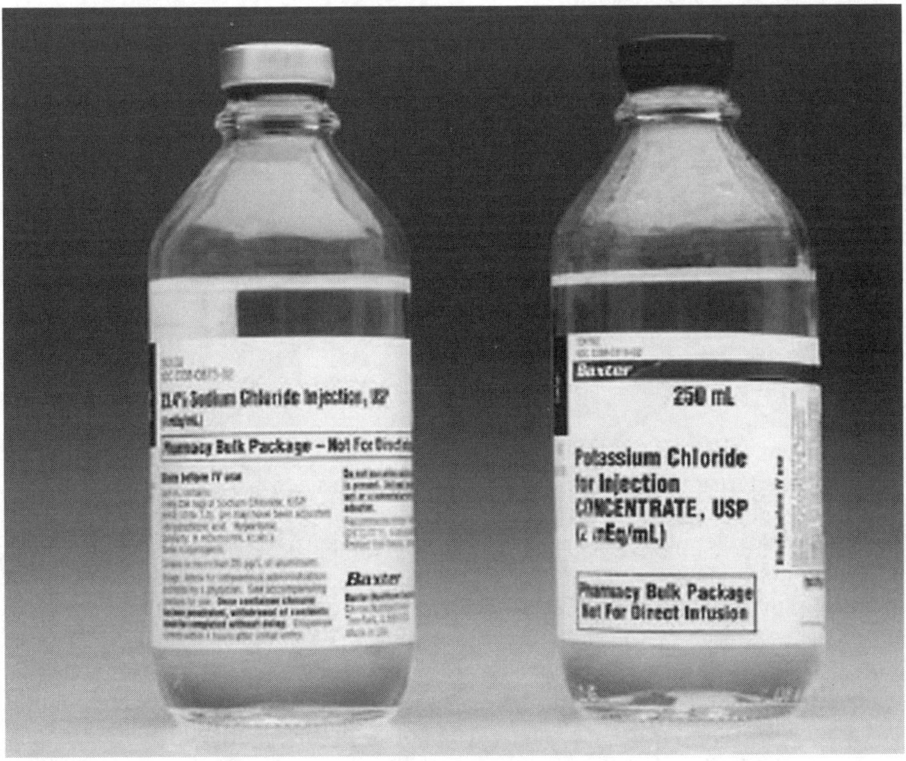

Figure 7.2 Manufacturer's vials of sodium chloride (NaCl) and potassium chloride (KCl)

Figure 7.3 The Calgary Health Region's safety framework

As part of the transformation of the Calgary Health Region, new policies were developed, including four specifically related to patient safety:

- reporting;
- disclosing;
- informing; and
- just and trusting culture.

The policy development process has been comprehensive. For these four safety-related policies, four policy working groups were established, with overlapping interdisciplinary membership. These policy working groups began by conducting an extensive search of the literature and a review of the available policies from other Canadian and international healthcare organizations, as well as other industries – including aviation. Group members discussed, analysed and debated, both within and among the policy working groups. The policy discussion documents were then drafted and redrafted. Consultations were undertaken on an ongoing basis with all levels of personnel, from those who worked on the wards and in the kitchens to those in top management. After multiple drafts, final documents were developed and approved, following which, development was undertaken of guidelines to accompany the policies. In addition, plans for implementation were developed with input from advisory working groups and educational resources. This complex and rewarding process took approximately 18 months, and is still ongoing, with respect to education, implementation and reflective discussion, including input from members of the public.

The above policy and guideline development process is detailed to illustrate that the Reporting Policy was just one of four major safety policies. Furthermore, development was not undertaken in isolation but as part of an overall safety management system.

Reporting policy

Regulatory implications

The aviation industry has been successful in improving world-wide aviation safety in part because of the International Civil Aviation Organization (ICAO). Established in 1944 in Chicago, when 52 nations signed the Convention on International Civil Aviation (the 'Chicago Convention'), today ICAO has 189 member states. With the exponential development of international civil aviation after World War II, ICAO developed international standards and recommended practices to facilitate and coordinate this development.

These international standards and practices include those first adopted in April 1952 for the investigation of aviation accidents. At that time, the Standards and Recommended practices for Aircraft Accident Inquiries were pursuant to Article 37 and designated as Annex 13 to the Convention on International Civil Aviation, Chicago, 1944. In addition, from 1994 on, serious incidents have been included in these standards. Since then, Annex 13 has provided an essential foundation for successful air safety investigations at both the international and domestic levels. Indeed, many countries have incorporated some of the 'best practices' into domestic legislation concerning accident investigation.

One of the investigatory principles embodied in Annex 13 dictates that certain 'records shall not be made available for purposes other than accident or incident investigation …' (ICAO, 1994). Similarly, reporting and investigating 'incidents and accidents' in healthcare in the Province of Alberta are protected by provincial legislation in the form of Section 9 of the Alberta Evidence Act:

> *(1) (a) Quality assurance activity means a planned or systematic activity the purpose of which is to study, assess or evaluate the provision of health services with a view to the continual improvement of*
>
> *(i) the quality of health services, or*
>
> *(ii) the level of skill, knowledge and competence of health service providers*
>
> *(2) A witness in an action, whether a party to it or not,*
>
> *(a) is not liable to be asked, and shall not be permitted to answer, any question as to any proceedings before a quality assurance committee, and*
>
> *(b) is not liable to be asked to produce and shall not be permitted to produce any quality assurance record in that person's or the committee's possession or under that person's or the committee's control.*

(3) Subsection (2) does not apply to original medical and hospital records pertaining to a patient.

According to the Act, information that is collected for the purpose of improving the quality of care cannot and shall not be released for the purpose of civil litigation (Alberta Evidence Act Queen's Printer). The experience of one of the authors (Dr Davies) over the past two decades is that this legislation has allowed healthcare providers to investigate and improve care in the Region in a manner that would not be possible without such legislation.

Previous reporting system

Although the previous reporting system had been use in the Region for several years, in more or less the same format, there were some limitations. Most reports were completed by nurses. Although doctors were required to 'sign off' a report under certain circumstances (such as a patient suffering severe harm), very few doctors filed a report. Reporting tended to be about severe harm only, with very few reports of close calls or what might happen to the 'next' patient. The form was designed to be all-encompassing and to include incorrect narcotic counts at the end of a nursing shift and problems related to Occupational Health and Safety (such as environmental mould). The process that the form was subjected to following its completion lacked confidentiality and allowed for copying and wide distribution. Reports were used by senior nurses for performance management of ward/unit nurses and occasionally in disciplinary decisions, with the number of 'Incidents' being added up to determine suitability for further employment. The overall concept of 'Incident Reporting' was confusing and had negative connotations. The reporting process supported a 'name, blame and shame' culture and many reports were not related to safety but to conflict, either between individuals or departments. The reporting process was neither timely nor efficient, with some forms appearing in the Office of Quality Improvement up to two years after the event. The form was paper-based on two-pages and often the second page (which outlined the investigation and resultant action taken) was left uncompleted. The reporter was required to tick off the appropriate categories and factors involved, but this was not easy as terms were duplicated, confusing and not systematic. The process was not fully supportive of the making of a decision to investigate the event or to gather more information. Nor could data be easily aggregated for the determination of organizational trends. With the old process, individuals involved in safety reviews were not able to reconcile clinical engineering reports about equipment with incident reports involving the same equipment. Also, the specific name of any medication involved in incidents was not collected, thus further impairing the ability of the safety office to identify regional trends. The reporting and investigation process also provided minimal feedback or dissemination of learning, at both the individual and organizational levels.

Despite these many flaws in the official Incident Report system, the Region was able to identify and correct many system safety deficiencies. Much of this success has been the result of the vigorous activity of the Regional Critical Incident Review Committee (RCIRC). As a Board of the Calgary Health Region designated Quality

Assurance Committee, the RCIRC has been able to act on reports, to undertake detailed, extensive and systematic investigations (Davies, 1996), and to assist with the generation of practical recommendations.

The new system

The new reporting system is the product of a complete 'overhaul'. Reporting in the Region is now:

- refocused;
- renamed;
- revised;
- revitalized;
- reformatted;
- reconfigured; and
- reconnected.

Refocused

Reporting now concerns what did or could happen to a specific patient or to any patient, in terms of whether the patient was harmed or nearly harmed, and also what saved the day. For example, if a patient falls and suffers a fractured wrist, reporting now centres on the fracture, rather than on the fall. This is also a change from the unstated but implicit 'Who did it?' Reporting is in turn refocused on the safety learning. Through reporting we hope to learn where the system safety deficiencies lie and what can be 'fixed,' for example, such as the loose handrail and the poor lighting, again, changing from the unstated but implicit 'Who's to blame?'.

Of course, this new focus requires a definition of harm, and the Calgary Health Region has adopted the definition from the College of Physicians and Surgeons of Ontario. Harm is defined as 'an unexpected or normally avoidable outcome that negatively affects the patient's health and/or quality of life, which occurs (or has occurred) in the course of the patient's illness' (CPSO 2003).

Renamed

A decade ago, Runciman noted that the term 'incident monitoring means different things to different people' (Runciman, 1996). Since then there has been much discussion of terms and definitions, including development of patient safety glossaries and dictionaries (*e.g.*, Canadian Patient Safety Dictionary 2003). The Region has chosen to rename Incident Reports as Safety Learning Reports (Morath and Turnbull, 2005) and to clarify terminology. For example, levels of harm have been determined and also correspond to a requirement for reporting that is either mandatory or encouraged. Healthcare workers must report when a patient is either fatally (Level 5) or severely (Level 4) harmed (see Table 7.1). Healthcare workers are encouraged to report all other levels of harm, hazardous situations and hazards.

Table 7.1 Levels of harm

5 =	fatally harmed
4 =	severe
3 =	moderate
2 =	minimal
1 =	none apparent at the time of reporting
0 =	nearly harmed ('near miss', close call)

One important ramification of mandatory reporting in the *Province* of Alberta is that these reports are not qualified under Section 9 of the Alberta Evidence Act. However, it is important to note that almost every case in which a patient suffered severe harm or was fatally harmed would also be reported to top management.

Revised

Reporting has been revised, such that incorrect narcotic counts are now dealt with directly by Pharmacy. Infectious disease reports and Occupational Health and Safety (OH and S) problems are tracked through Public Health and OH and S, respectively. Performance management issues are the domain of Human Resources and not the Office of Clinical Safety Evaluation.

Revitalized

Implementation of an electronic system will allow timely reporting, with healthcare providers encouraged to report at the time of discovery of the harm or hazard. The aim of the Office of Clinical Safety Evaluation will be to provide immediate electronic acknowledgement, with thanks, of the report and review of the report within one business day. By making the form reporter friendly and by requiring minimal data entry (on the part of the reporter or reviewer), reporting efficiency should also improve.

Reformatted

Reporting has been reformatted to allow confidential reporting, with feedback de-identified. The new system is also designed to allow web-based reporting, initially using the intranet and then, with experience and increasing organizational maturity, Internet. Telephone back-up will be maintained for system outages and will also be developed to enable patients, family members and the public to report. Reporters are asked to answer three simple questions and then to describe the event in narrative form. The questions are:

1. What are you reporting?
- Harm to a specific patient?

- A close call for a specific patient?
- A hazard or hazardous situation for any patient?

2. Where did this occur?
3. Did this involve:

- Medication?
- Equipment?

If the answer to either of the latter two questions is 'Yes,' then the questionnaire prompts for the name of the medication or the serial number and/or description of the equipment.

Reconfigured

The reporting system has also been reconfigured, in that a small number of trained investigators are now responsible for reviewing the reports, rather than a data entry clerk. In addition, the reviewers classify reports, instead of the reporter. Reports are now triaged as to the need for further investigation, an improvement over the previous, more classical method of 'word of mouth need'. Although this worked well in the past, close calls were rarely reported, often being seen as either related to 'good care' or to 'good luck', rather than being recognized as a sign of an underlying system safety problem.

Reconnected

The reporting system is now reconnected with the individual healthcare provider and the health system as a whole. Once the review is completed and recommendations generated, improvements to system deficiencies are undertaken and tracked. Feedback is provided to the reporter and to the entire system (where appropriate), with sharing of solutions.

Feedback is especially important. Reporters will receive an electronic 'Thank you!' on receipt of the report in the Office of Clinical Safety Evaluation. Reporters will also be provided with a reference number to enable them to track progress of the report, all the way to implementation of 'safety fixes'. The Ward or Unit Manager will receive regular aggregate data for the specific ward or unit and will also be able to make *ad hoc* requests for information across the organization. Top management will receive aggregate data from reports, but will also receive reports from a separate alerting system for serious events (as they do now).

Developing and implementing the new reporting system continues to have its challenges. Managers who currently own the old reporting process are very reluctant to give up 'control' of the Incident Report form. The Incident Report form, although dysfunctional from a system safety perspective, has rapidly become 'the form for all the processes that don't currently have a form'. These processes are known by various terms including:

- 'evidence';
- performance management;

- the 'Tell on other people' form;
- inter and intra-unit Communication tool (Re: Heads-up that last night's MD is coming round for a 'chat' …); and
- benchmarking document.

We need to consider all these challenges and other uses of the Incident Report form before implementation to realize all the advantages of this new system. However, we see addressing each of these challenges as part of a dynamic and ongoing process of continual system safety improvement. Finally, we look forward to the day when members of the public can report a safety hazard directly via the Safety Learning Report system, no longer needing to go through Patient Complaints or through a staff member, and helping to effect improvements that will contribute to safer healthcare in the Calgary Health Region.

Conclusion

Since the unexpected deaths in 2004 of two of the Region's patients, major system changes have been effected in the Region. These changes include review of the previous Incident reporting methodology and development of a new reporting policy and accompanying guidelines. Reporting has been refocused on the patient, Incident Reports renamed as Safety Learning Reports, and the form revised. The process has been revitalized to improve efficiency and reformatted to allow confidential reporting. Incident classification and coding has been reconfigured, with a small group of trained investigators replacing the reporters for these tasks. Reporting has also been reconnected – to the reporter, patient care managers, and top management. The challenges in changing from an old system to a new one have been recognized and are being addressed. These changes are all part of an ongoing process of continual system safety improvement.

Acknowledgements

We thank Linda Perkins RN MSc, Senior Consultant, Measurement and Evaluation, of Quality, Safety and Health Information, Calgary Health Region for her comprehensive contribution to the review of the safety reporting system.

References

Alberta Evidence Act, Queen's Printer, Province of Alberta <http://www.qp.gov.ab.ca/catalogue/catalog_results.cfm>, accessed 6 January 2006.

CPSO (2003), Disclosure of Harm Policy, The College of Physicians and Surgeons of Ontario, June 2003, http://www.cpso.on.ca/Policies/disclosure.htm. Accessed on 6 January 2006.

Davies, J.M. (1996), Risk Assessment and Risk Management in Anaesthesia,

Bailliere's Clinical Anaesthesiology, 1996, No. 10, 357–372.

Davies, J.M., Hebert, P. and Hoffman, C. (2003), The Canadian Patient Safety Dictionary, Ottawa: The Royal College of Physicians and Surgeons of Canada, 2003, http://rcpsc.medical.org/publications/index.php#other, Accessed on 6 January 2006.

External Patient Safety Review (Robson Report) – June 2004. http://www.calgaryhealthregion.ca/qshi/patientsafety/reports/patient_safety_reports.htm, accessed 6 January 2006.

Flemons, W.W., Eagle, C.J. and Davis, J.C. (2005), Developing a Comprehensive Patient Safety Strategy for An Integrated Canadian Healthcare Region, *Healthcare Quarterly*, 2005, No. 8 Special Issue, 122–127.

Health Quality Council of Alberta (HQCA) (2004). Review of Best Practices for Handling Potassium Chloride Containing Products in Hospitals, and the Preparation of Batch Amounts of Dialysis Solutions for Continuous Renal Replacement Solutions, Health Quality Council of Alberta http://www.calgaryhealthregion.ca/qshi/patientsafety/reports/patient_safety_reports.htm, Accessed 6 January 2006.

ICAO (1994), Aircraft Accident and Incident Investigation, *Annex 13 to the Convention on International Civil Aviation*, eighth edn, Montréal: International Civil Aviation Organization (ICAO).

ICAO (2006), The International Civil Aviation Organization, http://www.icao.org/, Accessed on 6 January 2006.

Johnston, R.V., Boiteau, P., Charlebois, K. and Long, S. (2004), Responding to Tragic Error: Lessons from Foothills Medical Centre, Canadian Medical Association Journal, 2004, No. 170, 1659–1660. [DOI: 10.1503/cmaj.1040713]

Morath, J.M. and Turnbull, J.E. (2005), *To Do No Harm: Ensuring Patient Safety in Health Care Organizations*. (San Francisco, CA: Josey-Bass Inc.).

Runciman, W.B. (1996), Incident Monitoring, *Bailliere's Clinical Anaesthesiology*, 1996, No. 10, 333–356.

Chapter 8

Using Aviation Insurance Data to Enhance General Aviation Safety: Phase One Feasibility Study

Michael Lenné, Paul Salmon, Michael Regan,
Narelle Haworth and Nicola Fotheringham
Monash University Accident Research Centre

The Aviation Safety Foundation Australasia commissioned the Monash University Accident Research Centre to undertake a study to assess the feasibility of using data held by aviation insurers in support of the development of an electronic database for storing accident and incident data. This chapter reports on some of the outcomes of the feasibility study, which included the following tasks: a review of the various sources of aviation safety data in Australia; a review of publicly available general aviation accident analyses; a review of the human error models that underpin the accident investigation process; and an analysis of a subset of claims held by the participating aviation insurers. While some inconsistencies in the insurance data were noted, further stages of the project aim to develop procedures to enhance the consistency of the insurance data such that it will contain sufficient information to permit a meaningful understanding of the causal and contributory factors involved in general aviation accidents and incidents.

Background

In 2003–2004 the Aviation Safety Foundation Australasia (ASFA) had in its business plan a project to investigate the collection, declassification, and analysis of General Aviation (GA) safety-related data that are held by aviation insurance companies with a view to supporting the development of ASFA's Research and Development (AVSAFE) database. ASFA commissioned the Monash University Accident Research Centre (MUARC) to undertake a study assessing the feasibility of achieving this aim. This chapter primarily reports on some of the outcomes from the feasibility study. The initial research was partitioned into a number of components, including: a) a review of publicly available analyses of GA accidents and incidents; b) a review of the models of human error that underpin the accident investigation process; c) a review of the various sources of aviation safety data in Australia; and d) an analysis of a subset of claims held by the participating aviation insurers. Due to the word limits of this chapter, components a) and b) are discussed only briefly in the following subsections, while components c) and d) are discussed in considerably more details in subsequent sections of the chapter.

Existing analyses of GA accident data

An examination of publicly available analyses of GA accidents and incidents was undertaken, with a focus on the data variables that were considered. In addition to providing an understanding of how aviation safety data are reported, it was important to review such reports to examine the recommendations for resolving the identified safety issues, where such recommendations were made. A definition of GA for the AVSAFE project was also derived from this review.

The publicly available analyses of GA accidents and incidents were primarily restricted to fatalities only. The reports reviewed shared some similarities in how data were reported. It is common to see reports that conclude that a substantial proportion of accidents (up to 85 per cent) are attributable to human error, and sometimes specifically attributable to pilot-error or maintenance error. However, the analysis of human error is not a focus of these current analyses of groups of accidents. This is an important issue for consideration in the context of the AVSAFE database.

The reports reviewed used consistent measures of exposure. Epidemiological studies of aviation accidents typically incorporate some measure of exposure (Li, 1994), such as flight hours documented during an investigation or self-reported flying hours. It is important to have a measure of exposure to interpret the relative risks associated with particular activities. In the reports on GA accident data reviewed during this project, the common measure of exposure was the total number of hours flown by GA per 100,000 hours.

The Bureau of Transport of Regional Economics (BTRE) publishes annually the number of hours flown in GA in Australia, defined as all non-scheduled flying activity in Australian-registered aircraft other than that performed by the major domestic and international airlines. All owners of VH-registered aircraft (with the exception of the Australian domestic and international airlines) are surveyed annually and asked to report hours flown by each aircraft in various categories of operation, as well as total landings per aircraft. The survey typically achieves a response rate consistently around 70 per cent, with estimates made for the remaining aircraft for which no response is received (BTRE, 2003). The survey results are merged with details from the civil aircraft register held by the Civil Aviation Safety Authority, which gives access to other information such as aircraft type, engine and fuel type, country and year of manufacture.

In this study it became apparent that there are some inconsistencies in the definitions of GA adopted by the various organizations involved in reporting of GA accidents. For example, the Nall Report includes only fixed-wing aircraft (AOPA Air Safety Foundation, 2004), while the U.S. National Transportation Safety Bureau incorporates additional categories, such as rotorcraft, gliders, balloons and blimps, as well as registered ultra light, experimental, or amateur-built aircraft (NTSB, 2004). It is thus important to establish a definition of GA that will be used for the AVSAFE research project. The definition of GA adopted is that used by the BTRE, namely that the GA sector is made up of all non-scheduled flying activity in Australian-registered aircraft (CASA VH-registered), other than that performed by the major domestic and international airlines. The major categories of flying are private, business, training, aerial agriculture, charter and aerial work (BTRE, 2005a, 2005b). In addition, the sport aviation segment of GA includes operations involving ultra light aircraft, gliders, hang gliders and autogyros.

Contemporary models of accident investigation

There is broad acknowledgement that human error is a contributory factor in a significant proportion of aviation accidents. For example, it is estimated that up to 85 per cent of all aircraft accidents have a major human factors component, and human error is now seen to be now the primary risk to flight safety (Matthews, 2005). Consequently, the importance of incorporating the analysis of human error into the accident investigation processes adopted by aviation insurers was acknowledged. Contemporary models of human error and accident causation in organizational systems take a systems approach to safety and human error (Reason, 1990). Systems-based approaches, such as the Swiss cheese approach presented by Reason (1990) purport that accidents are caused by a combination of human error at the so-called sharp end of system operation and inadequate or latent conditions (*e.g.*, inadequate equipment and training, poor designs, manufacturing defects, maintenance failures, ill-defined procedures, *etc.*) residing throughout the system. These latent conditions affect operator behaviour in a way that leads to errors being made. Systems approaches are particularly suited to accident investigation and analysis procedures for a number of reasons. Primarily, they facilitate the development of appropriate countermeasures that treat not only the errors made by operators, but also the latent conditions that lead to the errors being made in the first place. Without a systems approach, the typical outcome of accident investigation is the attribution of blame to the individual who made the error, and individual-based countermeasures, such as re-training, automation, training, discipline, selection and proceduralization, are developed. Systems approaches remove the blame culture that is typically associated with accident investigation in complex, dynamic systems and permit a comprehensive analysis of the errors and latent conditions involved in a particular accident. A number of systems approach-based accident investigation and analysis techniques have been developed, such as the Human Factors Analysis and Classification System (HFACS) (Wiegmann and Shappell, 2003). Such approaches, in particular HFACS, yield the type of data that is unique to this form of investigation, that is, they provide detailed information about the types of failure across different levels of system operation. Hence the data derived from such approaches provide the basis for an increased understanding of the causal factors involved in crashes. It was concluded that the data derived from such approaches could potentially be used to aid the development of countermeasures designed to reduce the occurrence of error-related GA accidents and incidents.

The capture of general aviation safety data

As stated earlier, a major component of this feasibility study was to identify the various sources of GA accident data to highlight the value of pursuing the aviation insurance data for a safety database. In Australia, the Australian Transport Safety Bureau is the organization responsible for the investigation and reporting of aviation safety occurrences, as defined by the Transportation Safety Investigation Act and associated Regulations. The ATSB collects data for fatal and non-fatal GA accidents and incidents, as do the aviation insurers, thus it is instructive to examine the data that are collected by the ATSB for differing levels of crash and injury severity. Hence this task involved an examination of the conditions under which the ATSB might investigate and the data sources yielded for safety occurrences that are both investigated and not investigated. The review indicated that the ATSB's primary focus is on fare-paying passenger safety and that all fatal accidents (aside from those related to sport aviation) are investigated.

The ATSB and the aviation insurers have different procedures for collecting data to meet their needs. In looking at the flow of information to the ATSB, it is important to note that there are requirements to report various safety occurrences to the ATSB. The conditions under which an individual must report a safety occurrence are outlined on the ATSB's website (ATSB, 2005). As required under the TSI Regulations, the owner, operator or crew of an aircraft must report an accident or serious incident to the ATSB as soon as practicable. Occurrences must be reported to the ATSB in writing in accordance with their status as immediately reportable events (including fatal and serious injuries, serious damage) and routine reportable events (including non-serious injury, minor damage). The report in writing to the ATSB is in the form of the Air Safety Accident and Incident Reporting (ASAIR) form. Thus, while the ATSB may not investigate all accidents and incidents, it still needs to be notified of all aviation occurrences so the information can be used in future safety analysis. The TSI Regulations available through the ATSB website list all reportable occurrences and responsible persons for reporting. For events that are classified as routine reportable (in accordance with the TSI Regulations), the data from the ASAIR form may be the only data the ATSB receives for that safety occurrence. Other events, likely to be immediately reportable, may be investigated, in which case considerably more data would be available.

It is important to note that while there are guidelines to determine which cases are investigated, there is flexibility in the interpretation of those guidelines depending on the circumstances of the case. Desk top investigations may also be conducted and may involve formal interviews and an investigation report. The flowchart presented in Figure 8.1 attempts to capture the potential data sources that might be available for GA accidents and incidents. The purpose of this flowchart is to broadly outline the type of data collected, by whom, and under what general conditions.

Broadly speaking, for operations covered by the BRTE definition of GA, the ATSB is likely to fully investigate fatal accidents. Although not illustrated in the flowchart, it is also possible that the ATSB may conduct a desk top investigation of accidents involving minor or serious injury. For accidents of this type, however, it is likely that the data yielded from the ASAIR form would be the data formally

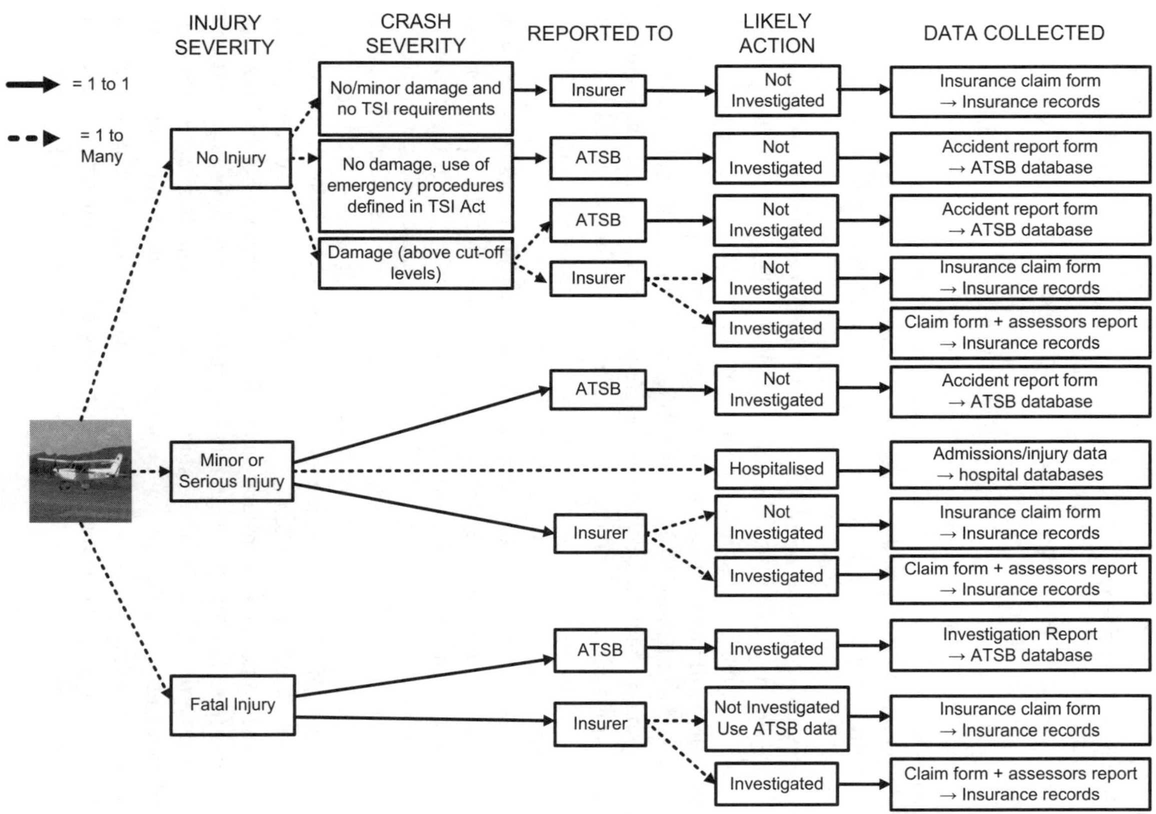

Figure 8.1 Potential data sources for accidents and incidents in general aviation (for operations covered in the BTRE definition of GA)

captured. For incidents involving no injury there are still requirements to report to the ATSB under certain conditions outlined in the TSI Regulations.

The aviation insurers also collect information for a range of claims. Initial discussions with the participating aviation insurers were held to gain an understanding of the conditions under which insurer-appointed loss adjustors would investigate a claim. This process was also informed by the detailed analyses of insurance claim files, discussed later in the chapter. Discussions with the insurers suggested that while claim forms should be completed for all claims, loss adjustors were appointed to investigate a wide variety of claims ranging from those involving no injury but minor damage through to minor, severe, and fatal injury accidents. It is the opinion of the authors that, if these data can be structured and harnessed appropriately, there is great potential to use it to enhance aviation safety.

In many areas of safety it is commonplace to examine the characteristics and contributory factors to better understand why an event occurred and to highlight areas for potential improvement. While there are variations across domains, it is widely reported that the ratio of non-fatal to fatal accidents is such that fatal accidents represent a very small proportion of all accidents. For example, Bird's near miss triangle, as presented by Jones, Kirchsteiger, Bjerke (1999), demonstrates that for each major accident there are a greater number of associated minor accidents (10), property damage incidents (30) and an even greater number of near misses (600). Heinrich, Peterson and Roos (1980); cited in Wierwille *et al.* (2002) also developed a triangle that demonstrates the relationship between near misses and fatal accidents in industrial settings. Heinrich and colleagues estimate that for every fatal accident, there are 10 major injury accidents, 100 moderate injury, 1,000 minor injury, and 10, 000 associated near miss scenarios.

In GA in Australia there are on average less than 10 fatal accidents per year (ATSB, 2004), so much time is needed to achieve sample sizes that permit robust statistical analyses of a wide variety of accidents and incidents. For these reasons safety analysts in many domains have studied the characteristics and causal factors for non-fatal accidents to provide a larger data set. In many areas the characteristics and casual factors for non-fatal accidents are similar to those for fatal accidents, and thus the study of non-fatal accidents can significantly aid the understanding and thus prevention of fatal accidents through countermeasure development and implementation. In other areas non-fatal causal factors are quite different from fatal accidents, suggesting that the two extremes of injury severity have quite different causes (*e.g.*, Haworth, 2003). Here the study of non-fatal and fatal accidents must be done independently. It is the opinion of the authors that the data collected by the aviation insurers potentially represents a valuable dataset of safety-related cases that could be used not only to understand the contributory factors in those accidents and incidents, but may also yield data to assist in the understanding and prevention of fatal accidents.

Examination of the data held by aviation insurers

An analysis of the GA accident and incident data currently held by the participating aviation insurers was conducted. Specifically, the research team was interested in identifying what information was collected, how often each data item was collected, and how the collection of data varied across injury severity. Due to the resource and time constraints associated with the project, only a subset of claims from each insurer involved were analysed. For this purpose, a surrogate database containing the data from the insurers' hard copy claim files was constructed. Three aviation insurers committed to this stage of the AVSAFE research project. The aim was to collect 25–30 cases from each insurer. MUARC requested from each insurer a complete sample of consecutive finalized claims over a defined period. The time period defined to produce the required sample was dependent upon the number of claims managed by each insurer, and thus varied. All claims were finalized from 2002 onwards. Key components of each claim file were the Insurance Claim Form and the Assessor Report. Briefly, the insurance claim form is completed by the assured and submitted to the insurer. The insurer then determines whether an assessor will be appointed to investigate the claim. The presence of other documents within each claim was also noted, including CASA maintenance forms and ATSB ASAIR forms. Claim files often contained considerable correspondence between the insurer, the insured, and the assessor. The contents of this correspondence was not noted as it would have been too resource intensive to analyse.

Initially, a subset of claims from three aviation insurers was examined to determine the completion rates for each item contained within the claims forms. The data from insurance assessors was then analysed. Data from the three insurers were then combined to form a unified database, which contained a total of 73 cases. The majority (88 per cent) of the cases involved no injury, 7 per cent involved minor to severe injury, and 5 per cent involved fatal injury.

The number of cases involving no injury far outweighed the number involving minor/serious and fatal injuries. A key issue therefore is how to best capture the non-injury data to maximize the potential for improving aviation safety. As discussed earlier, in many areas the characteristics and causal factors for non-fatal accidents are similar to those for fatal accidents, and thus the study of non-fatal accidents can significantly aid the understanding and therefore prevention of fatal accidents through countermeasure development and implementation. While 11 per cent of the sample involved injury to at least one person, there are a number of claims that did not involve injury but that have the potential to cause injury. The study of these cases could significantly improve the understanding of GA crashes. Injury is defined as a bodily lesion at the organic level resulting from acute exposure to energy (which can be mechanical, thermal, electrical, chemical or radiant) interacting with the body in amounts or rates that exceed the threshold of physiological tolerance (Krug, 1999). Furthermore, the most important correlates of pilot fatalities in GA are factors related to impact forces (Wiegmann and Taneja, 2003). While many claims involved incidents while taxiing, including wing clips and clipping signs, these incidents occurred at low speed and therefore involved reduced energy exchange and the potential risk for injury, while present, is relatively low. Incidents of this type are classified as non-safety-related for the purposes of the following discussions. This should not be read as suggesting that these incidents are not of great interest to the project, but rather as a means of differentiating between the potential injury severity. As such, safety-related claims in this chapter refer to those claims that involve injury or have the potential to cause injury. These claims form the basis of subsequent discussions.

A distinction was made between safety-related and non-safety-related cases to capture those incidents that could potentially contribute to an increased understanding of GA crashes. An analysis of the proportion of claims that were investigated by the aviation insurers was then conducted. Almost all of the claims defined as safety-related (95 per cent) were investigated by insurance assessors, whereas 61 per cent of non-safety-related claims were investigated. In the opinion of the authors, this finding suggests that if insurance data could be captured in appropriated formats, and almost all safety-related claims are investigated, a powerful dataset would be available for the identification of safety trends and for the setting of research agenda to target the development of potential countermeasures.

Next, to investigate the potential application of contemporary accident investigation and analysis approaches in the analysis of existing insurance data, an analysis of the eight crashes involving injury was conducted using the HFACS accident and incident investigation approach. In summary, it was concluded that there is potential for the application of HFACS-type approaches in the collection and analysis of such data. The analyses indicated that HFACS-type approaches are useful as they allow for the classification of both the unsafe acts that led to the incident and also the contributory factors involved, which in turn reduces instances of apportioning blame solely on the aircrew. However, it was also noted that significant development of current GA incident reporting procedures, and redevelopment of the error and latent conditions classification schemes employed within the HFACS framework, is required. In particular, it was concluded that the level of detail contained in the insurance data were insufficient in most cases, and that a portion of the HFACS approach may be inappropriate when used in a GA context, due to its origin from within the military and civil aviation domains. In conclusion, the authors recommend that a comprehensive analysis of GA accident data using the HFACS approach would yield a valuable data set highlighting the different active error types and the various latent or contributory factors involved in specific types of GA accidents.

Conclusion

Based on the findings of the research conducted, it was concluded that there are a number of shortcomings and inconsistencies in the manner in which aviation insurer-based GA accident and incident data are collected, classified, stored, analysed, and reported. The existing data collected by the aviation insurers does not, in the opinion of the authors, currently contain sufficient information to comprehensively understand the nature and causation of GA accidents and incidents. The existing information collected does, however, provide the basis for the most promising means of collecting quality data for the AVSAFE project. Based on the findings of the research conducted during this feasibility work, a series of recommendations were developed and presented to the sponsor.

ASFA has recently commissioned MUARC to continue the development of the AVSAFE database. The work to be conducted in 2006 will include the following: clarification of the data requirements for insurance claim forms and the development of a standard claim form template; the analysis of a much larger sample of aviation claims to report on safety patterns and to inform the modification of HFACS to non-fatal GA accidents and incidents; the development of a standardized template for insurer-appointed loss adjustors; and the development of the database. It is anticipated that insurance data collected from 2007 onwards will be fed into the AVSAFE database and subsequent analyses conducted to identify safety issues and to provide avenues for further research and strategic countermeasure development to promote the safety of GA operations.

Acknowledgements

The authors would like to thank the project sponsors, the Aviation Safety Foundation Australasia, and the organizations that supported it in the conduct of this project, namely the Australian Transport Safety Bureau, BHP Billiton, and the Civil Aviation Safety Authority.

The authors would also like to acknowledge assistance received from representatives from the following organizations during the Phase One Feasibility Study: the Australian Transport Safety Bureau, including Mike Watson, Dianne Coyne, Joy Sutton and Sylvia Loh; QBE Aviation, including Julian Fraser and Daniel Nash; Vero Aviation and AOA, including Peter Freeman and Grant Williams; and BHP Billiton, in particular Stan Medved. The authors would also like to thank representatives from ASFA, particularly Gary Lawson-Smith and Russell Kelly JP, for their continued support in undertaking this project.

References

AOPA Air Safety Foundation (2004), 2003 Nall Report. Accident Trends and Factors for 2002.

ATSB (Australia Transport Safety Bureau) (2004), General Aviation Fatal Accidents: How Do They Happen?, A Review of General Aviation Fatal Accidents 1991 to 2000, *Aviation Research Paper* B2004/0010, Canberra: Australian Transport Safety Bureau.

ATSB (Australia Transport Safety Bureau) (2005), Aviation Safety, http://www.atsb.gov.au/aviation/index.cfm, Accessed on March 2005.

BTRE (Bureau of Transport and Regional Economics) (2003), General Aviation 2003, Report published by the Bureau of Transport and Regional Economics and. Available at: http://www.btre.gov.au/statistics/aviation/general_aviation.aspx, Accessed on March 2005.

BTRE (Bureau of Transport and Regional Economics) (2005a), BTRE Transport

Statistics: General Aviation, http://www.btre.gov.au/statistics/aviation/general_aviation.aspx, Accessed on March 2005.

BTRE (Bureau of Transport and Regional Economics) (2005b), General Aviation: an Industry Overview (Report 111), Canberra: Department of Transport and Regional Services, Bureau of Transport and Regional Economics.

Haworth, N. (2003), How Valid Are Motorcycle Safety Data?, Paper presented at Road Safety Research, Education and Policing Conference, September 24–26, Sydney.

Jones, S., Kirchsteiger, C. and Bjerke, W. (1999), The Importance of Near Miss Reporting to Further Improve Safety Performance, *Journal of Loss Prevention in the Process Industries*, 12, 59–67. [DOI: 10.1016/S0950-4230%2898%2900038-2]

Krug, E. (1999), Injury: A Leading Cause of the Global Burden of Disease, http://www.who.int/violence_injury_prevention/index.html, Accessed on 16 March 2001.

Li, G. (1994), Pilot-related Factors in Aircraft Crashes: A Review of the Epidemiologic Studies, Aviation Space and Environmental Medicine, 65, 944–952.

Matthews, S. (2005), The Changing Face of Aviation Safety, Proceedings of the Safety In Action Conference, Melbourne, March, 2005.

National Transportation Safety Board (2004), U.S. General Aviation, Calendar Year 2000, Annual Review of Aircraft Accident Data NTSB/ARG-04/01, Washington, DC: National Transportation Safety Board.

Reason, J. (1990), *Human Error*, New York: Cambridge University Press.

Wiegmann, D.A. and Shappell, S.A. (2003), *A Human Error Approach to Aviation Accident Analysis, The Human Factors Analysis and Classification System*, Burlington, VT: Ashgate Publishing Ltd.

Wiegmann, D.A. and Taneja, N. (2003), Analysis of Injuries among Pilots Involved in Fatal General Aviation Airplane Accidents, *Accident; Analysis and Prevention*, 35, 571–577. [PubMed 12729820] [DOI: 10.1016/S0001-4575%2802%2900037-4]

Wierwille, W.W., Hanowski, R.J., Hankey, J.M., Kieliszewski, C.A. and Lee, S.E., Medina, A., Keisler, A.S., and Dingus, T.A. (2002), 'Identification and Evaluation of Driver Errors: Overview and Recommendations. U.S Department of Transportation, Federal Highway Administration', Report No. FHWA-RD-02-003.

Chapter 9

Grappling with Complexity: Analysing an Accident with the Theory of Constraints

Dmitri Zotov, Alan Wright and Lynn Hunt
Massey University

Safety recommendations are the end-point of an accident investigation, but many are not acted upon. Recommendations are more likely to be effective if they address underlying organizational or management deficiencies, rather than individual failings or local conditions. However, the difficulty has been, firstly, to identify such underlying deficiencies, and, secondly, to do something about them. The Theory of Constraints is a change methodology, designed to address these matters.

In this case study, an accident is analysed from the perspective both of the airline, and of the regulatory authority. Information from the investigation is put into the form required by the Theory of Constraints, and it is shown that a few core problems underlie all the undesirable effects observed. These leverage points are addressed, and the resulting few safety recommendations provide the necessary corrective action.

Introduction

In the Boeing 737 accident report of Manchester (AAIB, 1988), there were 31 safety recommendations. That sounds impressive, but they will not have been much use if nothing happens as a result. There were all sorts of useful ideas, like carrying smoke hoods in airliners, and internal sprinklers in future aircraft; but have many of them been implemented?

The end product of any investigation is one or more recommendations for improvement. Unfortunately, as Taylor (1998) showed, the reality is that very few recommendations are implemented. If effective recommendations are not made and implemented, the investigation has failed in its objective. The situation where many investigations are failures is both wasteful, and leaves potential hazards in place to strike again.

Why should this be? In part, it is because many accidents are very complex processes, and people have trouble grasping complexity in a written report (Johnson, Wright and McCarthy, 1995). We need a better way to present the picture. But also, the traditional way of making general recommendations, and leaving their implementation to the recipients (Wood and Sweginnis, 1995), seems a recipe for inaction. It might be more helpful if we had specifics, perhaps written jointly with those who need to act. Further, there is diffusion of the effort available, when there are a large number of recommendations.

Safety recommendations are about change. There are change methods used in business to improve deficient processes, and an accident can be considered to be a process (Hendrick and Benner, 1987) with an undesirable outcome. Also, many airline accidents stem from business problems – funding or resource constraints, or company structure. Could we bend one of these business improvement methods to our use?

Since the 1980s, a business improvement method called the Theory of Constraints (Goldratt, 1990; Dettmer, 1997) has become widely used (Mabin and Balderstone, 1998; Mabin and Balderstone, 2003). It uses graphical presentation to deal with complex situations, and isolates underlying 'core' problems, so that specific remedies can be devised. If the information from an accident investigation could be put in the prescribed format, it should be possible to devise changes which would make similar accidents unlikely, because the processes for devising remedies are well established.

A case-study approach is appropriate for exploratory studies such as this (Yin, 1994). For the subject of the study, an accident was needed where detailed information was available to form a Current Reality Tree – the situation at the time of the accident. From this, it should be possible to generate safety recommendations to form a Future Reality Tree – the situation we would like to bring about. The accident used for the study was the Ansett (New Zealand) Dash 8 controlled flight into terrain, near Palmerston North in New Zealand, because litigation had made the necessary information available.

The Current Reality Tree – the airline

The aircraft was on a VOR DME arc approach over the Tararua Ranges. On final approach to Palmerston North, an undercarriage leg hung up, and while the crew tried to lower it with the emergency procedure, the aircraft struck a hill. Four occupants were killed, and most of the rest were seriously injured.

Where to start? The various undesirable effects found during the investigation can be listed in a chronological array (Figure 9.1). An aircraft hitting a mountain is an undesirable effect, but so is a Ground Proximity Warning System (GPWS) that does not work properly (there were about 4 seconds warning, instead of the 17). Likewise, a crew distracted by an undercarriage that hangs up, and a pilot untrained in emergency procedures, are clearly undesirable effects. Notice that these are termed 'effects': they are not causes. They have come about because of underlying conditions, and we must now try to find those conditions.

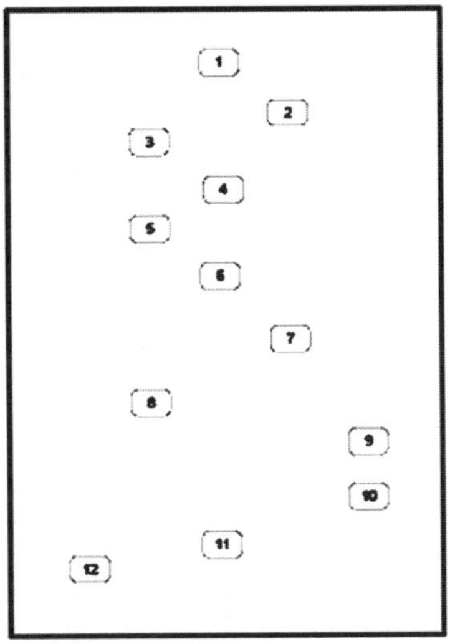

Key:

1. The aircraft struck the ground during an instrument approach
2. There was little warning before impact
3. The crew were distracted by the undercarriage malfunction
4. The undercarriage malfunctioned
5. The crew had not been trained to deal with an undercarriage malfunction
6. The undercarriage on the Dash 8 aircraft malfunctioned from time to time
7. Engineering attempts to fix the undercarriage had not provided a long-term solution
8. No simulator training was available for the crews
9. The company had been unprofitable long-term
10. The company was subsidised by its Australian parent company
11. The undercarriage mechanism was defective in design
12. Audits by the CAA had disclosed no warning of the accident

Figure 9.1 Undesirable effects in the Ansett Dash 8 accident

The Theory of Constraints suggests that most or all of the undesirable effects are linked in some way, so let us try to link some of them. Here is the string from 'design defect' to 'potential to strike the ground'.[1]

Generally, the linkages between the original effects will be 'long arrows,' and we have to find further information, to have necessary and sufficient conditions[2] at each stage. Figure 9.2 reads like this (from the bottom):

- If the design of the undercarriage latch is defective and the engineering instructions for rectification are ineffective, then the latch defect is not rectified
- If not, then undercarriage hang-ups occur, and if the instructions are still ineffective then the undercarriage hangs up repeatedly

1 By convention, the Current Reality Tree entities are written in the present tense, even though, as here, the accident is in the past.

2 The ellipses through which arrows pass denote a logical 'and'. Where there is no ellipse, there is an additive effect.

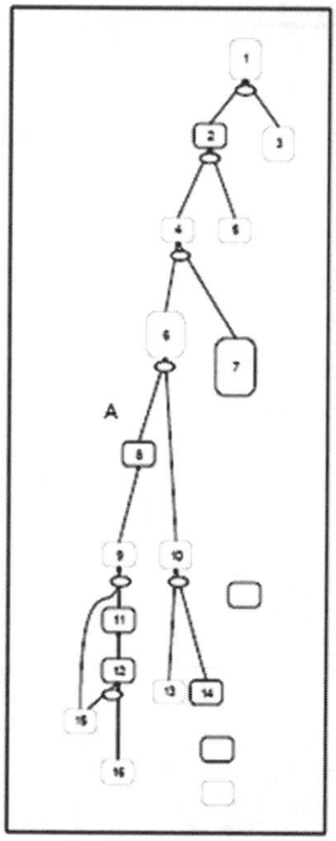

Key:

1. There is the potential for the aircraft to strike the terrain
2. There is the potential for undetected closure with terrain
3. There is little warning of terrain closure
4. There is the potential for crew distraction while attempting to perform undercarriage emergency procedures
5. The descent may intersect terrain short of the aerodrome
6. There is a high risk that the undercarriage emergency procedure will be performed incorrectly
7. Emergency procedures that are performed incorrectly may require the crew to take further corrective action
8. Crews encounter undercarriage malfunctions from time to time
9. The undercarriage latch malfunctions repeatedly
10. The crews are not trained to manage an undercarriage malfunction
11. Undercarriage latch malfunctions occur
12. The undercarriage latch defect is not rectified
13. The crews receive no simulator training
14. The crews receive no in-flight training in undercarriage malfunction
15. The engineering instructions for the rectification of the undercarriage latch are ineffective
16. The design of the undercarriage latch is defective

Figure 9.2 Connections between undercarriage latch design and potential to strike terrain

- If it does, then crews encounter hang-ups from time to time

Turning now to the crew training stream:

- If the crews receive no simulator training and they get no in-flight training in undercarriage malfunction then they are not trained to handle a hang-up.

Combining these two streams:

- If crews encounter hang-ups from time to time and they are untrained, then there is a high risk that the emergency procedure is performed incorrectly and

- If emergency procedures performed incorrectly may require further corrective action then there is the potential for crew distraction while attempting to fix a hang-up
- If the descent may intersect terrain short of the aerodrome then there is the potential for undetected closure with terrain
- If there is little warning of terrain closure then there is the potential for the aircraft to strike the terrain.

At this point, you might say 'but that's obvious ...' But that is just the point: it is obvious, and it cannot be disputed.

In exactly the same way, we can generate linkages for other sets of information.[3]

Unprofitability (Figure 9.3): Ansett was set up in opposition to Air New Zealand, who were well able to protect their market position by cross-subsidy from international operations. Ansett (NZ) never made an operating profit. The parent airline, Ansett

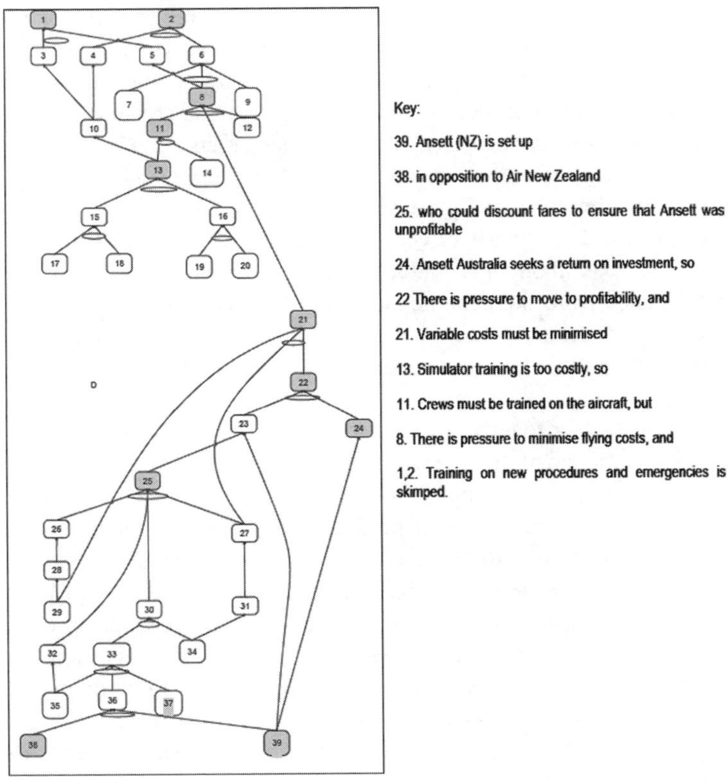

Figure 9.3 Financial stress and emergency training

3 Full details of the construction of the Current Reality Tree are shown in Zotov, Hunt, and Wright (2004).

Australia, wanted a return on investment, so there was pressure to move towards profitability, and since many of the costs of airline operation are fixed (fuel, landing fees, insurance and so on) variable costs had to be minimized. This led to the lack of simulator training, and little in-flight training: the co-pilot had never been shown the undercarriage emergency procedure, and there was no training on the newly-introduced Palmerston North arc approach (Figure 9.4), despite complaints from the crews.

It also led to deficient maintenance (Figure 9.5). The engineers decided that an unreliable undercarriage had no safety implications, so they would 'fix' the problem. The job card just said 'check [the up-lock latch] for wear' (the wear in question being a few thousandths of an inch, over perhaps a quarter of an inch), and the shop floor workers interpreted this as 'rub a thumbnail over it'. The increasing frequency of failures was not noticed (Figure 9.6).

Questions such as whether a defect had safety implications, and the increasing frequency of defects, were matters for the Safety Department (ICAO, 1984). So what was the Safety Department doing?

The Safety Department had previously been a one-man band, but much better than nothing at all. However, it was abolished 18 months before the accident, presumably to save money. Instead, it was announced that 'Safety is everybody's business' (Ansett (NZ) 1993). So the functions expected of a Safety Department were not performed, and the engineers were not alerted to the ineffective maintenance instruction, nor were the training staff advised of a need to introduce recurrent training in a recurring emergency, nor were the crews alerted to the risk of trouble-shooting undercarriage failures on final approach (see Figure 9.7).

In addition to all of these effects, the GPWS did not produce the expected warning. There was about 4 seconds warning when there should have been 17 seconds. Evidence from the designer was that this was due to severe corrosion of the radar altimeter aerial (Morgan, 2001), and it appears that this may have resulted from the drain hole in the radome becoming blocked when the radome was painted (Figure 9.8).

These networks link together, to form the Current Reality Tree (Figure 9.9). The dashed lines near the top of the tree represent potentiation. At this point, we no longer have necessary and sufficient conditions for the next step. Instead, we have a hazardous situation, wherein the occurrence of random and ordinarily benign conditions (for example, low cloud) can tip the situation from potential accident to actuality.

The Current Reality Tree has a definite origin, in competitive pressures, and it should be no surprise that this shows a fundamental conflict between profitability and safety. The Theory of Constraints argues that, if one can address such fundamental problems, then – since all the undesirable effects stem from them – the entire problem is resolved. Unfortunately, the conflict in this case stemmed from political dogma, that unfettered competition is a Good Thing. Political dogma is likely to be intractable. The need for profitability led to:

- Engineering economies
- Faulty rectification
- No Safety Manager

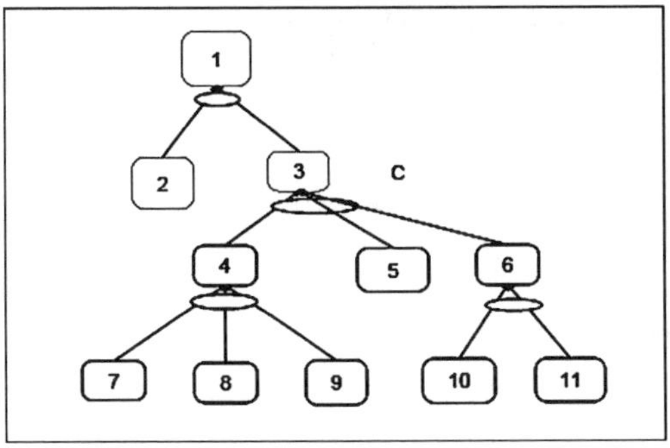

Key:

11. Minimum altitude for holding pattern appears to apply to start of approach

8. Crews get no training on new approach

3. There is the potential to join the final approach too high

1. The descent path may intersect terrain short of the aerodrome

Figure 9.4 Lack of continuation training – approach procedure

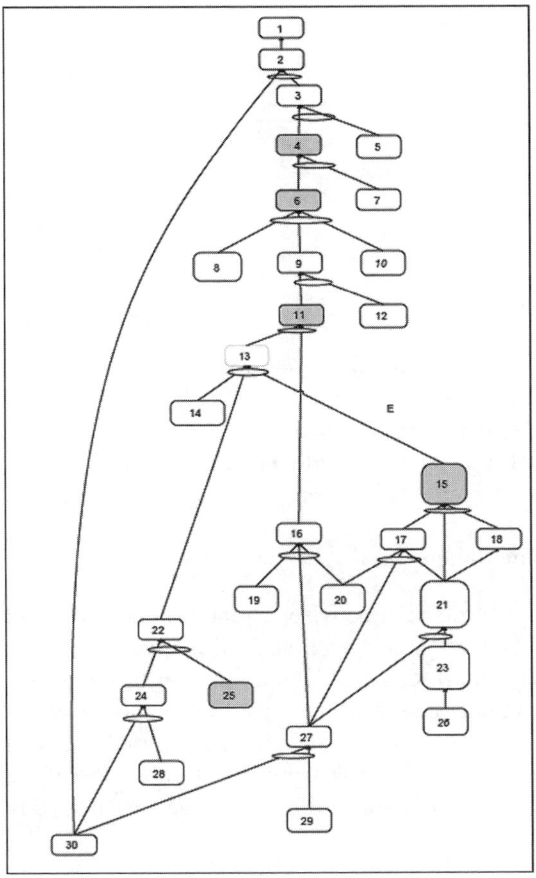

Key:

25. There is pressure to cut maintenance costs, and

15. Engineers decide that U/C hang-ups have no safety implications, so

13. Replacement parts are not bought.

11. Instead, the Engineers 'fix' the problem:

6. The job instruction is to 'check for wear'.

4. The shop-floor workers 'run a thumbnail over it'.

1. Hang-ups recur

Figure 9.5 Maintenance aspects

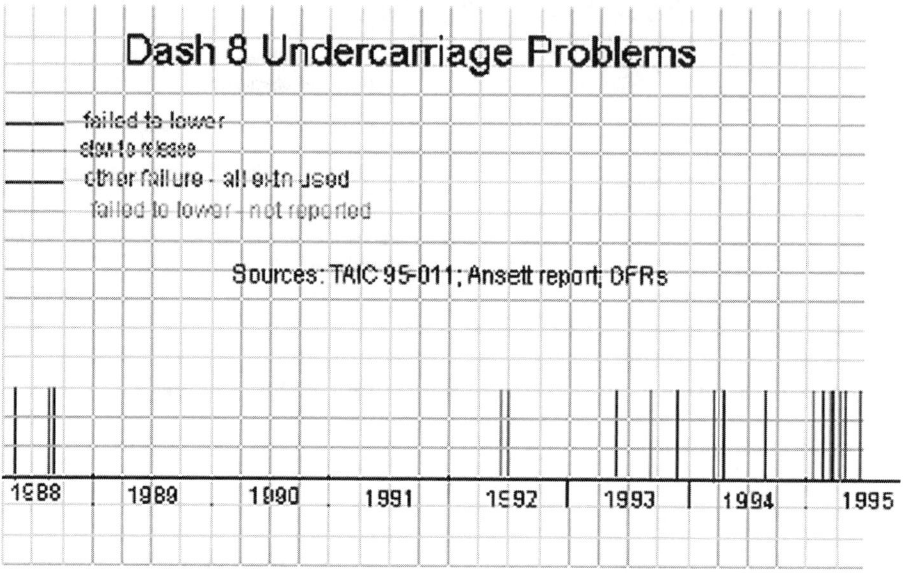

Figure 9.6 Undercarriage failures

- Training economies
- No recurrent training
- No emergency procedures training.

And, unknown to anyone,

- The GPWS was unserviceable.

The core problems – those from which multiple adverse effects are:

- The initial faulty design of the undercarriage latch
- The absence of any individual responsible for safety oversight
- The engineers' decision that undercarriage hang-ups had no safety implications and
- The recurring undercarriage malfunctions.

Faulty design happens, from time to time. Likewise, erroneous decisions will always be with us. The essential thing is to deal with such difficulties in such a way that the potential for harm is minimized, and this is a function of the Safety Department. The recurring malfunctions stemmed not only from a faulty work-sheet – again, something that happens from time to time – but also from the lack of safety oversight which should have detected what was happening. It looks, therefore, as though the absence of a Safety Manager is something which we can profitably address.

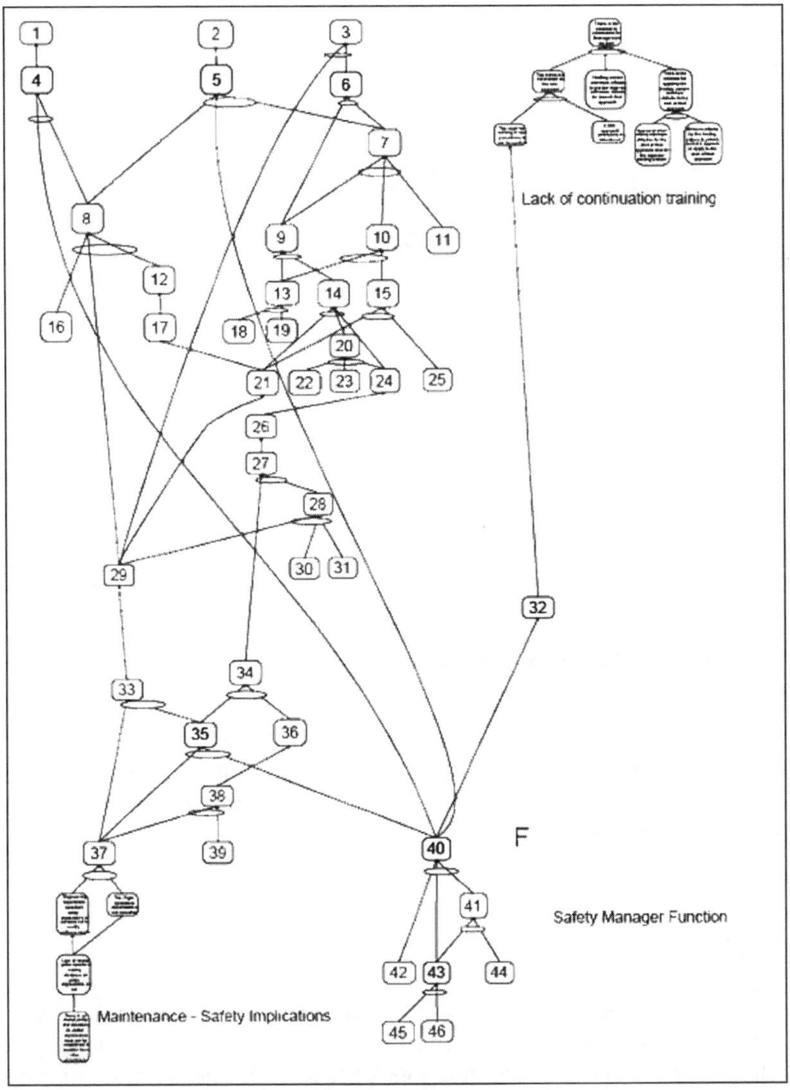

Figure 7. Absence of Safety Manager

Key

43. The Safety Manager position has been abolished, so
40. No individual is responsible for safety oversight and risk management, so
35. No review of Engineers' decision
4. Engineers are not alerted to the faulty instruction
5. Training staff are not advised to institute recurrent hang-up training
6. Crews are not alerted to the risk of distraction
32. No review of training for new procedures

Figure 9.7 Absence of safety manager

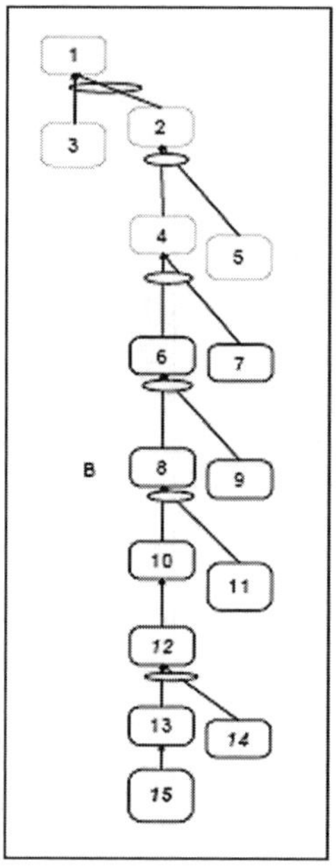

Key

13. The radar altimeter radome is painted over, and

12. Paint clogs the drain-hole, so

4. The radar altimeter aerial becomes severely corroded, and

1. There is little warning of terrain closure

Figure 9.8 Shortness of GPWS warning

It does not look much like a Swiss cheese, does it? The failed GPWS was clearly a failed defence, but otherwise defence in depth was lacking. Rather than pathogens, the core problems (highlighted in red) might be compared with tumours, spreading their tentacles throughout the body to set off new tumours, and needing to be excised if the body is to survive.

The Current Reality Tree – the CAA

If all this was going on in the airline, what was the Civil Aviation Authority (CAA) doing? As a result of 'reforms', it was strapped for cash and staff, so it had decided to abandon old-fashioned surveillance in favour of 'auditing,' meaning seeing that the airline had proper systems in place (NZCAA, 1995). The net result was that the CAA had no idea what was really happening. Some of the undesirable effects are shown in Figure 9.10.

The undesirable effect 'Audits are ineffective in averting accidents' is the end-point of a sequence of undesirable effects relating to auditing. Those relating to this particular accident include:

- Auditors are unaware of the absence of a Safety Manager (although the CAA had been advised)
- Auditors are unaware of deficient crew training
- Auditors are unaware of recurring malfunctions (although the Airworthiness Department was 'watching' the situation).

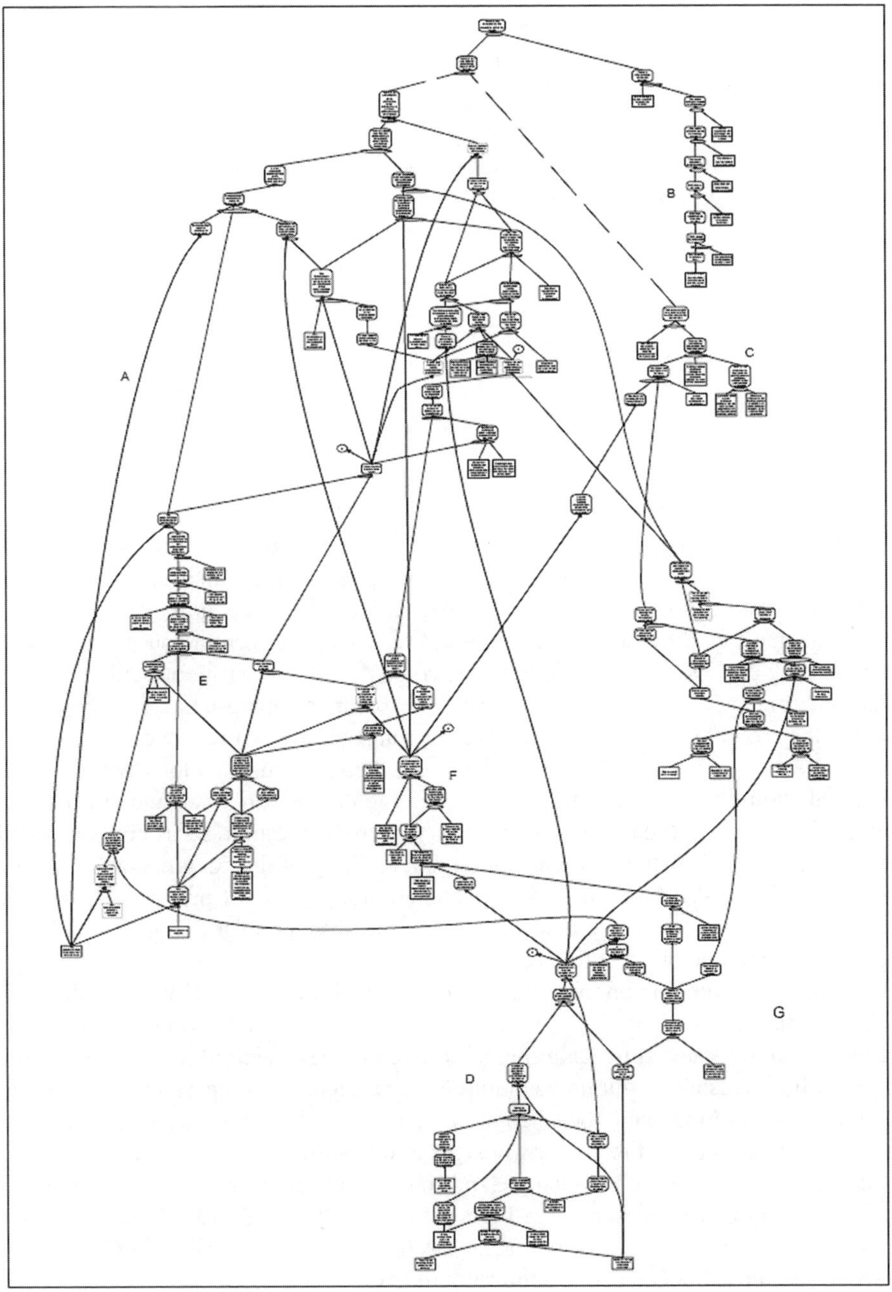

Figure 9.9 Ansett Dash 8 current reality tree

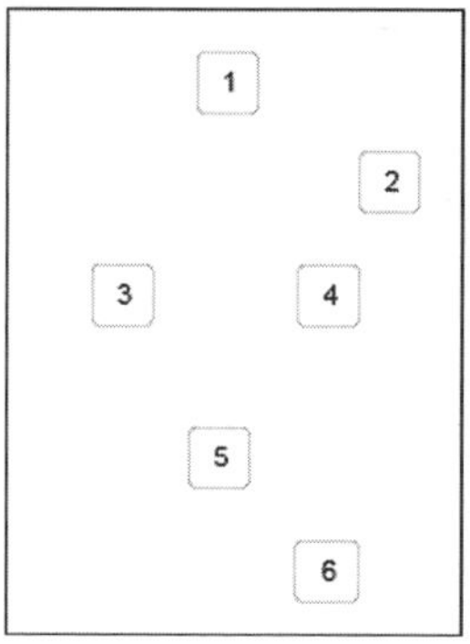

Key

1. Audits do not avert accidents
2. Auditors are unaware of the absence of a Safety Manager
3. Auditors are unaware of recurring malfunctions
4. Auditors are unaware of deficient crew training
5. Regulatory authority does not perform surveillance of airline operations
6. Some airlines operate unsafely

Figure 9.10 Undesirable effects at the CAA

And, arguably,

- CAA does not perform surveillance of airline operations.

From these undesirable effects, a second Current Reality Tree, showing the situation at the CAA, can be generated. The auditors will be unaware of abnormal events if they do not review crew reports, as well as procedural documentation. Also, the auditors did not examine physical reality by performing surveillance and competency checks, which could have observed defects recurring or being rectified. The auditors were therefore unaware of the recurring hang-ups. Had they been aware, they might have enquired into the risk management of these events, *e.g.*, what training the aircrew had received to deal with them. Enquiry as to why the events recurred should then have led the auditors to the deficient maintenance instruction which did not detail the required inspection. Emergency training was not performed in accordance with airline documentation, but auditing of the documentation, alone, could not detect the deficient training. The auditors were not prompted to review the training in undercarriage emergencies, so the absence of training to handle the recurring emergency was not detected.

Risk management is one function of the Safety Department. However, although the CAA had been advised that Ansett no longer had a Safety Manager, the CAA auditors did not know this. Audit preparation could have alerted the auditors to the change, but pressure to minimize audit cost, and therefore preparation time, led to deficient preparation, and the change went unnoticed. Surveillance should have brought the absence of the Safety Manager to the auditors' notice. In the event, the auditors were unaware of the change, and so of the absence of risk management.

These factors are addressed in Figure 9.11, which illustrates why audits alone could not keep the CAA informed as to the true state of the airline. Had the CAA been aware, possible corrective actions included:

- Insisting that emergency training be performed
- Requiring correction of the defective maintenance instruction and
- Requiring the Safety Manager position to be reinstated.

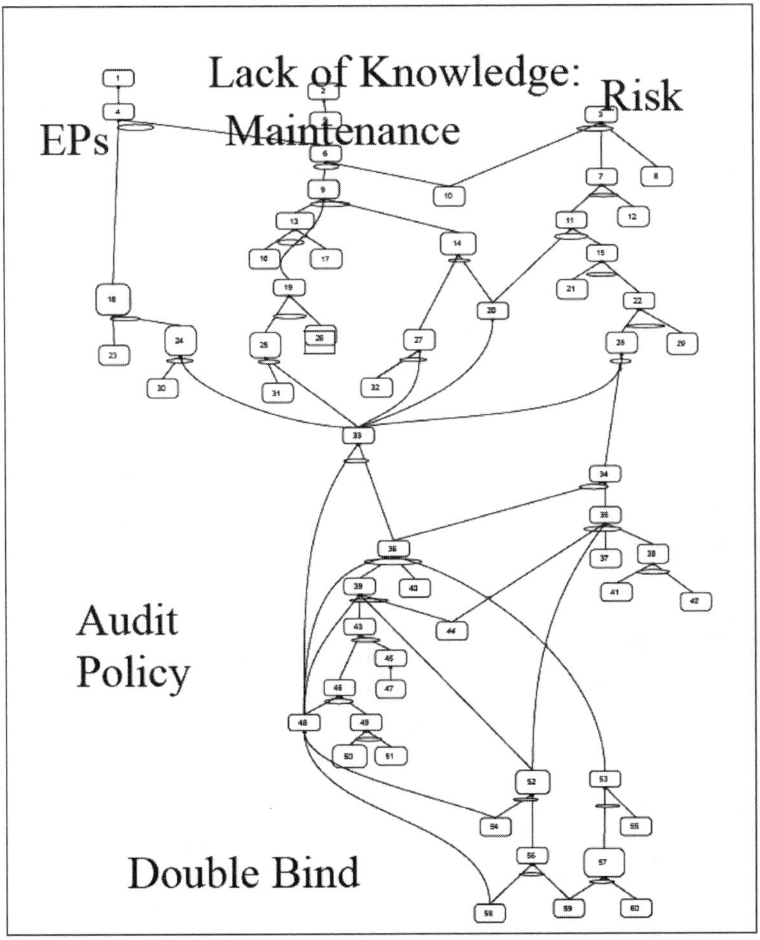

Key:
59. The CAA's function is to assure a safe airline system, so
53. effective oversight is necessary, but
52. this has to be done within a limited budget, and
48. the CAA has insufficient qualified staff to perform required surveillance, so
36. the CAA decides to conduct audits only.
20. The auditors do not examine physical reality, so

19. they are unaware of a general problem with the undercarriage, and
18. do not detect the lack of emergency procedure training, and
5. do not ask why undercarriage maintenance is unsuccessful.
11. Auditors have no indication that company systems have changed, and
7. believe there is still a Safety Manager, and
1, 3, 4. Absence of flight crew training, faulty maintenance instructions and absence of risk management are undetected.

Figure 9.11 Absence of mindfulness at the CAA

Collectively, these actions would have made the accident unlikely.

An airline under financial stress will try to minimize costs, possibly to the detriment of safe performance. Ansett's actions in not buying replacement undercarriage parts, not training its crews in emergency procedures, and abolishing the Safety Manager position, can be seen as a response to financial pressures. So, it would be reasonable to increase the level of safety oversight of airlines under financial pressure. However, the CAA had been deprived of direct knowledge of airlines' financial positions, as financial viability was considered to be a matter for market forces. Since non-viable airlines may cut corners, increased oversight is desirable, but absent financial information, greater depth of oversight is unlikely, so unsafe conditions arising from that stress are less likely to be detected (Figure 9.12).

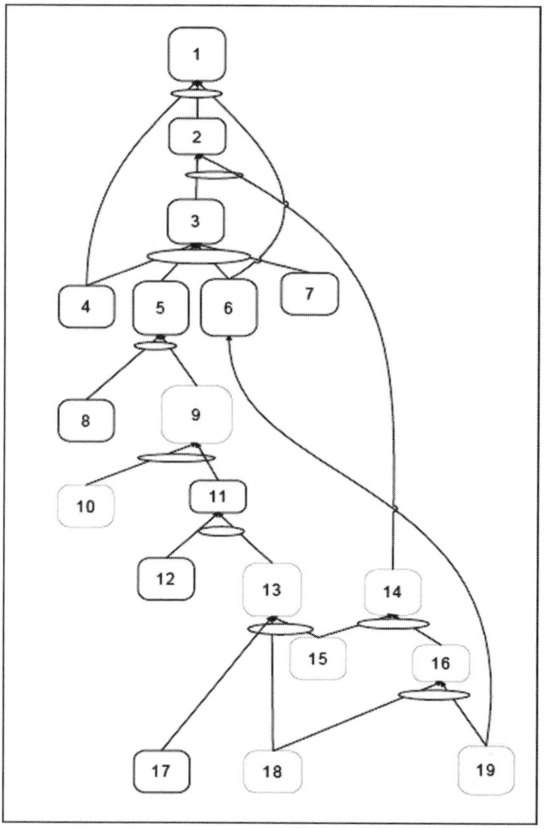

Key:

18. CAA does not require airlines to demonstrate viability, so
14. CAA is unaware of airlines' financial state, and
12. Non-viable airlines are allowed to remain in business, but
10. Safe operation is expensive, so
3. Greater depth of auditing would be advisable, but
2. this does not happen, so
1. Unsafe conditions arising from non-viability are unlikely to be detected.

Figure 9.12 Airline financial status effects

When Figures 9.11 and 9.12 are combined, we get the Current Reality Tree from the perspective of the CAA (Figure 9.13).

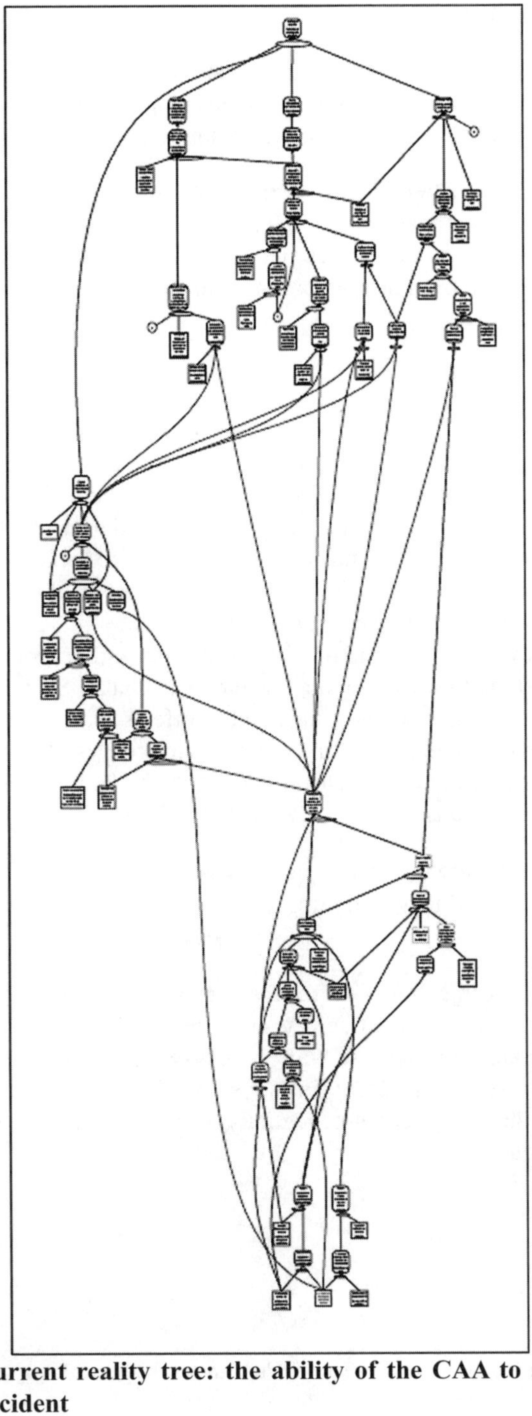

Figure 9.13 Current reality tree: the ability of the CAA to avert the Ansett accident

The core problems here are:

- Insufficient qualified staff for required surveillance
- Greater depth of oversight of non-viable airlines does not occur and
- Audits are defined as reviewing the documentation of safety systems.

Insufficient staff is a matter for appropriate funding. Discovering the viability of airlines could be as simple as reading the airlines' returns to the Companies Office. However, the dominant problem was the definition of the oversight role as being confined to review of the documentation. This was entirely a matter within the province of the CAA.

The Future Reality Tree – Ansett

The next part – producing the Future Reality Tree – should have been easy. Either the core problems could be addressed individually, or even better, we could discover an underlying feature common to the core problems, and address that. With hindsight, it is obvious why these approaches did not work as intended; in relation to the problems at Ansett the Theory of Constraints methodology is designed to help businesses to solve their problems, and in this case produced ideas like 'find a niche market where Air New Zealand won't compete,' or 'fold your tents and go gracefully'. These are not recommendations that Government bodies can make. So a third method was used, replacing the (non-financial) undesirable effects with the opposite desirable effects, together with the necessary injections[4] to make these true.

One set, dealing with the Safety Department, is shown in Figure 9.14.

The essential injections are:

- Mandate the Safety Manager appointment
- Mandate Safety Manager training and
- Audit the Safety Department performance.

Arising from these injections, we find that recurring defects are reviewed, to locate the causes; safety implications of defects are considered; and safety implications of proposed savings are reviewed. As with the Current Reality Tree, we can form all the sets into a Future Reality Tree[5] (Figure 9.15). The square-cornered 'injections' form the basis of the safety recommendations. The Future Reality Tree is formed by combining the sectors:

- Crew training
- Safety management
- Maintenance

4 This terminology is used in Theory of Constraints literature. In psychological literature the 'injections' would be termed 'interventions'.

5 Details of the construction of the Future Reality Trees are available in Zotov, Wright and Hunt (2005), at www.aavpa.org.

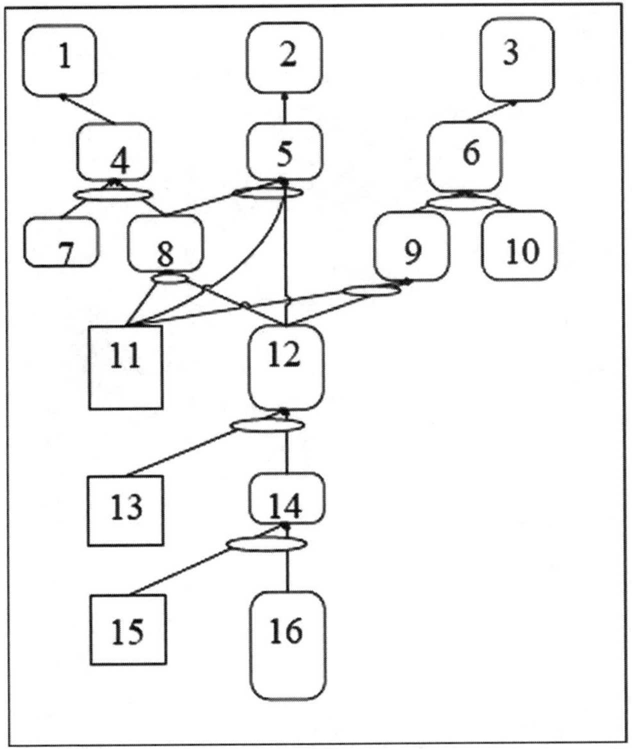

Key:

1. Information on recurring defects is reported to the Chief Engineer
2. Crews receive training in anticipated emergency procedures
3. Implications of proposed savings are reported to department heads and to CEO
4. Recurring defects are reviewed and causes found
5. Safety implications of defects are considered
6. Proposals for savings which have safety implications are reviewed
7. Defects recur
8. Defects are reported to the Safety Manager
9. Proposals for savings are reported to the Safety Manager
10. Savings are proposed which may have safety implications
11. The CAA audits the performance of Safety Management functions
12. The Safety Manager is able to perform the Safety Management role
13. The CAA mandates training for the Safety Manager
14. A Safety Manager is appointed
15. The CAA mandates the appointment of a Safety Manager
16. A nominated individual is needed to be responsible for safety oversight and risk management

Figure 9.14 Future reality tree: safety manager functions

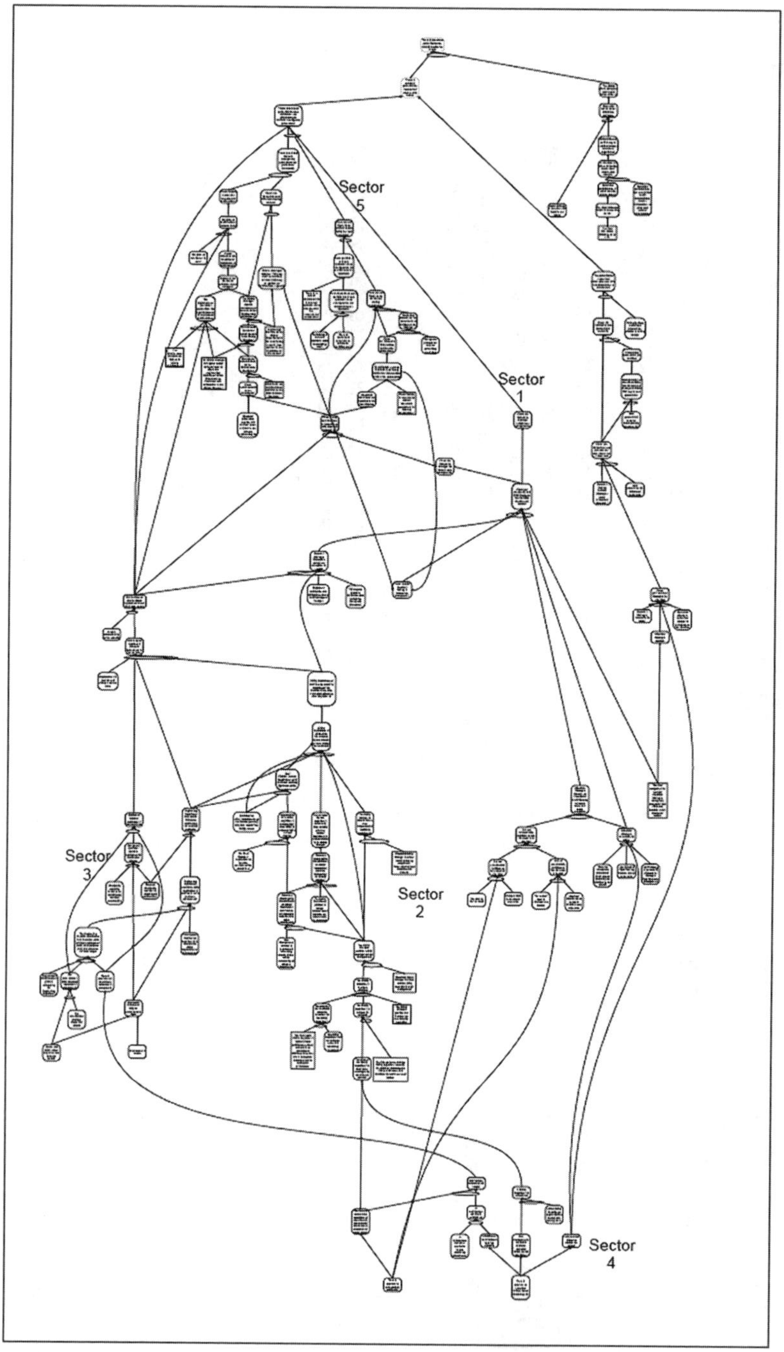

Figure 9.15 The Ansett future reality tree

- Pressures on the airline
- Crew distraction and
- GPWS maintenance.

When those areas have been addressed, there is negligible potential for a CFIT accident.

When constructing the CAA Future Reality Tree, there are no constraints on the type of recommendation, and because of the shape of the Current Reality Tree (Figure 9.13) we can address a fundamental problem. At the very base of the tree there is a conflict diagram: the CAA would like to deploy all necessary resources to do the job properly, but has to work within a limited budget, and cannot do both. If we can break this conflict, then since all the undesirable effects stem from it, the whole tree will collapse, and it can be re-built as we would like it to be.

Every arrow in a Current Reality Tree represents assumptions. Here are some of the assumptions in the CAA conflict diagram (Figure 9.16).

Consider the lower line of assumptions:

- The statement that safety oversight of an airline comprises surveillance, audit, and review of safety management, is non-controversial.
- Likewise, it is certain that the public expects the Government to assure the safety of airlines, as shown by the demands for public inquiries after major accidents.

Turning to the upper line:

- The resource intensive nature of surveillance, and the lengthy time required for audit preparation, are established facts.
- If the first set of assumptions cannot be broken, then the CAA is indeed in a bind: it is required to assure airline safety, but cannot have the resources to do so effectively. These assumptions must therefore be examined in detail:
- The Government policy is 'user pays': 'given'.
- Safety oversight costs must be charged directly to airlines: This assumption stems from the management buzz-speak of referring to those the CAA interacts with – airlines, pilots and engineers – as 'clients'. Clients are those for whom 'services' are performed, whether they should wish it or no. Clients are then charged for services. This is certainly one mode of charging, but may not be the only one.
- Given the composition of the CAA Board ('CAA Board Appointments' 1994), comprising senior members of the aviation industry, airlines could protest effectively at the level of charges; the result was that audits are required to be 'cost-effective'.

There may be an alternative approach to funding. The question is, 'who is the user'? Who benefits when the CAA acts to prevent airlines taking short-cuts to enhance their finances? Who benefits if the CAA makes the airlines spend more money on simulators and training, instead of on attractive lounges? Surely not the airlines.

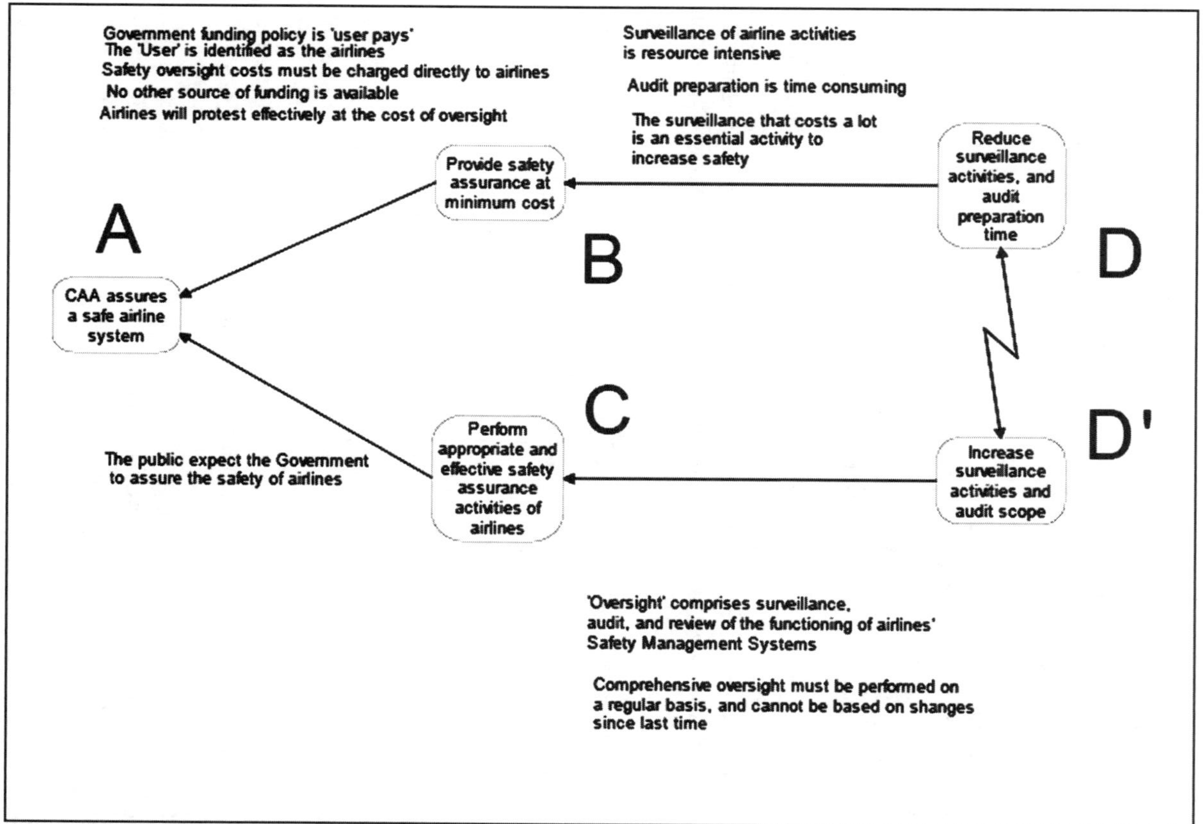

Figure 9.16 Conflict resolution diagram from CAA CRT

The answer, of course, is the travelling public, who are more likely to step off the aircraft in one piece as a result of the CAA's actions. So, if the public is the 'user,' could there be a direct levy on the travelling public, to fund CAA activities, and would it cause a public outcry? Yes, and no, in that order. Until 1990, the (NZ) Air Services Licensing Authority was funded by a ticket levy, and no one even noticed. Indeed, the American FAA is substantially funded by this means. Because numbers are large, the required levy would be small: a $2 levy would leave the CAA awash with cash.[6] All other funding could be foregone.

If funded in this way, there would be no constraint on safety oversight activities: that which is needed could be done. The conflict between the need to deploy resources, and the inability to do so, would disappear.

Such a change at the bottom of the Current Reality Tree causes major effects. Removing the constraint on safety oversight removes all of the undesirable effects, one after another, until the whole tree is like the Cheshire cat, with nothing left but the grin. There is nothing left to transform, so the Future Reality Tree must be generated in another way. We can do this by establishing Desirable Effects – the things that are needed to achieve mindfulness about the state of an airline – and work out how to bring them about. Consider the design of safety oversight (Figure 9.17).

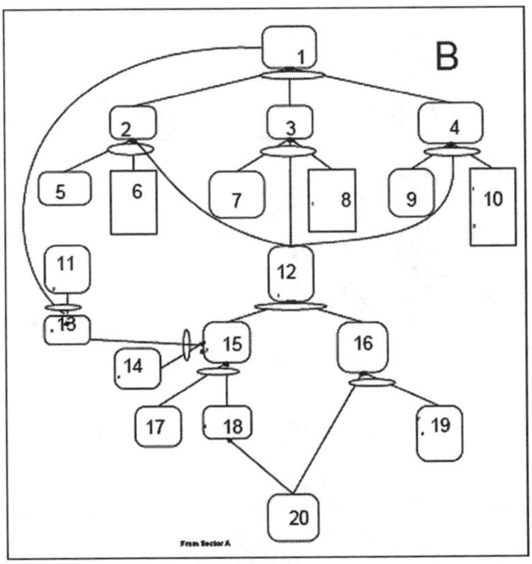

Key:
18, Qualified staff can be recruited
12. CAA has sufficient resources for all aspects of safety oversight
5. ICAO Manual requires surveillance
7. A review advised need for auditing
9. Flight Safety Foundation advised oversight of Safety Management Systems
2, 3, 4. CAA performs appropriate surveillance, auditing and safety oversight.

Figure 9.17 From sufficient resources to design of safety oversight regime

The various sectors generated in this way are amalgamated as before, to form the new Future Reality Tree for the CAA (Figure 9.18).

The object, however, is not to generate Future Reality Trees, but safety recommendations, and these come from the injections which make the Future Reality trees valid.

6 Figures for New Zealand, at the time of the accident, are not readily available, but they should be in proportion for those from Australia. Figures for Australia, 2004: 90 million passenger-sectors; @$2 = $180 million.

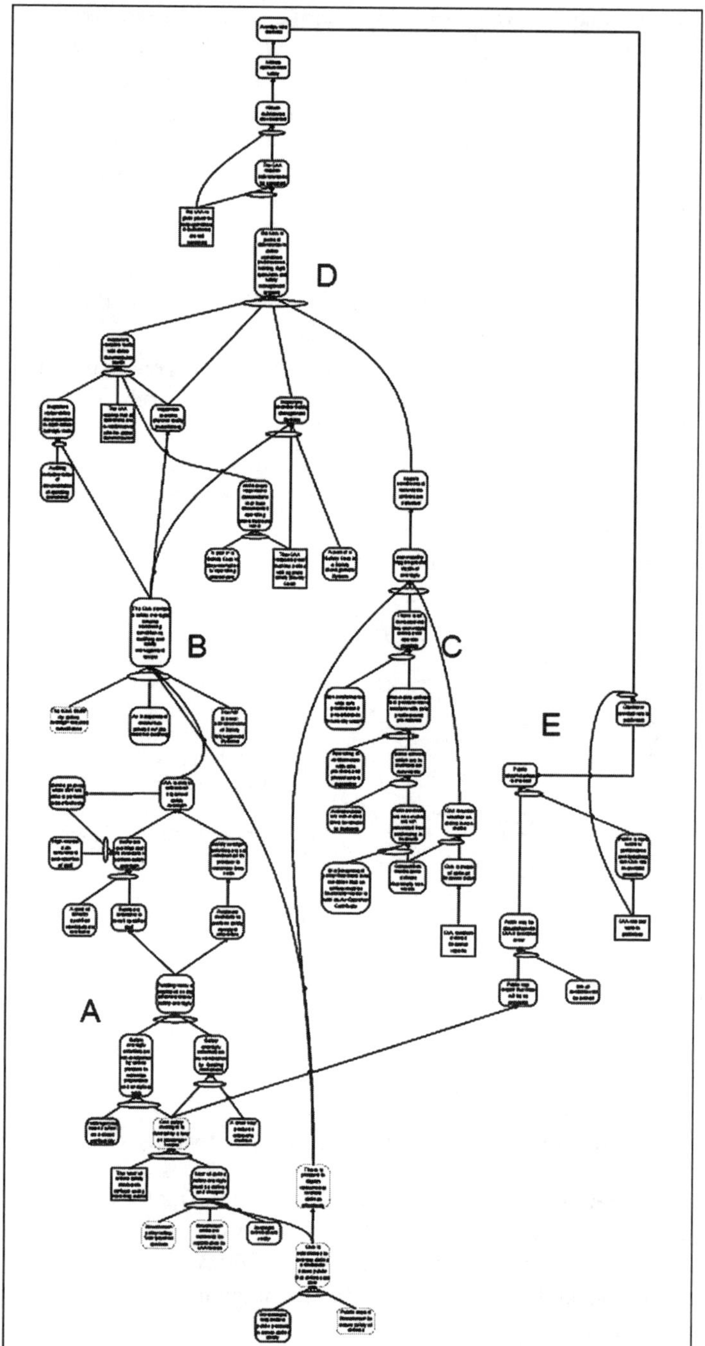

Figure 9.18 Future reality tree for the CAA

Arising from the injections in the Ansett Future Reality Tree, the safety recommendations are:

- The CAA perceives that the Safety Department increases the safety of air travel.
- The Safety Manager must be an Approved Person.
- The CAA audits the Safety Department performance.
- The CAA perceives the need for simulator training of airline pilots.
- The Maintenance Controller must be trained in safety management.
- There is a formal requirement for a missed approach when emergencies occur on approach.
- Radar altimeter outputs are to be tested *in situ*.

None of these should be controversial (Safety Management Systems are about to be mandated in Australia, in CASR 119, and are shortly to be the subject of an ICAO Manual Doc 9859) and all have been demonstrated to reduce the potential for the recurrence of accidents similar to that involving the Ansett Dash 8.

The safety recommendations arising from the CAA Future Reality Tree are:

- The 'User' of CAA services is defined as the travelling public.
- Airlines must produce a Safety Case.
- Airline operations must be acceptably documented, and conducted in accordance with documentation.
- The CAA seeks power to stop an airline operating outside its safety case.
- The CAA monitors airlines' financial returns.
- The CAA's role and work is publicized.

Notice how many recommendations are subsumed into a Safety Case regime – something already required for most hazardous industries in Australia, except aviation. This is an area that needs to be addressed, in Australia and elsewhere, but the need might not have been brought out by the conventional process of writing recommendations to address undesirable effects. The recommendations are all addressed to the CAA, not Ansett. Safety costs money, and lots of it, and there was no realistic possibility that Ansett could generate sufficient funds, so recommendations to them would have been futile. Forcing an airline to reinstate its Safety Department, train the crews, and do proper maintenance, might well accelerate the airline's demise. However, the airline might not kill anyone on the way.

Conclusion

This case study has illustrated the complexity of an apparently simple accident, and shown how that complexity can be displayed in comprehensible form. The information could be put into the format required for analysis using the Theory of Constraints methodology. The few injections required to transform undesirable Current Reality into desirable Future Reality form the basis of safety recommendations which should result in far-reaching reform. The Theory of Constraints methodology should be a useful tool in major accident investigations.

References

AAIB (1988), 'Report on the accident to Boeing 737-236 G-BGJL at Manchester Airport on 22 August 1985 (Aircraft accident report 8/88)', London: HMSO.

Ansett, (NZ), (1993), Ansett New Zealand Flight Operations Policy Manual (vol. Section 6 – Flight Safety Programmes). Christchurch: Ansett (NZ).

CAA Board Appointments (1994), *New Zealand Wings*, 12.

Dettmer, H.W. (1997), *Goldratt's Theory of Constraints: A Systems Approach to Continuous Improvement*, Milwaukee: Quality Publishing.

Goldratt, E.M., (1990), *What is this Thing Called the Theory of Constraints and How Should it be Implemented?*, Croton-on-Hudson: The North River Press.

Hendrick, K. and Benner, L. (1987), *Investigating Accidents with Step*, New York: Marcel Dekker.

ICAO (1984), *Accident Prevention Manual*, first edn, Montréal: International Civil Aviation Organization.

Johnson, C.W., Wright, P.C. and McCarthy, J.C. (1995), Using a Formal Language to Support Natural Language in Accident Reports, *Ergonomics*, 38, 1264–1282.

Mabin, V. and Balderstone, S. (1998), *A Bibliographic Review of the Theory of Constraints*, Victoria: Wellington University.

Mabin, V.J. and Balderstone, S.J. (2003), The Performance of the Theory of Constraints: a Review of Published Reports of Theory of Constraints Applications, Paper presented at the Theory of Constraints Conference 2003: Profitting from strategic constraint management, Kuala Lumpur.

Morgan, J. (2001), Corroded Aerial Possible Cause of Warning Fault, *Dominion*, 7.

NZCAA (1995), Lower Hutt: New Zealand Civil Aviation Authority.

Taylor, A.F. (1998), Airworthiness Requirements: Accidents Investigation and Safety Recommendations, Paper presented at the 29th international Seminar of the International Society of Air Safety Investigators, Barcelona.

Wood, R.H. and Sweginnis, R.W. (1995), *Aircraft Accident Investigation*, Casper: Endeavour.

Yin, R.K. (1994), *Case Study Research and Design Methods*, Thousand Oaks: Sage.

Zotov, D., Hunt, L. and Wright, A. (2004), Analysing Systemic Failure with the Theory of Constraints, *Human Factors and Aerospace Safety*, 4, No. 4, 321–354.

Zotov, D., Hunt, L. and Wright, A. (2005), Grappling with Complexity: Finding Core Problems from Safety Information, Canberra, Australia: Proceedings of the Australian Aviation Psychology Seminar, 25–26 October 2005. Available from www.aavpa.org.

Chapter 10

Drought in Safety Management – More Tools, Less Theory

Steve Tizzard
Civil Aviation Safety Authority (CASA), Australia

I leap at any opportunity to talk on flight safety as the price of ignorance – in terms of horrendous human suffering and financial costs. How often have you heard the expression – 'it was harder in my day'. Older professionals and trade persons alike often use this expression. I had always considered the expression stemmed from functions previously done by hand, but now by machine. For example, the butchers graduated to power saws; the bakers graduated to mechanical dough mixers; and the candlestick makers, well I have never met one to find out.

Normalization of deviancy

A sociologist by the name of Dr Diane Vaughn wrote a book in the late eighties and popularized and may have even coined the term normalization of deviancy – the tendency of an organization over time to let its course drift, normalizing that deviant new course each time by accepting the deviation until it becomes norm. When we look at where we go from here in the safety system, one of the biggest risks we face is the tendency to normalize deviancy.

Incremental changes, incremental acceptance of less and less safety in each decision, is not associated to a rationale for standing up and saying: 'Wait a minute, we've got a major problem, and this is clearly going to lead to an accident.' Nevertheless, if we let those increments occur, we end up with normalization of deviancy and eventually we end up with disaster.

In my view this means that an original standard, procedure or experience level is reduced one or more times, often for economic reasons, without due regard for the weakening of flight safety. Here are some examples:

Solo general flying hours and solo navigation hours for private pilot licences (PPL) have reduced (see Table 10.1), and the minimum hours for a PPL were static at 40 hours from 1947 until 1965.

Australia reverted to a 40-hour PPL syllabus in about 1992. But what no one has ever done to this day is to reduce the syllabus content. One must ask if the minimum time to train for the task is reduced. What are the resultant safety implications stemming from this decision?

Table 10.1 Solo general and navigation hours, 1947–present

	1947–1958	1958–1965	1965–1992	1992–present
Solo GF	22	20	15	5
Solo Nav	3	3	7 + 5ICUS	5

Source: CASA

Table 10.2 Minimum *ab initio* instructional hours to be a Chief Flying Instructor

	1946–1955	1955–1965	1965–1980	1980–present
Required hours	250	1,000	500	200

Source: CASA

Ab initio instructional hours to be a Chief Flying Instructor have also seen a similar decline (see Table 10.2).

Let me reinforce this problem of declining experience with a personal anecdote. About 35 years ago I was approved as a part time Chief Flying Instructor (CFI) and an Approved Testing Officer (ATO) – 'the department,' as the Civil Aviation Safety Authority (CASA) predecessor was affectionately known then, were reluctant to appoint me because I was too young and inexperienced. I was 28 years old, had 4,500 flying hours under my belt. I simply did not fit the mould of what was expected. However, my case was fought hard by a sympathetic examiner of airmen on the grounds of over 1,000 hours combat flying during the Vietnam conflict; graduate of the world-renowned Central Flying School in England; and, two tours of duty in Papua New Guinea with check and training experience. I was then the Standardisation Officer of the Royal Australian Air Force Basic Flying Training School. Today a lad or lass could actually demand approval for the same appointment with less than 1,000 hours flying experience under 20 years of age and have no commercial experience.

As an ATO in that era, I could tell a candidate if I had given a fail assessment. After the test and the assessment of the applicant's progress records, license and logbook went to the department. There, an examiner of airmen made the final decision on whether the applicant passes or fails. The department also conducted an every fifth flight test to check on training standards.

Today approximately 900 ATOs conduct most tests with few (if any) checks on the ATO's performance by the regulator. If normalization of deviancy was not interrelated to other social issues, such as watering down flight safety, the problems would be more easily correctable. For example, flight tests are often virtually demanded on minimum allowable hours rather than standards. Here are some of the rules Bill Gates is said to have given during a recent high school speech:

- Life is not fair – get used to it;

- The world will not care about your self-esteem;
- If you think your teacher is tough wait until you get a boss;
- Your school may have done away with winners and losers – life has not;
- Be nice to nerds – chances are you will end up working for one;
- If you can read this thank a teacher; and
- If you are reading it in English thank a soldier.

A friend sent me an email recently that I was going to delete. But it contained too many flight safety messages to ignore it. Here is what it said, in part:

> The Obituary for Mr Common Sense
>
> He was with us for many years but his birth record was lost in bureaucratic red tape. He will be remembered for such things as knowing when to come in out of the rain and that life is not always fair. His health rapidly deteriorated when well meaning but overbearing regulations were put in place including reports of a six-year-old boy charged with sexual harassment for kissing a class mate and a teacher fired for reprimanding an unruly student. He finally lost the will to live when *The Ten Commandments* became contraband; churches became businesses; and criminals received better treatment than their victims. *Common Sense* was preceded by his parents, truth and trust. His wife discretion, his daughter responsibility and his son reason.

The above is indicative of the world we live in. So we must become even smarter with flight safety and not let the 'new age' complexities of life beat us at our own game. When I fail someone during a flight test it is because they did not meet the standard. I am not trained to know if they had a deprived upbringing. Trainee pilots we see today are understandably part of the new generation. Their schooling probably contained much information on their rights, and what the world owed them, none of which gets to the basics of aviation discipline. Avoid overusing buzz word generators when convincing staff, superiors, regulators and even politicians that you really are concerned about flight safety. Experience has taught me that.

Negatives are all I have given you so far, about flight safety at the lower end of the market. Flying training is not rocket science. Australia still follows the teachings of Robert Smith Barry (1886–1948), following his major works which are now some 85 years old. Yet the message is not getting through. Here are some examples of recent comments on flight tests:

- 'I hate conducting CPL flight tests – you have to take over on most landings because the poor kids have only gone on navigation exercises for the last few weeks and forgotten how to land.'
- 'The kid didn't use the rudder trim once throughout the (CPL) test. Afterwards I pointed to the control and asked what is this. Answer: part of the ventilation control system – sir.'
- 'In an overseas CPL validation flight test, an applicant "forgot" pre take-off checks and enters the runway without adequate lookout. He obviously failed but the ATO, out of curiosity, allowed the applicant to climb to 3000 feet overhead. The ATO simulates engine failure. The applicant's only checks

were "landing light on, undercarriage down" and ignores the 10,000 feet and 8,000 feet runway below, finally attempting to land on a lawn bowling green in the centre of the town.'

The above three examples of unacceptable performance in a flight test tend to be written off as anecdotal evidence but during a recent three-year overseas stint, I flew with many young Australian flying instructors, all of which needed extensive ground and flight retraining before they were permitted to operate.

The CASA CEO, Mr Bruce Byron, has a strong training background and is determined to see flying training, the foundation and cornerstone of all flying, is on track. He has formed the Flying Training Industry Development Panel chaired by aviation legend Mr John Willis OAM and some experienced industry personnel with CASA support staff. So far, Version 1 of a Revised Flight Instructor manual has been produced. The previous edition was some 30 years old.

Additionally, Flight Operations Inspectors with strong training backgrounds have been appointed as Flying Training Roles Specialists and will soon be highly visible in flight testing roles. Already some one-third of all instructor ATOs (about 60 in total) have attended an intense live in training course on how instructor should conduct flight tests.

There may have been a 'drought in safety management' in flying training but the rains are coming. It will be particularly heavy on organizations and individuals that do not shape up, readily acknowledging that many if not most do it right. If a code of ethics was ever needed in the flying training industry, the time to have one is now. A suggested outline would be a good start:

- Code of Ethics;
- Personal (Resource, Candour, Devotion, Curiosity, and Independence);
- Corporate;
- Comradeship (Patriotism and Loyalty);
- Airmanship (Coolness, Control, Endurance, and Decision).

Costs of aviation accidents

Is there a great cost in flight safety – certainly not – the real cost is in having accidents – an Australian Transport Safety Bureau Report[1] reveals that in a recent 10-year period there were 196 fatal general aviation accidents that resulted in 379 fatalities. Depending on which source is used, each death in a VH registered aircraft costs our nation between one and two million dollars. Talking on the mid-range figure these 196 fatalities have cost the nation just under one billion dollars with an incalculable cost in human suffering and sorrow.

1 *Preliminary Analysis of Fatal General Aviation Accidents in Australia.*

Conclusion

In conclusion – flight safety management built on past experiences should not be forgotten.

Reference

Author (2004). General aviation fatal accidents: how do they happen? ATSB Aviation Research Paper B2004-0010. *Australian Transport Safety Bureau.* Canberra: ATSB.

Chapter 11

Aerial Agriculture Accidents 2000–2005: The Human Factors and System Safety Lessons

Geoff Dell

Protocol Safety Management Pty Ltd and the Safety Institute of Australia

The accident rate in Aerial Agriculture has been unacceptably high in Australia for several years. Insurance records show there have been at least sixty-one (61) serious accidents since 2002 that resulted in direct costs of over $A12 million. Thirty-three (33) of those occurred in the 18 months to June 2005 and cost over $A9 million. This chapter analysed the common factors leading to these accidents based on information reported to the insurers by the airframe owners at the time of the events. It also addressed common factors relating to human performance and safety management practices based on interviews with insurers case management personnel. Some of these factors related to what has often been referred to as 'basic airmanship,' but in fact revealed underlying failures in safety management, training and operational surveillance. This analysis also provided warning signals for other arms of industry where the operational imperatives may be perceived to dictate actions contrary to accepted risk control practices. Some recommendations for effective intervention are also offered.

Introduction

Insurance records show that at least sixty-one (61) serious incidents that resulted in claims have occurred in the three years from mid-2002 to mid-2005 with a cost over $A12million. Thirty-three (33) of those have occurred in the past 18 months costing over $A9million. This contrasts with the three-year period 1994–1996 when insurance records show 30 serious accidents occurred in aerial agriculture operations. There seems little doubt the industry will come under considerable pressure if the increased accident rates are not arrested. Insurance premiums will climb sharply as the underwriters try to recover costs and the insurance companies may become more and more reluctant to cover the risks and underwrite the operations.

While most would agree that there is a need for rapid intervention, a better understanding of the failures that have led to the accidents and clear perception of any underlying safety management problems faced by the industry will be necessary if effective intervention is likely to be achieved.

This enquiry was conducted to attempt to make use of the existing insurance claims data and knowledge within the insurers to inform the intervention debate and any subsequent corrective action.

Methods

The data supporting this chapter were drawn from insurance claim files of QBE Aviation, the largest underwriter of aerial agriculture operations in Australia. A retrospective analysis was carried out of reports and correspondence compiled by the insurance investigators and assessors at the time the events occurred. In addition, case managers familiar with the events were interviewed to establish as complete a description of each occurrence as possible.

Not withstanding, the insurance data were almost devoid of causation information beyond the active failures which occurred immediately before the events. There was little or no data on the organizational circumstances or other latent systems failures which may have been precursors to the events.

Accordingly in this study, analysis of the insurers data, for the most part, was limited to examination of the active failures, that is the mistakes or omissions, of the pilots involved. To analyse these human factors aspects of the data, the relevant Human Factors Analysis and Classification System (HFACS) categories were used (see Wiegmann and Shappell, 2003).

Modern safety investigation doctrine based on the works of Reason (see for example Reason, 1990 and Reason, 1997), mandates analysis of the contribution to causation of latent failures and accident pre-conditions in organizational and other management arrangements prevailing at the time of the events.

However, since the insurers claims and other data reviewed in this study contained little or no such information, an alternative method of investigating possible latent organizational and management systems failures was employed using an opinion survey by means of interviews (see Scheuren, 2004) of a sample of pilots with aerial agriculture experience.

A structured interview method was applied addressing the following issues:

- Check and training;
- Surveillance of operations;
- Support organization;
- Operational arrangements;
- Aircraft and systems.

Findings

QBE Aviation's records included a total of sixty-one (61) aerial agriculture aircraft accidents that had occurred in the period May 2002 to October 2005. However, reliable causation information was available for only 44 of those events. Records on the other 17 accidents were limited to a basic event description and detailed claims cost information for each case.

Analysis of the forty-four (44) events for which causation information was available, found that forty-one (41) or 93 per cent could be attributed, at least in part, to human performance failures by the respective pilots and only three (3) could be attributed to equipment failures.

The three equipment failures were a tyre puncture on take-off, a control column pistol grip detaching on take-off and a broken throttle cable on approach. These failures were most likely outside the pilots' direct control and were excluded from further analysis since there was insufficient data available to determine whether causation related to human performance shortcomings by other personnel or not.

The 41 events which can be attributed, in part, to active failures of the respective pilots can be grouped into eight categories; wire strikes, failed to get or stay airborne on take-off, landing accidents, controlled flight into terrain, taxying accidents, loss of control or stall in procedure turns and fuel exhaustion (see Table 11.1).

Table 11.1 Pilot human factors accidents

	Number	Per cent
Wire strikes	13	33
Failed to get/stay airborne	9	22
Landing accidents	7	17
CFIT	4	10
Taxying accidents	3	8
Loss of control/stall in procedure turn	2	5
Fuel exhaustion	2	5
Failure to re-configure aircraft correctly pre-flight	1	2

Thirteen of the human factors accidents involved wire strikes; the most common (33 per cent) type of occurrence; nine involved aircraft which failed to get or stay airborne on take-off (21 per cent); seven were excursions from the airstrip or ground loops on landing (18 per cent); four involved controlled flight into terrain (10 per cent). There were two loss of control or stall in procedure turn, two in-flight fuel exhaustion accidents and two taxying accidents, and two failure to re-configure the aircraft correctly pre-flight (3 per cent each) incidents.

Analysis of the 13 wire strike accidents revealed that in 10 of the events the pilot knew of the presence of the wires and had actively avoided the wires on previous runs. However, on the accident runs the pilots failed to avoid the wires. HFACS analysis of these 10 failures revealed that in every case (all 10) the pilots probably became spatially disorientated due to perceptual errors. In two of the cases the spatial disorientation was coupled with inappropriate manoeuvres (see Table 11.2) related to decisions to fly under wires where there was insufficient clearance for the aircraft to safely transit under the wires, decision errors under the HFACS criteria.

Table 11.2 Wire strikes

Group	Error type		Precondition	
Pilot knew of wires (10)	Perceptual error	Spatial disorientation (10)	Adverse mental state	Loss of situational awareness (10)
	Decision error	Inappropriate manoeuvre (2)		
Pilot did not know of wires	Violation	Failed to adhere to brief (1)	Adverse mental state	Misplaced motivation
				Haste

In at least one of the other three wire strike events, the pilot was not aware of the presence of the wires. In this case, analysis of the incident revealed that the pilot elected to fly at very low level on a route which was outside the area of the pre-operation survey and briefing. The route taken was shorter than that which was briefed. Under the HFACS classifications this was considered a violation, since the pilot failed to comply with the briefing which was probably driven by a misplaced motivation and haste to return to base for the next load, both adverse mental state error preconditions under the HFACS framework.

Analysis of the nine take-off accidents where the aircraft either failed to get airborne or stay airborne, procedural non-compliance appeared to be the most common failure on behalf of the pilots involved (see Table 11.3). In five cases the pilots failed to dump the load when the aircraft failed to perform as expected on or immediately after take-off and in a further case the pilot failed to reject the take-off when the aircraft failed to accelerate in the take-off roll. In one other case, the pilot made a poor decision by electing to take-off downwind and the aircraft failed to get airborne before it struck the fence at the end of the field. Most likely amongst the preconditions for all these errors were the predominant factors of a misplaced motivation to get the job done with a minimum of delay, heavy workload and fatigue.

In one case the pilot misjudged the width of the airstrip which was too narrow for the spray boom to fit on both sides resulting in the aircraft ground looping during the take-off roll. Again, misplaced motivation to quickly get the job done appeared to have been a precondition for this error of judgement.

In the three of the take-off accident cases, the aircraft were overloaded for the prevailing conditions, clearly violations under the HFACS framework. Misplaced motivation to quickly get the job done again appeared to have been a precondition for these violations.

Table 11.3 Failed to get/stay airborne

Error	Error type		Precondition	
Failed to dump (5)	Decision error	Procedural non-compliance – wrong response to situation	Adverse mental states	Misplaced motivation
				Task saturation
				Fatigue
				Channelized attention
Failed to reject take-off (1)	Decision error	Procedural non-compliance – inappropriate manoeuvre	Adverse mental states	Haste
Down wind take-off (1)	Decision error	Poor decision	Adverse mental states	Misplaced motivation
				Haste
Misjudged width of airstrip (1)	Perceptual error	Misjudged	Adverse mental states	Distraction
				Haste
Overloaded on take-off (3)	Violations	Overloaded	Adverse mental states	Misplaced motivation
				Haste

Table 11.4 shows the analysis of the landing accidents. In two cases the pilots carried out inappropriate manoeuvres by deciding to land in standing water in one instance and to land downwind in the other. Preconditions for both these errors were again misplaced motivation to get the job done as quickly as possible.

Another landing accident resulted from the pilot incorrectly responding to a situation where the engine power apparently fluctuated during a run and he closed the throttle and carried out a precautionary landing in which the aircraft tipped over. One of the preconditions for this event seems to have been a loss of situational awareness on behalf of the pilot who committed to the course of action seemingly without consideration of alternatives.

There were also two landing over-run accidents which resulted from pilots misjudging conditions and landing with too much speed to stop before the end of the strip. In both cases the pilots had low time in aerial agriculture.

Table 11.4 Landing accidents

Error	Error type		Precondition	
Landing in standing water (1)	Decision error	Procedural non-compliance – inappropriate manoeuvre	Adverse mental state	Misplaced motivation Haste
Downwind landing (1)	Decision error	Inappropriate manoeuvre	Adverse mental state	Misplaced motivation Haste
Precautionary landing after power fluctuation (1)	Decision error	Wrong response to situation	Adverse mental state	Loss of situational awareness
Failed to correctly reconfigure aircraft between operations resulting in product covering windshield on approach (1)	Skill error	Omitted step in procedure	Adverse mental state	Distraction Haste
Loss of control after touchdown, ground loop and hit fence (1)	Skill error	Poor technique	Poor supervision	Low pilot ag experience
Overshot on landing and struck fence or object (2)	Perceptual error	Misjudged speed and distance	Poor supervision	Low pilot ag experience

Three of the four controlled flight into terrain (CFIT) incidents related to pilots electing to fly at low level on the lee side of hills or trees while transiting to or from the area of operations and being unable to remain airborne in the downward moving air on the lee of the obstacle (see Table 11.5). Preconditions for these events seem to have been a loss of situational awareness possibly due to workload and desire to get the job done quickly.

Another CFIT accident occurred when a pilot with limited aerial agriculture experience flew the aircraft into the ground during night spraying operations. The pilot seemingly misjudged the aircraft's altitude during the operation. It is apparent the pilot was stressed and distracted by personal issues and had low aerial agriculture experience.

Table 11.5 Controlled flight into terrain

Error	Error type		Precondition	
Flew in lee of hills/trees too low to recover from down draught (3)	Decision error	Inappropriate manoeuvre	Adverse mental states	Loss of situational awareness
				Task saturation
				Haste
Flew into ground during night spraying (1)	Perceptual error	Misjudged altitude	Adverse mental states	Stress
			Poor supervision	Distraction
				Low pilot ag experience

The causation of the two loss of control/stall in procedure turn events was similar (see Table 11.6). Both probably resulted from a loss of situational awareness on behalf of the pilots during a critical phase of the flight, a procedure turn conducted at the end of a run to line up for the next. One event resulted from an inadvertent increased bank angle which led to a stall while the pilot's attention was focused on lining up for the next run and more aggressive handling of the aircraft. The other resulted from a breakdown in the pilot's visual scan and distraction with checking the product level during the turn which led to the failure to provide terrain clearance.

As Table 11.7 shows, the two fuel exhaustion accidents appeared to have similar causation. The pilots failed to give appropriate attention to fuel management. Both were low time pilots in aerial agriculture. However, there was insufficient information available to gain an understanding of why these pilots were unable to satisfactorily manage their fuel and take action to return to the airstrip before the fuel state became critical.

Table 11.6 Loss of control/stall in procedure turn

Error	Error type		Precondition	
Stall and spin in procedure turn (1)	Skill error	Failed to prioritize attention	Adverse mental states	Channelized attention
		Break down in visual scan		
	Violation	Flew over aggressively – reduced turn radius	Adverse mental states	Channelized attention
Loss of control in procedure turn, pilot checking chemical level (1)	Skill error	Failed to prioritize attention	Adverse mental states	Loss of situational awareness
		Breakdown in visual scan		Distraction

Table 11.7 Fuel exhaustion

Errors	Error types		Preconditions	
A/C crashed due to fuel exhaustion (2)	Skill error	Failed to prioritize attention (on fuel)	Poor supervision	Pilot low aerial agriculture experience

The three taxying accidents were different in outcome, but there was some overlap in error causation (see Table 11.8). Two of the events resulted from the pilots becoming distracted: one by a car adjacent to the strip which led to the pilot rapidly applying the brakes that compressed the nose gear oleo and caused the propeller to strike the ground; the other by task saturation and loss of situational awareness associated with taxiing the aircraft on a steep airstrip resulting in the aircraft taxying into a fertilizer truck. Authorization of the operation from the steep airstrip by the company was a contributory factor which was a precondition to this accident. The third taxying accident was another propeller strike, this time during engine run-up, in which the nose gear oleo compressed and the aircraft lurched under brakes. While pilot technique was a possible contributory factor, there was insufficient information to positively establish the pre-conditions leading to this event.

Table 11.8 Taxying accidents

Error	Error type		Precondition	
During taxi pilot applied brakes aggressively and the prop struck the ground (1)	Skill error	Poor technique – brake use	Adverse mental state	Distraction by car adjacent to strip
Aircraft taxied into fertilizer stack on steep airstrip (1)	Perceptual error	Spatial disorientation	Adverse mental state	Loss of situational awareness
				Task saturation
			Supervisory violation	Authorized ops from steep strip
Propeller struck ground during engine run-up under brakes (1)	Skill error	Poor technique	Not known	

There was one event which resulted from an error during the pre-flight inspection (see Table 11.9). The pilot failed to re-fit the fuel cap after refueling the aircraft and during the subsequent flight, fuel and vapour from the open cap was ignited by the hot engine exhaust. There was insufficient information to determine the preconditions which led to this pre-flight error.

Table 11.9 Failure to re-configure aircraft correctly pre-flight

Error	Error type		Precondition
Pilot failed to replace fuel cap – subsequent in-flight fire	Skill based error	Procedural non-compliance – failed to replace cap	Not known

Findings from pilot interviews

Check and training

It was apparent that many aerial agriculture operators had virtually no pilot standards surveillance or checking arrangements in place. Also there was very little recurrent pilot training taking place, in many of these operations.

It seems that many of those operators engaged pilots on a needs basis in a similar fashion to any small business hiring a tradesman. So long as the pilots had the appropriate license endorsements, they were assigned to flying work with the operator trusting that the prior training and licensing processes had delivered a pilot with all the necessary knowledge and skills to achieve a satisfactory outcome from both the safety and operational perspectives.

One obstacle to effective check and training was perceived to be the small size of many aerial agriculture companies. Sometimes, the pilot was a one-person-company who tried to me*et all* operational task requirements with a minimum of support.

In contrast, there was consensus that a few operators were trying to establish and maintain effective check and training activities. However, these companies found this challenging because of the seasonal and competitive nature of much of the business which made effective prior planning and preparations problematic. The combination of high customer demand and pilot availability meant that there was little scope available during the season to spare, and often well intentioned pre-season plans fell into disrepair as the operators reacted to changing industry and commercial demands.

Surveillance of operations

There was consensus that surveillance of actual operations by the Regulator seemed to be inconsistent at best, with some pilots being unable to recall any time the Regulator had shown interest in their activities after they had achieved their aerial agriculture endorsement.

Some pilots had experienced their operations being scrutinized by their management from time to time. However, the number of operators who had formal arrangements in place for regular surveillance of application operations in the field was thought to be very small.

It was apparent that most pilots had experienced some surveillance by their companies or by client companies when they were newly employed or contracted by the organization. However, the frequency of surveillance dropped off rapidly as the season developed and was almost non-existent once a pilot was considered to be experienced.

Organizational support

The pilots felt there were wide variations in the level of support provided by the organizations who engaged their services. Some of the larger operators were thought to make every attempt to support the pilots and engender a safety first culture.

However, all the pilots indicated that there were many operators who seemed to accept no accountability for flight safety and believed it was solely the responsibility of the pilot. These operators were perceived to be continually pushing for productivity gains without due consideration of the impact on safety of operations.

Operational arrangements

All pilots could recall, at some time during the past 5 years, carrying out operations where they were not entirely happy with the circumstances. The standard of the airstrips were sometimes at the extremes of acceptability for surface condition, gradient, length or width.

Due to the seasonal nature of the operations which sometimes had narrow windows of opportunity, time pressure was a constant in the minds of most. This sometimes resulted in all activities occurring with a sense of haste which most of the pilots felt required them to constantly and consciously guard against the urge to shortcut the routines like procedures and checklists.

All the pilots reported being sometimes concerned about the accuracy of the information provided in the pre-operations briefings. Sometimes conditions at the airfield or the site to be sprayed or dusted were different from expectations and all indicated they habitually inspected the sites from the ground where-ever possible, although there were limits to the ability to achieve this successfully.

There was also a common belief that the farmers or crop owners took little or no responsibility for safety. While most co-operated with the operators and provided requested information, the accuracy of the information was sometimes a problem.

Aircraft and systems

All the pilots in this study felt technological developments were for the most part helping the pilots achieve their aims, especially in the area of product delivery. However, the new technology did have the tendency to increase pilot workload and their use had to be carefully integrated into operations to ensure pilots were not distracted at critical times.

There was a feeling amongst all the pilots that the newer and heavier aircraft were more robust and more error tolerant, which gave the pilots more confidence in the level of operational safety. Certainly, there was a feeling that the newer aircraft were less vulnerable to wire strikes because of their sturdier construction and had more crash-worthy design features which increased pilot survivability over earlier designs if an accident occurred.

Discussion

That aerial agriculture is a high risk activity is beyond question. The literature has many references to high accident rates in the industry in Australia and overseas (see for example ATSB (2005) and CDC (2004). Indeed, the high accident rates in aerial agriculture have been identified as a problem for many years. In 1996, the Australian

Aviation Underwriting Pool data (AUP, 1996) showed a consistent occurrence rate in Australia at least since 1985, as Table 11.10 taken from the Pool's newsletter 'Insight' suggests.

Table 11.10 Aerial agriculture accidents 1985–1993

Year	Accidents per 100k flying hours
1985	20.07
1986	28.90
1987	22.15
1988	24.47
1989	28.30
1990	23.59
1991	22.69
1992	31.32
1993	23.54

The 'Insight' article lists seven clear groups of accidents based on the 1985–1993 data:

- Aircraft stalled on turn 15 per cent
- Aircraft flew into (rising) ground 9 per cent
- Laden aircraft failed to gain height on take off 15 per cent
- Aircraft struck object (excluding wires) 12 per cent
- Fuel mismanagement 6 per cent
- Aircraft struck wire in flight 27 per cent
- Aircraft taxying 9 per cent.

The similarities with outcomes of the 2002–2005 period reported here would seem to be obvious and while there is no way to measure the effectiveness of any previous interventions, it seems likely that the industry's past attempts to stem the flow have failed.

Reliance on human performance

The outcome of this study confirms that of the earlier authors that the aerial agriculture accident problem seems to be predominantly a human performance issue. Ninety-three per cent of the accidents in the 2002–2005 period related to pilot human factors. Although this would not be a surprise to many in the industry, a *prima facie* case could be made that the methods of managing safety in aerial agriculture to date have not been error tolerant. Perhaps not so surprising when the nature of the aerial agriculture operation is considered and the notion of the 'single error accident' in aerial agriculture seems very plausible indeed.

However, it is interesting that the accident rates in other high risk work activities that also rely heavily on human performance, such as industrial diving, appear to be significantly lower than that of aerial agriculture. Although Australian comparison data are difficult to find, accident rate data in Lincoln (1997) suggested pilots operating in Alaska were more than 12 times more likely to be involved in fatal accidents than were industrial divers.

What then is the difference between the approaches to safety management in aerial agriculture and those other high risk, high reliance on human performance industries?

Back to basics: the way forward for aerial agriculture

The principles of effective safety management in successful high risk industries have changed little in over 50 years. There's no silver bullet. Across all these industries, what has been proven has been the systematic analysis of the hazards, careful consideration of the effective ways to control those hazards and the diligent application of programs to consistently implement those controls.

In the area, other sections of the aviation industry, such as airline and military air operations, have been at the fore, the principles and practices of system safety having been introduced in the early 1960s, through the pioneering work of the late C.O. Miller, the then head of the Bureau of Aviation Safety at the NTSB.

At the core of these system safety methods have been practices that have underpinned flight standards in the best performing operators. These include, but are not limited to:

- consistent pilot screening and selection;
- ongoing standards surveillance with routine checks and recurrent training;
- the use of high fidelity simulators, especially for rehearsal of new sequences, such as route checks and field endorsements as well as critical and emergency procedures; and
- line-orientated flight training, rehearsing the normal sequences until they become conditioned responses.

Add to these the more recent developments such as crew resource management, real time surveillance using quick access recorders you have the recipe that has delivered consistently high flight standards in the military and airlines for the past couple of decades. The data from the events studied and the pilot interviews in this study would suggest that aerial agriculture is yet to consistently adopt these basics.

Ironically, most of the system safety achievements in the other industries have been through copying these methods and principles that were originally founded in aviation. That many of these have not yet been embraced by aerial agriculture is indeed a paradox.

Doug Edwards (1997) in his book *Fit to Fly, Cognitive Training for Pilots*, suggested that

> Denial which is the refusal to acknowledge the obvious; the preference for self justifying fantasy is a pathogen, a danger factor, which allowed individuals to 'gloss over' evidence of a problem with potential disastrous effects on safe operations.

Of course, Edwards was referring to pilots failing to acknowledge the onset of hazardous flight circumstances, but the same can be applied to aerial agriculture operators who fail to acknowledge that high incident rates are a direct result of poor management practice and who underemphasize the importance of proven safety management principles.

Of course, in many industries, like aerial agriculture, commercial pressures and market realities make the task difficult. The economics of success can be a difficult balance in cash strapped operations. However, diligent application of system safety principles is even more important when there is pressure on the business, when operations are most vulnerable. It seems that this is particularly true of aerial agriculture when the vagaries of seasonal work, the ups and downs of primary industry and the continual economic pressure are considered.

Leadership and due diligence: the only effective strategies

In his book *Safety Culture and Risk: The Organizational Causes of Disasters* Andrew Hopkins suggested the

> focus on organizational practices places responsibility for [safety] culture squarely on senior management, for it is the leaders of an organization who determine how it functions, and it is their decision making which determines, in particular, whether an organization exhibits the practices which go to make up a culture of safety (Hopkins, 2005).

To this, the *Australian Standard* on Safety Management Systems guidelines document (AS/NZS 4804:1997), adds that leadership entails appropriate financial and human resource allocation, setting accountabilities, ensuring decisions are followed through and effective performance assessment.

In aerial agriculture, this is especially the role of the senior management of the operators, the company owners, the AOC holders and the Chief Pilots. Really, no one else in an organization have sufficient influence to effect change. How much effort is required by individual operators and chief pilots will depend on how much of the system safety doctrine they have yet to embrace.

However, it is clear that continued management abdication of responsibility for safe flight operations to the aerial agriculture pilot, as seems to have often been the situation in the past, is not going to deliver the change in industry safety performance required.

According to Hopkins (1995), leaders create and change safety cultures, by 'what they systematically pay attention to', 'anything from what they notice and comment on to what they measure, control, reward and in other ways systematically deal with'. 'So leaders drive safety and set standards and cultures', says Hopkins.

Effective accident prevention in aerial agriculture in future clearly requires the adoption of proven aviation practices and strong safety leadership. Like it or not, it is the role of the operators and chief pilots to take the lead. The alternatives spell industrial oblivion if the insurers refuse to underwrite the business in future.

Responding to the issues arising from the accidents

Wire strikes

Data show that wire strikes have been the most common type of aerial agriculture accident since at least 1985 consistently accounting for about one-third of human factors related accidents. While this statistic alone suggests fundamental shortcomings in methods adopted by the industry to address the problem in the past, more concerning is the statistic that almost all of the wire strikes involve wires that the pilot knew were there and had previously avoided, but due to a range of well-known human error conditions, such as excessive task load, distraction, fatigue, *etc.*, the pilots failed to avoid the wires.

Earlier efforts to address this issue have focused on methods to identify and plot the location of the wires, such as the Freeman's Power Line Location Guide formerly published by the Australian Civil Aviation Authority, and to emphasize the seriousness of the wire strike issue during agricultural pilot training programs.

However, none of these methods are failsafe in that they do not control for the primary problem, that of pilots failing to avoid known wires. Since most wires are very difficult to see from the air, it is not surprising that once a pilot has been distracted or performance has been degraded by those other factors mentioned above, the aircraft hit the wires without any warning that an accident is imminent.

System safety principles would suggest this circumstance is unacceptable where a simple operator error can seemingly directly lead to a catastrophic accident. There would appear to be a clear need for interventions to make the wire strike methods more error tolerant. Some wires in known high risk areas have been fitted with high visibility markings to aid pilot identification of wires in the path of the aircraft. However, these have not been universally adopted, presumably due to the large number of wires and the costs involved installing the markers. No doubt many aerial agriculture pundits would vigorously argue it's not a practical solution.

However, the same short focus argument used to be put in other high accident rate domains, such as the argument in the 1960s that it would be too cost prohibitive to fit seat belts to all cars, or to fit rollover protection to farm tractors. Yet the solutions approach taken in those jurisdictions can also be applied to aerial agriculture.

This author believes there is a clear need for CASA to regulate for high visibility markings for all power lines located on properties on which aerial agriculture operations are planned. Like the seat belts and rollover protection examples, the regulator should set a deadline for compliance with a moratorium period to allow compliance to be achieved without undue impact on businesses. Further, compliance should be the responsibility of the property owner who wishes to engage the aerial agriculture services.

Like everything else in safety, it will not be a panacea and it will take many years for power line markings to be the norm rather that the exception. However, the other jurisdictions show these intervention methods can work effectively without undue economic hardship on business. Indeed, the introduction of the legislation alone will probably drive design and introduction of more simple and innovative power line marking methods.

Apparent need for Airmanship 101

Several of the accident categories, such as 'failed to get airborne', 'landing overruns', fuel exhaustion suggest a need for a return to emphasis on basic airmanship. However, such suggestion also fails to address that these errors were made by pilots who had previously qualified for the operations, had been selected for the task by those engaging their services and who had gained a certain level of experience and proficiency. They were all mostly intent on getting the job done at the time of the accidents.

This author believes these accidents were also the result of a lack of integrity in the safety management methods applied by the industry and symptoms of over-reliance on the pilots to maintain performance at a high level under what would elsewhere be considered conditions adverse to safe flying.

The almost total lack of an effective ongoing flight standards, recurrent check and training and surveillance system, reported by the pilots interviewed in this study, could be considered a direct precursor to all these supposed basic airmanship failings.

Aerial agriculture needs to take a leaf from the successful airlines books if the accident rates are to be effectively and consistently stemmed. Operators' management practices must be changed to include the core systems safety flight standards practices mentioned above, especially introduction of routine use of simulators for recurrent check and training.

Modern enhancements such as quick access recorders should be introduced to aerial agriculture operations to enhance operations surveillance capability across the industry. This is another area where regulatory intervention could precipitate a step change in accident prevention in the industry.

Conclusions

Aerial agriculture accidents rates appear to have increased in Australia in recent years. Ninety-three per cent of aerial agriculture accidents can be attributed, at least in part, to errors and violations on behalf of the pilots involved. Wire strikes account for around one-third of those events. Apparent failings in basic airmanship by pilots involved in many of the accidents belie the causative impact of the underpinning non-error tolerant accident prevention strategies in the past which relied heavily on pilot performance with little support from the organizational arrangements.

Aerial agriculture is yet to adopt many of the successful flight standards methods proven in other sections of the aviation industry, in particular the lack of flight simulation which would enable rehearsal of critical flight sequences and safe, cost-effective pilot recurrent training.

There is a clear role for regulatory intervention in the areas of wire strike prevention, through mandatory marking of wires on properties planning the use of aerial agriculture, and of operations surveillance through fitment of quick access recorders to facilitate introduction of contemporary flight standards surveillance activities.

Changes in safety culture across the industry will be necessary if the accident trends are to be reversed and it will be the clear role of the operators and chief pilots to lead the way and make the changes necessary.

Unless the system safety shortcomings of the aerial agriculture industry are addressed, the industry will face the burden of significant insurance premium increases as the insurers attempt to recover costs, or if the worsening accident trends continue, the prospect of the insurers refusing to underwrite the business may become a reality.

References

AS/NZS 4804:1997, 'Occupational health and safety management systems – General guidelines on principles, systems and supporting techniques,' *Standards Australia*, Sydney.

ATSB (2005), Risks Associated with Aerial Campaign Management: Lessons from a Case Study of Aerial Locust Control, Canberra: Australian Transport Safety Bureau.

AUP (1996), Aerial Agriculture – A Risky Business, *Insight Newsletter*, Australian, Melbourne: Aviation Underwriting Pool.

CDC (2004), Work Related Pilot Fatalities in Agriculture – United States, 1992–2001, in Centre for Disease Control, http://www.cdc/mmwr/preview/mmwrhtml/mm5315a4.htm

Edwards, D. (1997), *Fit to Fly, Cognitive Training for Pilots*, Brisbane: CopyWrite Publishing Company.

Hopkins, A. (1995), Due Diligence and Safety Management Systems, Presentation to the Prosecution Developments – Directors' and Officers' OHS Liability and the Use of Crimes Act/Manslaughter in the Industrial Context Seminar, Melbourne: Centre for Employment and Labour Relations Law.

Hopkins, A. (2005), *Safety Culture and Risk: The Organisational Causes of Disasters*, Sydney: CCH Australia.

Lincoln, J. (1997), The Cold Facts: Occupational Diving Injuries and Fatalities in Alaska, Proceedings of the Diving Safety Workshop, Fairbanks: University of Alaska.

Reason, J. (1990), *Human Error*, New York: Cambridge University Press.

Reason, J. (1997), *Managing the Risks of Organisational Accidents*, Aldershot: Ashgate.

Scheuren, F. (2004), What is a Survey?, American Statistical Association, http://www.amstat.org/sections/srms/.

Wiegmann, D. and Shappell, S. (2003), A *Human Error Approach to Aviation Accident Analysis*, Aldershot: Ashgate Publishing.

Chapter 12

Learning from Accidents and Incidents

Joanne De Landre
Safety Wise Solutions Pty Ltd

Miles Irving, Iaen Hodges and Bruce Weston
RailCorp

In late 2004 RailCorp adopted a new safety occurrence investigation method – ICAM (Incident Cause Analysis Method). ICAM breaks new ground in the area of incident investigation which, in the past, has tended to focus on intentional or unintentional acts of human error – those things that people did or did not do – that lead to an incident or accident. ICAM is a holistic method, which aims to identify both local and wider systemic factors within the entire organizational system that contributed to the incident. Through the analysis of this information, ICAM provides the ability to identify what really went wrong and to make recommendations on what needs to be done to prevent recurrence. It is directed towards building error-tolerant defences against future incidents.

The use of ICAM at RailCorp has not only provided a framework for the investigation of specific incidents but also ensured that the broader organizational issues are identified and addressed via corrective actions. This chapter will briefly introduce the principles of ICAM before discussing an incident displaying the application of ICAM and the subsequent learning. The incident resulted in an investigation that addressed deficiencies through corrective actions to actively target prevention of recurrence and reduction of risk, not only for this specific incident but also for the broader safety management systems at RailCorp. The key learning from the incidents ensure that the broader rail industry can learn and benefit from the findings.

Introduction

The principal objective of incident investigation is to prevent recurrence, reduce risk and advance health, safety and environmental performance. The incident investigation methodology that an organization uses should not only achieve this objective, but also be a practical, intuitive and consistent methodology that can be integrated into existing management systems to achieve the best possible outcome. This chapter will commence with a brief description of the incident investigation methodology selected by RailCorp – the Incident Cause Analysis Method (ICAM); an incident will then be discussed which demonstrates how the principles underlying the ICAM model not only identify the basic cause of incidents, but also ensure that all deficiencies that contributed to an incident are identified and addressed.

The Reason Model

The principles of ICAM stem from the work of organizational psychologist and human error expert, Professor James Reason and his modelling of organizational accidents. Reason and his colleagues from the University of Manchester in the United Kingdom developed a conceptual and theoretical approach to the safety of large, complex, socio-technical systems, of which aviation and mining are excellent examples. Reason defines organizational accidents as those in which latent conditions (arising mainly from management decisions, practices or cultural influences) combine adversely with local triggering conditions (weather, location, *etc.*) and with active failures (errors and/or procedural violations) committed by individuals or teams at the front line or 'sharp end' of an organization, to produce an accident (Reason, 1990, 1997).

A fundamental concept of ICAM is acceptance of the inevitability of human error. Human factors research and operational experience has shown that human error is a normal characteristic of human behaviour, and although it can be reduced, it cannot be completely eliminated (Helmreich and Merritt, 2000). An organization cannot change the human condition, but they can change the conditions under which humans work, thereby making the system more error tolerant (Reason, 2000).

Applying ICAM

ICAM is an analysis tool that sorts the findings of an investigation into a structured framework. An adaptation of the Reason Model is shown in Figure 12.1. The ICAM model organizes incident causal factors into four elements:

Absent or failed defences

Defences are those measures designed to prevent the consequences of a human act or component failure producing an incident. Defences are equipment or procedures for detection, warning, recovery, containment, escape and evacuation, as well as individual awareness and protective equipment. These contributing factors result from inadequate or absent defences that failed to detect and protect the system against technical and human failures. These are the control measures which did not prevent the incident or limit its consequences.

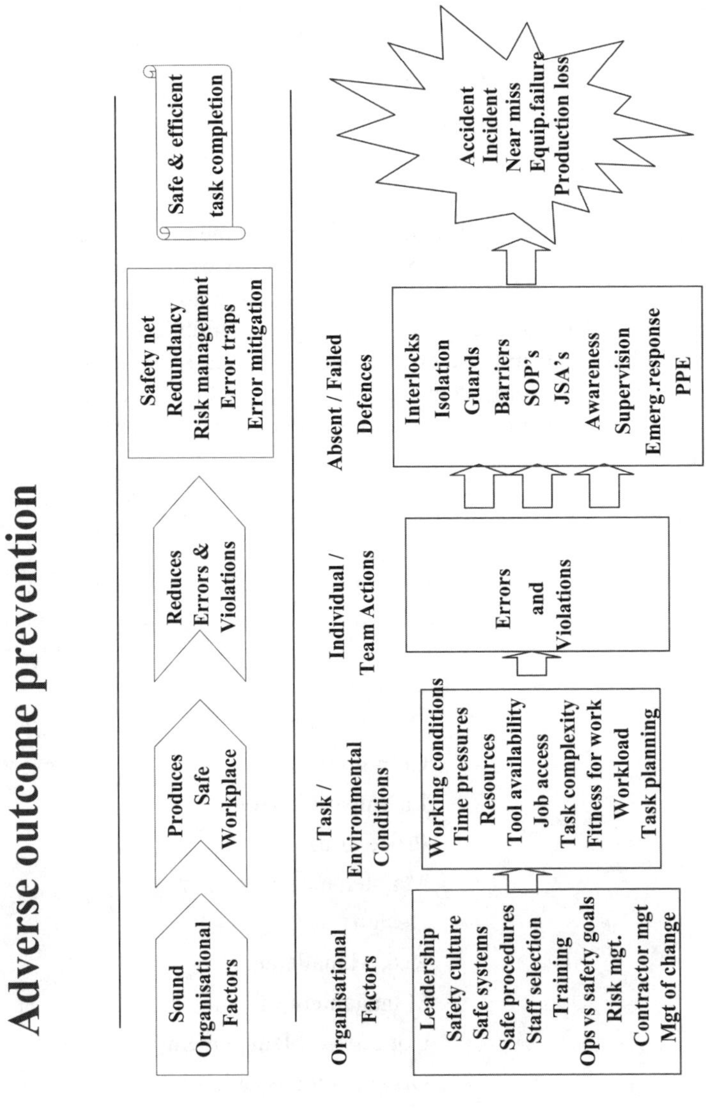

Figure 12.1 ICAM model of incident causation

Individual or team actions

These are the errors or violations that led directly to the incident. They are typically associated with personnel having direct contact with the equipment, such as operators or maintenance personnel. They are always committed 'actively' (someone did or did not do something) and have a direct relation with the incident. For most of the time, however, the defences built into our operations prevent these 'human errors' from causing harm. Individual/team actions within the ICAM model are initially categorized into intended or unintended actions and then categorized as slips, lapses, mistakes or violations.

Task/environmental conditions

These are the conditions in existence immediately prior or at the time of the incident that directly influences human and equipment performance in the workplace. These are the circumstances under which the errors and violations took place and can be embedded in task demands, the work environment, individual capabilities and human factors. Deficiencies in these conditions can promote the occurrence of errors and violations. They may also stem from an organizational factor type such as risk management, training, incompatible goals, or organization, when the system tolerates their long term existence.

Organizational factors

Table 12.1 ICAM Organizational Factor Types

HW	Hardware
TR	Training
OR	Organization
CO	Communication
IG	Incompatible Goals
PR	Procedures
MM	Maintenance Management
DE	Design
RM	Risk Management
MC	Management of Change
CM	Contractor Management
OC	Organizational Culture
RI	Regulatory Influence
OL	Organizational Learning

Source: Safetywise Solutions

Organizational Factor Types (OFTs) produce the adverse task/environmental conditions, allow them to go unaddressed or undermine the system's defences. These are the underlying organizational factors that produce the conditions that affect performance in the workplace. They may lie dormant or undetected for a long time within an organization and only become apparent when they combine with other contributing factors that led to the incident. These may include management decisions, processes and practices. ICAM classifies the system failures into 14 Organizational Factor Types (OFTs) as seen in Table 12.1.

Specific objectives of ICAM

The specific objectives of incident investigations using the ICAM process are to:

- Establish the facts;
- Identify contributing factors and latent hazards;
- Review the adequacy of existing controls and procedures;
- Report the findings;
- Recommend corrective actions which can reduce risk and prevent recurrence;
- Detect organizational factors that can be analysed to identify specific or recurring problems;
- Identify key learnings for distribution.

It should be noted that it is not the purpose of an ICAM investigation to apportion blame or liability.

Diversity of ICAM

ICAM has been used in the investigation of incidents and accidents in the aviation, rail, road, mining, marine and petroleum sectors. It has been successfully applied in a range of countries including Australia, Canada, Chile, Indonesia, Pakistan, Papua New Guinea, Peru and South Africa, migrating readily to these different industrial contexts, work domains and cultural settings as a practical investigative tool (Gibb, Hayward and Lowe, 2001).

A case study from RailCorp will be used to demonstrate the practical application of ICAM and display how all deficiencies were identified and addressed. The incident involved an empty Intercity Train entering the Lawson Springwood section where track machines were being block worked – effectively resulting in two 'trains' being in one section at the same time. This was at variance with RailCorp Network Rules and Procedures and a potential existed for a collision between the Intercity Train and the track machines. Further details can be seen below in the Incident Description, and the contributing factors will be commented on as the chapter progresses (a glossary of terms can be found below the incident description).

Case study incident description

At 0204 hours on Friday, 13 May 2005, Run No. W700 (an Intercity Train) entered the Lawson Springwood section where two Track Vehicles (forming Train M378) were being manually block worked through the Lawson to Springwood section.

The Signaller had cleared a Shunting Signal, intending to shunt W700 forward from the Up Main line and then back into the Up Refuge Loop. The Signaller did not inform the Driver of W700 of the intended unplanned move, believing it to be the same Driver as on the previous night when the same movement was carried out. The Driver of W700, expecting to travel from Lawson towards Springwood (as per the Special Train Notice or STN), responded to the Shunting Signal as if it was a 'running signal' and commenced to travel through the Lawson to Springwood section at up to normal line speed.

The Signaller observed the movement of W700 beyond Lawson Yard Limits (as displayed on his panel) and realized that the train had exceeded its limit of authority. The Signaller could not initially contact the Driver on the Train Control Radio as the train radio was not 'logged on'.

The Signaller then contacted the Train Controller (in Sydney) and reported the incident. The Train Controller confirmed, by reference to the information displayed on his monitor that W700 was not 'logged on'.

After this conversation with the Train Controller, the Signaller contacted run number 1871 (a Down freight service), on the 450.050 MHz open channel Radio system (known as the WB radio) and asked the crew to warn W700 to stop as the two trains passed each other. The crew of run number 1871 altered the lead locomotive's marker lights to red. By so doing they attempted to alert the crew of W700 that something might be wrong. The crew of W700 observed the red marker lights, but other than reporting (at interview) a heightened state of awareness, they did not take any action in response. The Signaller was able to contact W700 shortly afterwards via the MetroNet Train Radio and instructed the Driver to bring the train to a stand.

Signal log shows that the minimum separation between W700 and M378 approximated to five minutes. However, there was a potential for a collision between the Intercity Train, travelling at line speed (the Datalogger shows that W700 travelled at a maximum speed of over 75 km/hour) and the rearmost Track Machine if it had, for any reason, stopped in the section.

Once M378 had cleared the section, W700 was authorized to continue to Springwood where it was met by the local Network Operations Superintendent (NOS) whom the Train Controller had called out to assist with investigating the incident.

Glossary of terms

- Block Worked (manual Block Working): a method of special working which ensures sole occupancy by manually maintaining the block between rail traffic movements.
- Limit of Authority: a location to which rail traffic may travel under a proceed authority.

- Panel or 'mimic board': displays the track layout and location of trains within the Signaller's area of control.
- Running Signal: a fixed signal placed near a running line to authorize and control running movements.
- Running Line: a line (other than a siding) which is used for through-movement of trains.
- Section: a portion of track between two controlled signalling locations.
- Shunting Signal: a fixed signal provided to authorize and control shunting movements.
- Shunt: to move trains for purposes other than for through movements.
- STN – Special Train Notice: issued for additional (or alterations to) train movements.

Investigation perspective

A comprehensive safety investigation not only looks at how an incident occurred, but also looks at why it occurred. Most importantly, the investigation recommends corrective actions that can be taken across the organization to prevent such incidents happening again. An incident investigation should be seen as a safety improvement process. It is not the purpose of ICAM to apportion blame or liability.

Contemporary research by leading experts in human error such as Reason, Pariès and Helmreich all stress that if an organization's investigative technique is limited to the factors found at the 'sharp end' of the incident scene, they will dramatically limit the lessons to be learnt from safety investigations.

ICAM is designed to ensure that the investigation is not restricted to the errors and violations of operational personnel. It identifies the local factors that contributed to the incident and the latent hazards within the system and the organization (Gibb and Landre, 2002).

What happened versus why it happened

What happened in any incident can be inferred from incident descriptions. However, after establishing what happened, the more important questions to ask are why did it happen and how can a recurrence be prevented? The ICAM framework guides investigators in finding the answers to these two questions.

Details of the contributing factors identified in the case study are included below. Figure 12.2 includes the ICAM Analysis Chart for the incident organizing the causal factors into the four ICAM elements of:

- absent/failed defences;
- individual/team actions;
- task/environmental conditions; and
- organizational factors.

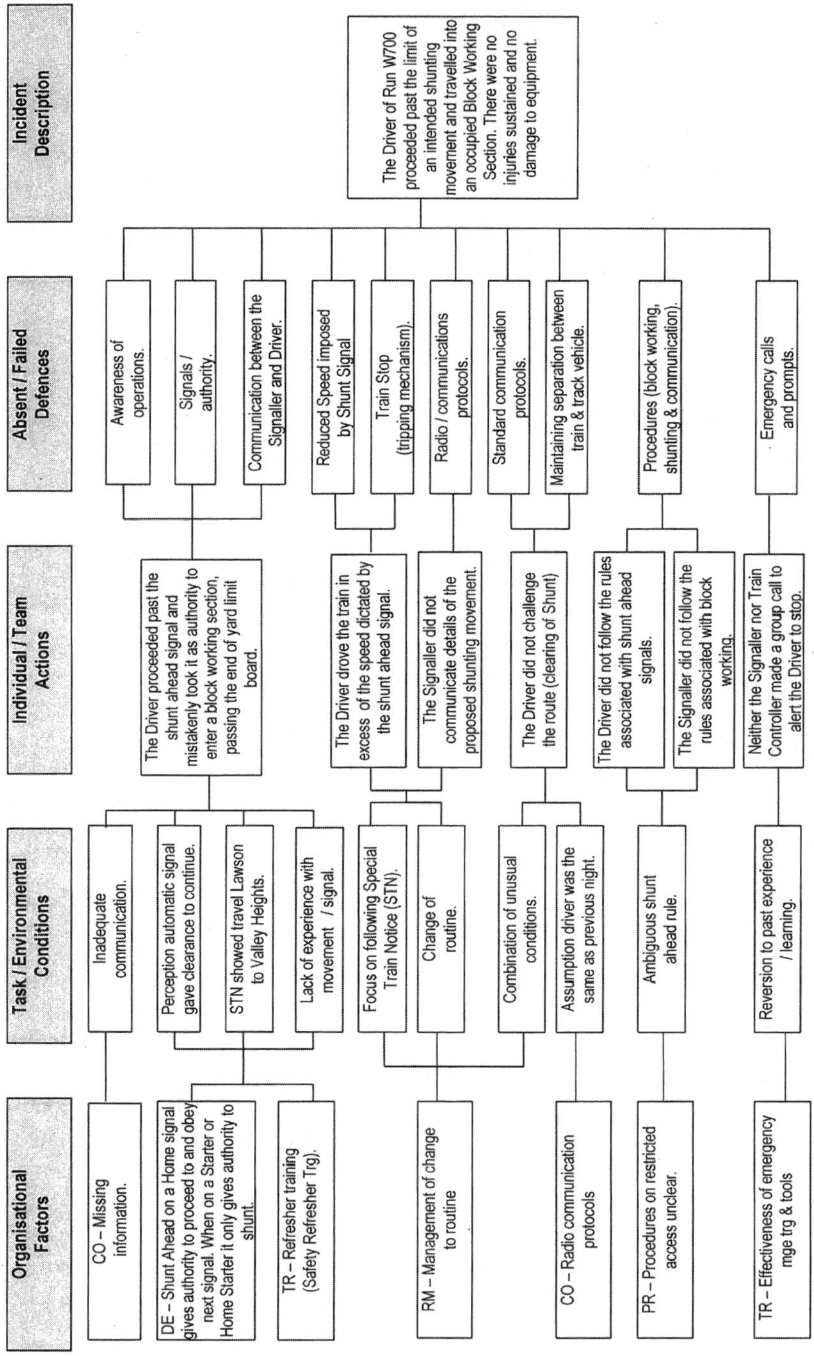

Figure 12.2 ICAM analysis chart

Case study: basic cause and contributory factors

The investigation report into the case study found that the basic cause of the incident was a lack of communication between the Signaller and Driver of Run No. W700 about the proposed movement from the Down Main to the Up Main at Lawson. Some of the other contributing factors found by the Investigation Team included:

- There were altered working conditions and a change in routine.
- There was a focus on following the Special Train Notice (STN) by the Driver.
- The Signaller thought the Driver was the same Driver who had performed the movement the previous day.
- There was a perception by the Driver of Run No. W700 that the shunting signal gave clearance to proceed through the section.
- The Driver of Run No. W700 had limited experience with the movement and signal.
- The rules relating to the meaning of shunting signals rules was ambiguous.

The ICAM Chart (Table 12.2) provides further detail on the contributing factors.

Investigation recommendations

Apart from establishing the details of what happened in specific incidents, the fundamental aim of an investigation is to prevent future incidents by identifying and correcting system deficiencies. Simply focusing on the actions of the Driver or Signaller would not achieve this objective. While recommendations may arise from a specific incident it is important for all operators to recognize the benefit of learning from other's mistakes and realizing that the implementation of such learning could benefit their operations and prevent an incident.

A required outcome of applying the ICAM process is the formation of clear recommendations to address deficiencies in system processes. The investigation team must make recommendations which address all absent or failed defences and all organizational factors identified as contributing factors. The investigation into the RailCorp incident resulted in several Recommended Safety Actions being proposed. Some of these are outlined in Table 12.2 below.

Business case for ICAM

An important force driving the application of ICAM within industries from an organizational view is the increasing commercial costs of accidents and incidents and their effect on company profits and financial viability, as well as the social and environmental impact of occurrences. Organizations are becoming increasingly aware of the direct and indirect costs of occurrences, many of which are the result of human factors. The ability to use a human factors based safety management program such as ICAM to reduce the incident rate, particularly, for recurrent incidents, will result in a cost benefit for the organization. Therefore, the actual cost of implementing ICAM and training staff in the methodology is viewed as minimal compared to the potential savings which result with an incident reduction rate.

Consequently, prevention and error management programs based on system safety and human factors concepts not only improve safety within the workforce, but are also commercially attractive. As Professor Patrick Hudson, a leading world expert in system safety and human factors has argued, a company safety department or system should be seen as a profit centre, not as a cost centre (Hudson, 1998).

Table 12.2 Case study and recommended actions

Organizational factor or absent/ failed defence	Recommended safety action
Driver and Signaller did not communicate as provided for in NTR 420. Shunting and Marshalling trains and NGE 204 Network Communications and OSP16 Shunting in RailCorp yards and Maintenance Centres	Reinforce the rules and procedures that relate to communication requirements in Safeworking Refresher Training (SRT). Review whether this training is adequate – Continue to audit communications to ensure an acceptable standard is both achieved and maintained
Driver did not act in accordance with NSG 606 in three respects: • 'A shunting signal authorises a movement at restricted speed past that signal,' • 'Drivers and track vehicle operators must proceed as if the line is already occupied' and •'… a shunting signal is an authority to proceed up to and not beyond (in this case END YARD LIMIT) sign'	Ensure that the requirements of NSG 606 and associated rules and procedures are included in forthcoming SRT – carry out a gap analysis to identify if there are training needs for train crew in this area
There is a divergence between NSY512 (Block Working) and NWT 304 (Track Occupancy Authority) and also NPR 701 (Using a Track Occupancy Authority).	Review NSY 512 and also NWT 304 and NPR 701 can be better aligned so as to avoid possible confusion
There is a divergence between NSY 512 (Block Working) and NSG 606 (Responding to Signals and Signs).	Review NSY 512 and NSG 606 to examine if they can be better aligned so as to avoid possible confusion.

Organizational factor or absent/ failed defence	Recommended safety action
Metronet radio on W700 was logged on to the Katoomba radio area (047) at Lawson rather than Spingwood's area (046). Under normal circumstances this would have resulted in the Springwood Signaller not having W700 'on his system' while the Train Controller would have W700 displayed on his.	Emphasize to drivers the importance of, and the requirement to, log on to the correct area.
Although the Train Radio on W700 was logged on (to area 047) the Train Controller was unable to 'see' W700 on his system	Siemens have identified a fault in the Metronet Train Radio system and will make the necessary changes to system software
Neither the Signaller nor the Train Controller attempted a 'Group Call' nor did they attempt to contact the Train via its 'Stock Number'	Carry out a gap analysis to identify training needs of both Signallers and Train Controllers in regard to the Metronet Train Radio System

Source: Safetywise Solutions and RailCorp

Conclusion

ICAM has been adopted by many businesses as their designated incident and accident investigation technique. The model has been used successfully in a diverse range of industries and countries and has been found to be a very practical and easy to apply investigative tool. ICAM exceeds the regulatory requirements for accident and incident investigation within many industries and is consistent with many organization's commitment to continual improvement in performance and safety by addressing the latent conditions and hazards in the system that produce human error.

The application of ICAM at RailCorp for incident investigation has provided the organization with a framework which enables identification of systemic health, safety and environmental deficiencies, assists investigation teams to identify what really went wrong, ensures recommendations are focused on what needs to be done to prevent recurrence and is directed towards building 'error-tolerant' defences against future incidents.

References

DeLandre, J. and Bartlem, S. (2005), Learning from Accidents and Incidents, Proceedings of the Queensland Mining Industry Health and Safety Conference, Townsville, Queensland.

Gibb, G., Hayward, B. and Lowe, A. (2001), Applying Reason to Safety Investigation:

BHP Billiton's ICAM, Proceedings of the Townsville, Queensland: Queensland Mining Industry Health and Safety Conference.

Gibb, G. and De Landre, J. (2002), Lessons from the Mining Industry, Paper Prepared for the Human Error, Safety and Systems Development (HESSD) Conference, Newcastle, NSW.

Helmreich, R.L. and Merritt, A.C. (2000), Safety and Error Management: The Role of Crew Resource Management, in *Aviation Resource Management* eds Hayward, B.J. and Lowe, A.R., Aldershot, UK: Ashgate.

Hudson, P.T.W. (1998), Safety Cultures in Aviation, Keynote Address, Vienna: European Aviation Psychology (EAAP) Conference.

Reason, J. (1990), *Human Error*, New York: Cambridge University Press.

Reason, J. (1997), *Managing the Risks of Organizational Accidents*, Aldershot, UK: Ashgate.

Reason, J. (2000), Human Error: Models and Management, *British Medical Journal*, 320, 768–770. [PubMed 10720363] [DOI: 10.1136/bmj.320.7237.768].

Chapter 13

Managing Road User Error in Australia: Where Are We Now, Where Are We Going and How Are We Going to Get There?

Paul M. Salmon, Michael Regan and Ian Johnston
Monash University Accident Research Centre

Recent research indicates that driver error contributes to up to 75 per cent of all roadway crashes. Despite this estimate, very little is currently known about the errors that road users make, or about the conditions within the road transport system that contribute to these errors being made. Additionally, the use of formal methods for collecting, analysing and using error-related data to enhance safety within the road transport domain has previously been neglected. This article describes the work conducted to date as part of an overall research program of which the main aim is to develop a novel error management program designed to promote the tolerance of road user error within the Australian road transport system. In conclusion, a novel methodology for a proof-of-concept pilot study on road user error and latent conditions at intersections in Victoria is presented.

Introduction

Human error has historically been implicated as a contributory factor in a high proportion of the accidents and incidents that occur in most complex sociotechnical systems. For example, within the civil aviation domain human error has been identified as a causal factor in around 75 per cent of all accidents, and is now seen as the primary risk to flight safety (Australian Civil Aviation Authority, 1998). Similarly, in rail transport, human error was identified as a contributory cause of almost half of all collisions occurring on UK Network Rail between 2002 and 2003 (Lawton and Ward, 2005). Within the health care domain, the US Institute of Medicine estimates that between 44,000 and 88,000 people die each year as a result of medical errors (Helmreich, 2000). In road transport, recent research has indicated that human or driver error contributes to as much as 75 per cent of all roadway crashes (Hankey, Wierwille, Cannell, Kieliszewski, Medina, Dingus and Cooper, 1999; cited in Medina, Lee, Wierwille and Hanowski, 2004).

Despite these estimates, there has been only a paucity of human error-related research conducted in road transport to date. Consequently, very little is currently known about the different errors that road users make, or about the conditions within the road transport system that contribute to these errors being made. In addition, the use of error management approaches within road transport systems worldwide has previously been neglected. This article describes the findings derived from the first three phases of an overall research program, the main aim of which is to develop a novel error management framework that can be used to promote error tolerance within the Australian road transport system. The research conducted to date included a review of the human error-related research conducted in other domains and in road transport, a review of the error management approaches previously employed in complex sociotechnical systems, the development of a conceptual framework for an error tolerant Australian road transport system, and the design of a pilot study on road user error at intersections in Victoria.

Human error

Human error is a complex phenomenon and people, regardless of skill level, ability, training and experience, make errors every day. Human error is formally defined as 'a generic term to encompass all those occasions in which a planned sequence of mental or physical activities fails to achieve its intended outcome, and when these failures cannot be attributed to the intervention of some chance agency' (Reason, 1990).

Whose fault is it anyway?

The complexity of the construct has led to the development of two incongruent theoretical perspectives on human error: the person approach and the systems perspective approach. The person approach focuses upon the identification and classification of the errors that operators make at the so-called 'sharp-end' of system operation (Reason, 2000). According to the person approach errors arise from aberrant mental processes such as forgetfulness, loss of vigilance, inattention, poor motivation, carelessness, negligence, and recklessness (Reason, 2000). When using the person approach, human error is treated as the cause of most accidents; the systems in which people work are assumed to be safe; human unreliability is seen as the main threat to system safety; and safety progress is achieved by protecting systems from human unreliability through automation, training, discipline, selection and proceduralization (Dekker, 2002). The systems perspective approach, however, treats error as a systems failure, rather than an individual operator's failure and considers the combined role of latent or error causing conditions (*e.g.*, inadequate equipment and training, poor designs, inadequate supervision, manufacturing defects, maintenance failures, ill-defined procedures, *etc.*) and human errors in accident causation. Human error is seen to be a consequence of the latent conditions residing within the system and is therefore treated as a symptom of problems within the system.

The systems perspective model of human error and accident causation proposed by Reason (1990) is the most influential and widely recognized model of human error. The Swiss cheese model (as it is more commonly referred to due to its resemblance to a series of layers of Swiss cheese) considers the interaction between latent conditions and errors and their contribution to organizational accidents. According to the model, systems comprise various organizational levels that contribute to the production of system outputs (*e.g.*, decision makers, line management, productive activities and defences). Each of the levels has various defences (*e.g.*, protective equipment, rules and regulations, engineered safety features, *etc.*) in place. These are designed to prevent accidents and safety compromising incidents. Holes or weaknesses in the defences created by latent conditions and errors create 'windows of opportunity' for accident trajectories to breach the defences and cause an accident. Accidents occur when the holes line up in a way that allows the accident trajectory to breach each of the defences that are in place. On most occasions, accident trajectories are halted by the defences at the various levels in the system. However, on rare occasions, the holes or windows of opportunity line up to allow the accident trajectory to breach all of the defences, culminating in an accident or safety-compromising incident. The Swiss cheese model is presented in Figure 13.1.

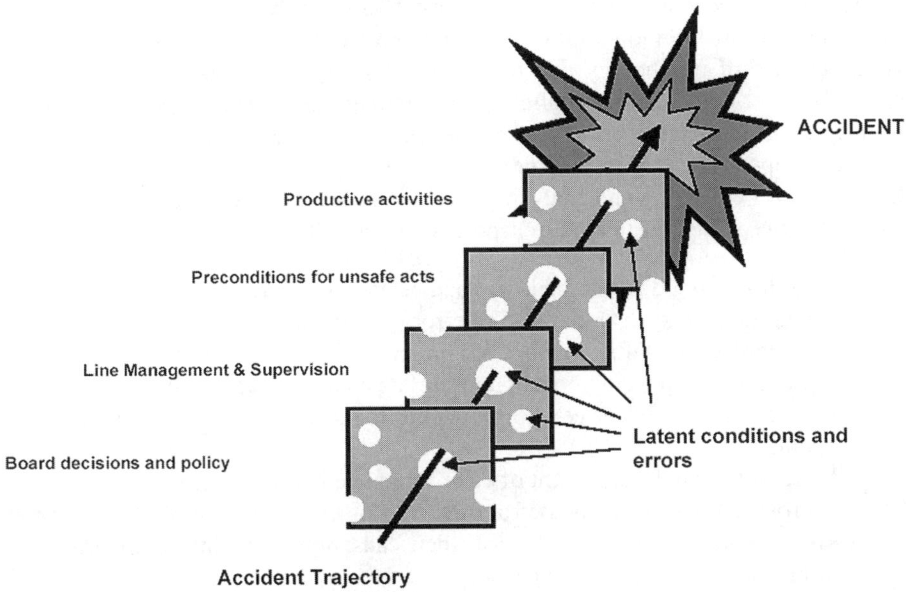

Figure 13.1 The Swiss cheese model of error and accident causation

Error management and Swiss cheese

Error management approaches are used to combat the problem of human error in most complex, sociotechnical systems. A review of the error management approaches used previously in such systems was conducted. The literature indicated that such programs are employed in most safety critical systems to identify, remove, reduce or manage the errors that are made by operators and the conditions that lead to them. Typically a combination of error management-related techniques is used to collect data regarding the nature of, and factors surrounding, error occurrence and to develop appropriate error countermeasures. It was concluded that the key component of error management programs is the collection of specific information on the different types of human errors made in a particular system. Such information is then used, amongst other things, to inform the design and development of error tolerant systems, and the development of countermeasures, remedial measures and strategies designed to eradicate or reduce error and its contributing conditions. The literature also indicated that there is a plethora of different error management-related techniques available, including accident investigation, human error identification, human reliability analysis, incident reporting schemes, specific error management approaches (*e.g.*, REVIEW, TRIPOD-DELTA, Reason, 1997) and error management training programs.

It is the opinion of the authors that the Swiss cheese model is particularly suited for error management purposes within the road transport domain. The model is a simplistic, intuitive approach to human error and provides a useful insight into the interaction between latent conditions and errors and their role in accident causation. The simplicity of the model allows it to be applied by practitioners with little or no experience of psychology and human factors theory or methods. Further, the model is generic and can be applied to any domain, and countermeasures derived from systems perspective-based analyses are aimed at the entire system, and not just individual operators. It was concluded that adopting the Swiss cheese model as a framework for collecting and analysing human error-related data and for developing error tolerance strategies in road transport is attractive for the following reasons:

- In addition to the errors made by road users, it considers the latent conditions throughout the system that lead to errors being made;
- It recognizes the fallible nature of humans and accepts that errors will occur;
- It promotes the development of appropriate countermeasures that are designed to treat both the latent conditions within the system and the errors made by operators;
- It promotes the development of error tolerance within systems;
- It promotes a shift in focus from the role of the individual road user to the system-wide failures involved in accident causation, removing the apportioning of blame to individual road users.

The literature also indicated that the systems perspective is beginning to receive attention worldwide within road transport. For example, in the World Health Organization (WHO) report on road traffic injury prevention, it is acknowledged that human behaviour is governed not only by the individual's knowledge and skills, but also by the environment in which the behaviour takes place (Rumar, 2004) and that indirect influences, such as road design and layout, vehicle nature and traffic laws and enforcement affect behaviour (WHO, 2004). According to the WHO report, evidence from a number of highly motorized countries demonstrates that a systems approach to road safety leads to a significant decline in road deaths and serious injuries (WHO, 2004).

Prototype models of road user error and contributing conditions

From the work conducted during Phases one and two of this research program, a prototype model of road user error and contributing conditions (*i.e.*, latent conditions) was developed (Salmon, Regan and Johnston, 2005b). The model is presented in Figure 13.2 and highlights the interaction between contributing conditions and road user behaviour that, in some cases, leads to errors being made by road users. According to the model inadequate conditions from the following five categories of contributing conditions impact road user behaviour in a way that can potentially lead to road user errors being made:

- Road infrastructure, *i.e.*, inadequate conditions residing within the road transport system infrastructure, including road layout (*e.g.*, confusing layout), road furniture (*e.g.*, misleading signage), road maintenance, (*e.g.*, poor road surface condition) and road traffic rules, policy and regulation related conditions (*e.g.* misleading or inappropriate rules and regulations)
- Vehicle, *i.e.*, inadequate conditions residing within the vehicles that are used within the road transport system, including human-machine interface (*e.g.*, poor interface design), mechanical (*e.g.*, brake failure), maintenance (*e.g.*, lack of maintenance), and inappropriate technology-related conditions (*e.g.*, mobile phone usage)
- Road user, *i.e.*, the condition of the road user involved, including road user physiological state (*e.g.*, fatigued, incapacitated), mental state (*e.g.*, overloaded, distracted), training (*e.g.*, inadequate training), experience, knowledge, skills and abilities (*e.g.*, inadequate skill), context-related (*e.g.*, driver in a hurry) and non-compliance related conditions (*e.g.*, unqualified driving)
- Other road users, *i.e.*, the contributing conditions caused by other road users, including other driver behaviour, passenger effects, pedestrian behaviour, law enforcement and other road user behaviour related conditions and
- Environmental, *i.e.*, the environmental conditions that might affect road user behaviour, including weather conditions, lighting conditions, time of day and road surface-related conditions.

Contributing conditions from one or more of the five categories described above can potentially impact road user behaviour, both in terms of cognitive behaviour such as perceiving, planning, hazard perception, situation awareness achievement and maintenance and decision-making, and physical behaviour such as vehicle control tasks. In most cases the effect of these conditions on road user behaviour is only minimal, and most road users can cope with the conditions and still perform the required activity safely. In other cases, the effect of the conditions may be greater, but the road user in question is able to cope with the conditions (due to factors such as skill level, ability and experience) and perform the required activity safely and without error. However, in some instances, the effect of the contributing conditions

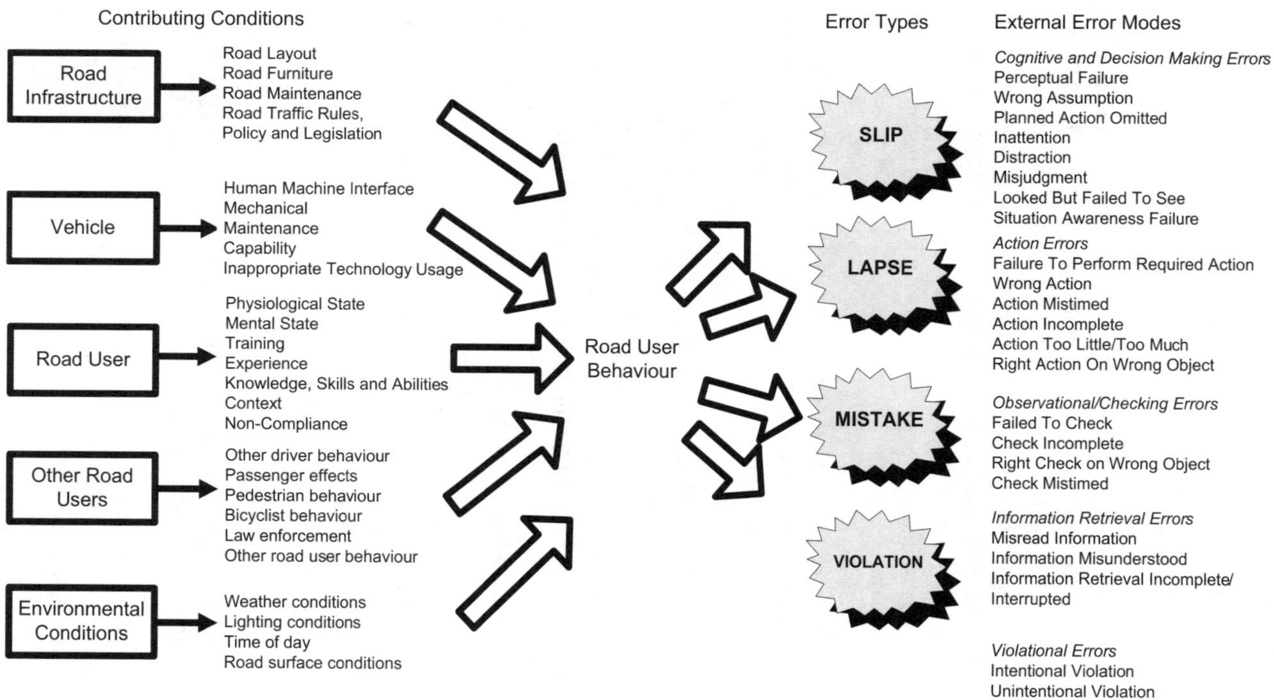

Figure 13.2 Prototype model of road user error and contributing conditions

is sufficient enough to cause the road user in question to make an error of some sort. The errors can either have no impact on the driving task, be recovered by the driver and have no impact, lead to an accident or safety compromising incident, or lead to further errors being made.

Framework for error tolerant road transport system

The second phase of this research program involved the development of a conceptual framework for promoting error tolerance at intersections in Victoria and throughout the Australian road transport system as a whole (Salmon, Regan and Johnston, 2005b). This involved identifying potential human error-related applications from other domains that could be used in road transport as part of an error management program. The underlying aim of the framework was to define a range of methods that would be appropriate for collecting and analysing human error-related data within the Australian road transport system, and also a range of methods that could be used to collate the data in a way that could inform the development of error management approaches and strategies designed to reduce, eradicate or manage road user error, and its contributory conditions. The framework was proposed both as a way of increasing our understanding of road user error, and for enhancing error tolerance at intersections and in the general road transport system. The proposed framework (presented in Figure 13.3) contains the following human error-related applications:

- Application of existing human error theory in the road transport domain;
- Collection of error-related data at specific road sites;
- Development of road user error and latent or error causing condition classification schemes;
- Development of a human factors-orientated road transport accident reporting tool;
- Development of a road transport incident or near-miss reporting system;
- Development of road user error and latent condition databases;
- Development of a valid human error identification technique for the road transport domain;
- Development of a road transport specific error management technique;
- Development of error tolerance strategies (*e.g.*, infrastructure and in-car technology design) and policies (*e.g.*, legislation and advertising campaigns) designed to increase error tolerance and/or mitigate error and its consequences within the road transport domain;
- Development of error management driver training interventions.

The internal structure of the framework can be decomposed into the following four levels: data collection (provides the techniques necessary for the collection of error and latent conditions data); data analysis (used to analyse the data collected); error identification (this is used to identify, both qualitatively and quantitatively, the different types of latent conditions residing in, and errors made in, the road transport system); and error tolerance strategies (involves the development and application of strategies designed to increase error tolerance within the road transport system

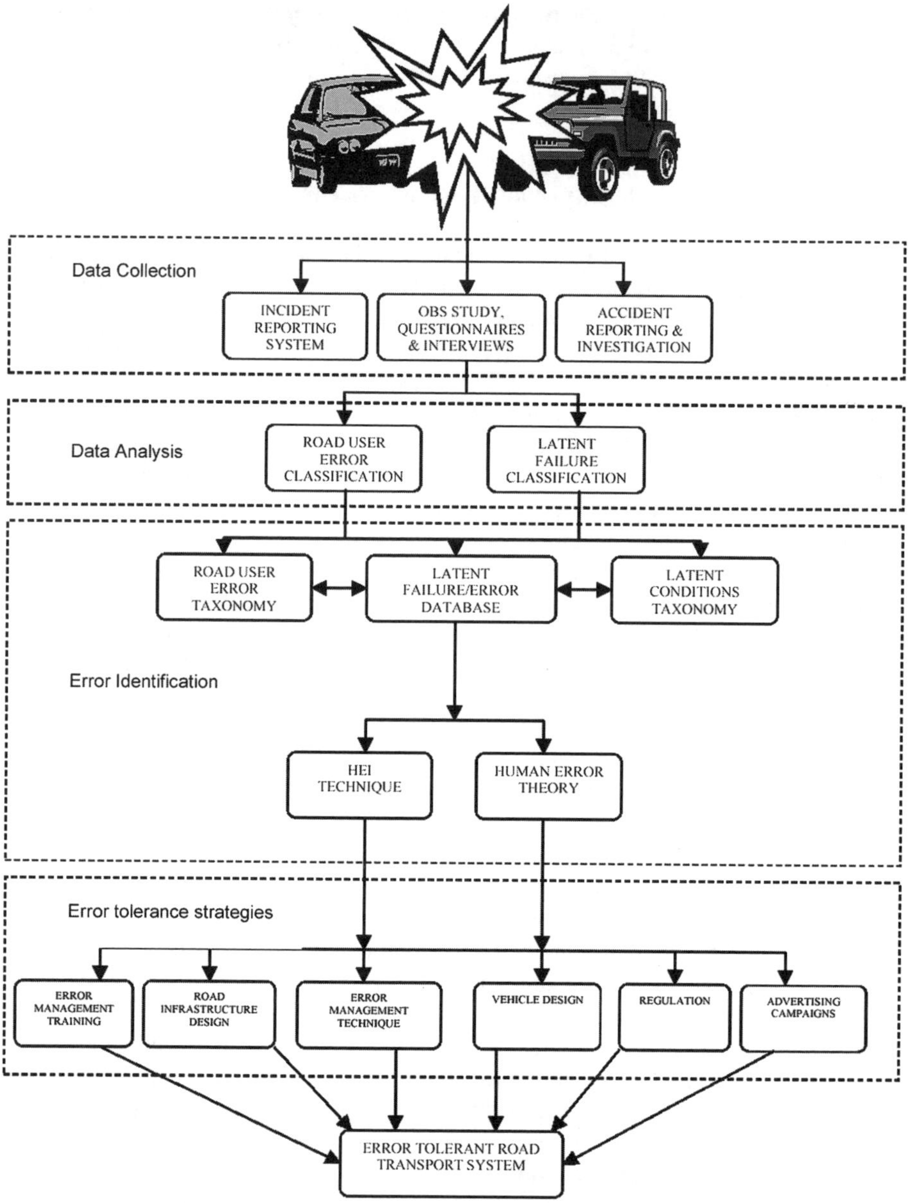

Figure 13.3 A framework for an error tolerant road transport system

and countermeasures designed to mitigate the latent conditions and errors identified previously).

Pilot study on road user error and contributing conditions at intersections in Victoria

The final two steps (Phases three and four) of this research program involve the design (Phase three) and conduct (Phase four) of a proof-of-concept pilot study on road user errors and latent conditions at intersections in Victoria. It is proposed that the pilot study be used to test and refine a novel methodology for collecting human error data within road transport and also to refine and validate the prototype model of road user error and contributing conditions presented in Figure 13.2. The proposed Phase four pilot study has the following three main aims:

- To collect data on the errors that are made by road users at intersections in Victoria and on the latent conditions within the road transport system that contribute to the errors being made (referred to as contributing conditions from this point onward);
- To collect data on various features associated with the errors made, such as the road user involved, the consequences of the errors, any recovery strategies used in the event of the errors being made and the tasks being performed when the errors are made; and
- To test, validate and refine a novel methodology for collecting human error-related data within the Australian road transport system.

Implications for data requirements

The aims of the pilot study described above have a number of implications for the data requirements associated with the study. Specifically, the data collected needs to cover not only the errors themselves, but also a range of factors surrounding the errors and also information related to the scenarios in which the errors are made. Based on the aims of the pilot study specified above, for each of the errors recorded, a series of data requirements were specified. The data requirements are presented in Figure 13.4.

An initial proof-of-concept pilot study design was proposed on the basis of the data requirements presented in Figure 13.4. The methods to be used in the proposed pilot study were extracted from the error tolerance framework (see Figure 13.3) and a brief review of the methods available for studying driver behaviour and collecting error-related data. The proposed pilot study methodology is presented in Figure 13.5.

A brief summary of each component step of the proposed methodology is presented on the following page:

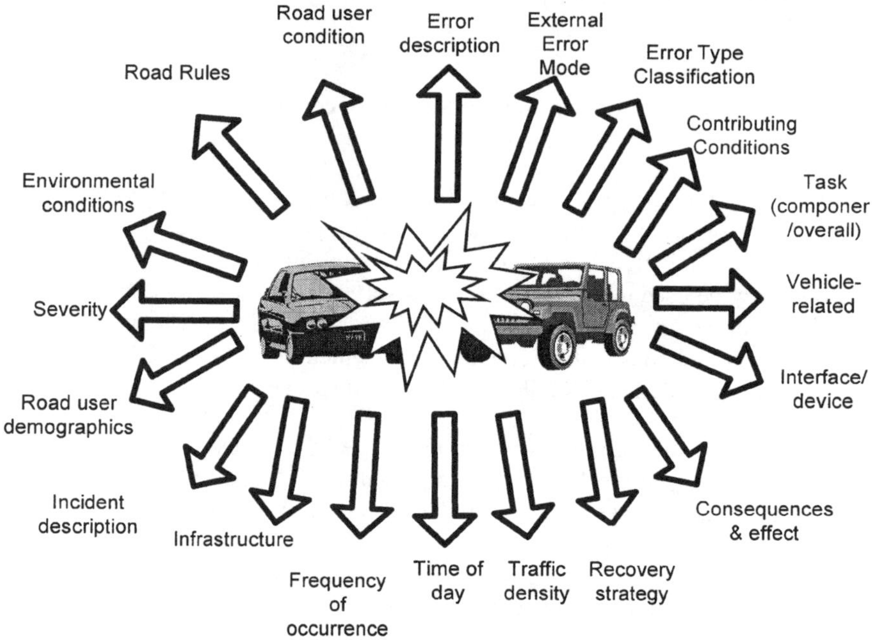

Figure 13.4 Data requirements for Phase four pilot study

- Selection of appropriate intersection sites. The first step of the pilot study methodology involves the selection of appropriate intersection sites on which to focus the Phase four pilot study. It is proposed that a range of intersection sites will be used for the study, including a T-intersection, a roundabout, a crossroad intersection, and a railway level crossing intersection. The intersection sites will be chosen on the basis of the VicRoads BlackSpot intersection ranking scheme and their suitability for observational study.
- SHERPA Error Predictions. Once the intersection sites are chosen, a panel of subject matter experts will use a modified Systematic Human Error Reduction and Prediction Approach (Embrey, 1986) to predict the errors that road users could potentially make while negotiating the intersections in question. The modified SHERPA analysis will involve developing hierarchical task analyses for each intersection site and then considering a range of contributing conditions (taken from the prototype model of road user error) for each component task step and then identifying any potential errors that could be made by road users whose behaviour is affected by the contributing conditions in question. It is proposed that, to validate SHERPA as an intersection error prediction tool, the errors predicted by the expert panel will be validated using the results of the pilot study. To validate the error predictions, a signal detection paradigm (Baber and Stanton, 1996) will be used. To do this, the predictions made by the expert panel will be compared to the errors that are observed during the

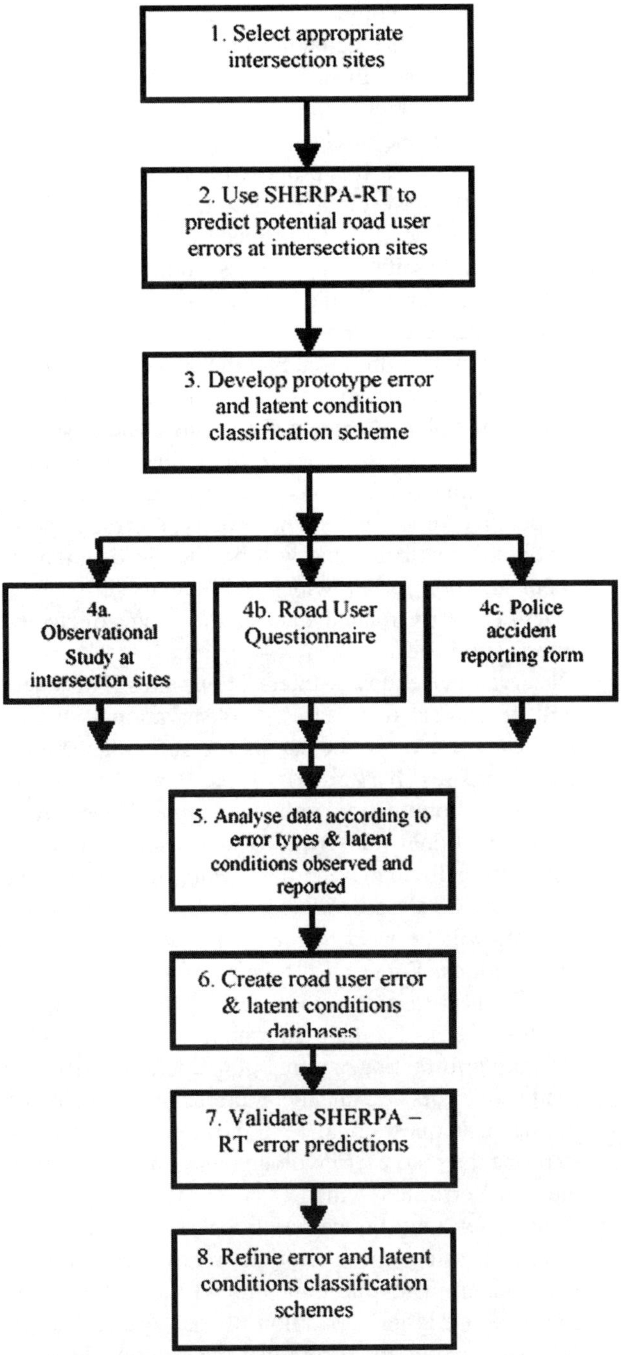

Figure 13.5 Proposed pilot study methodology

observational study component of the study. This allows validity statistics to be computed using the signal detection paradigm.

- Prototype Road User Error and Latent Conditions Classification Schemes. Prototype road user error and latent conditions classification schemes will be developed for use in classifying and analysing the data that is collected during the pilot study. It is proposed that these classification schemes will be developed on the basis of the findings derived from Phases one and two of this research and will be further informed by existing road user error classification schemes, error classification schemes from other domains, appropriate human error theory and through the conduct of walkthrough analyses conducted at the chosen intersection sites.
- Data collection. It is proposed that the data collection phase of the pilot study will comprise the following three different error data collection procedures:
  - Victorian Police Accident Reporting. The first component of the data collection phase will involve the collection of error-related data during the reporting of accidents occurring at intersections in Victoria within a specified time frame. The prototype error and contributing condition taxonomies as presented will be used to develop an additional accident reporting *pro forma* which will be used by Victoria Police officers to collect error-related data at the scene of accidents which occur at intersection sites.
  - Observational Study at Intersections sites. The selected intersection sites will be subject to a series of observational studies. The observational studies will involve the use of site-surveillance and follow up road user questionnaires. It is proposed that three data collection techniques will be used in conjunction with one another. These are 1) video surveillance, 2) observational transcript, 3) road user interview/questionnaire. Either conventional video recording equipment (*e.g.*, digital camcorder) or existing CCTV cameras (*e.g.* from the VicRoads Traffic Management Centre) will be used to record footage at each intersection during the observations. For each error-related incident, an incident master sheet will be used to create an observational transcript of the incidents. The master sheet will contain a narrative description and sketch of the incident, location information, a rating of incident severity, and a description of the road user errors and latent conditions involved. Where possible, a follow-up road user questionnaire will be administered to the road users involved to probe the road user involved regarding the contributing conditions that led to the error being made.
  - Road User Questionnaire/Survey. A road user questionnaire will be designed using the prototype road user error and latent condition taxonomies. The questionnaire will be administered via post or the Internet to a large population of road users. The questionnaire will ask drivers to recall any errors that they had made, or had seen being made in the recent past. A series of questions will then be used to probe the respondent for information regarding the nature of the error and the contributing conditions involved.

- Data Analysis. The data collected during the pilot study will be analysed to identify the factors associated with each of the errors recorded presented in Figure 13.4.
- Road User Error and Contributing Conditions Database Development. It is proposed that the data collected during the pilot study will be used to develop an initial database of road user error and contributing conditions at intersections. The database will contain information related to each of the different error types and contributing conditions identified. The database will be used to identify trends within the error data and also to inform the development of error tolerance strategies and error countermeasures. Further, the database will be used to inform the development of the larger study on error and latent conditions within the whole road transport system.

Conclusion

The overall aim of this project is to develop a novel error management program that ultimately leads to significant enhancements in the overall error tolerance of Australian road transport system, reductions in the errors made by road users and the eradication of the conditions that lead to errors being made in the first place. The first step in the development of such a program requires that human error-related data be collected and analysed to increase our knowledge and understanding of road user error in Australia and the conditions that lead to it. Only then can further steps be taken to develop error tolerance strategies and error countermeasures. The first three phases of this research program were conducted in order to develop a pilot study protocol for collecting these data. A prototype model of road user error and contributing conditions that will be used to classify and analyse the error data collected was developed. Further, a conceptual framework for an error tolerant Australian road transport system was proposed. The proposed framework contains both the methods with which to collect and analyse error-related data and also a number of error management approaches designed to reduce, eradicate or manage road user error and its contributory conditions. Finally, a study design based on the conceptual error tolerance framework was proposed for an initial proof-of-concept pilot study on road user error and latent conditions at intersections in Victoria. The conduct of the pilot study will take place during the next phase of this research program.

Acknowledgements

The authors wish to acknowledge that the research described within this article was funded by the Australian Transport Safety Bureau (ATSB) and the Monash University Baseline research program. The authors would also like to thank Dr Marcus Matthews from the ATSB for reviewing the project reports on which this article is based.

References

Baber, C. and Stanton, N.A. (1996), Human Error Identification Techniques Applied to Public Technology: Predictions Compared with Observed Use *Applied Ergonomics*, **27**, 119–131. [PubMed 15677051] [DOI: 10.1016/0003-6870%2895%2900067-4]

Civil Aviation Authority (1998), *Global Fatal Accident Review 1980–96* (CAP 681), London: Civil Aviation Authority.

Dekker, S.W.A. (2002), Reconstructing Human Contributions to Accidents: The New View on Human Error and Performance, *Journal of Safety Research*, **33**, 371–385. [PubMed 12404999] [DOI: 10.1016/S0022-4375%2802%2900032-4]

Embrey, D.E. (1986), SHERPA: A Systematic Human Error Reduction and Prediction Approach, *Paper Presented at the International Meeting on Advances in Nuclear Power Systems*, Tennessee: Knoxville.

Helmreich, R.L. (2000), On Error Management: Lessons from Aviation, *British Medical Journal*, 2000, No. 320, 781–785. [DOI: 10.1136/bmj.320.7237.781]

Lawton, R. and Ward, N.J. (2005), A Systems Analysis of the Ladbroke Grove Rail Crash *Accident Analysis and Prevention,* **37**, 235–244. [PubMed 15667809] [DOI: 10.1016/j.aap.2004.08.001]

Medina. A.L., Lee, S.E., Wierwille, W.W. and Hanowski, R.J. (2004), 'Relationship between infrastructure, driver error, and critical incidents', In *Proceedings Of The Human Factors and Ergonomics Society 48th Annual Meeting*, 2075–2080.

Reason, J. (1990), *Human Error*, New York: Cambridge University Press.

Reason, J. (1997), *Managing the Risks of Organizational Accidents*, Burlington, VT: Ashgate Publishing Ltd.

Reason, J. (2000), Human Error: Models and Management, *British Medical Journal*, **320**, 768–770. [PubMed 10720363] [DOI: 10.1136/bmj.320.7237.768]

Salmon, P.M., Regan, M. and Johnston, I. (2005a), Human Error and Road Transport: Phase one – Literature Review, Melbourne, Australia: Monash University Accident Research Centre.

PART 3
Normal Operations Monitoring and Surveillance Tools

Part 3 will explore methods of safety surveillance and monitoring behavior in the operational environment. It will emphasise the essential elements of monitoring normal operations as gleaned from the success factors surrounding LOSA (Line Operations Safety Audits)-type programs in aviation and rail. It will also recommend these practices as applicable to both medical and road domains, or for that matter any safety-critical industry. New dimensions in error management training will be covered with specific focus on cognitive elements, interpersonal skills and performance.

Chapter 14

Performance and Cognition in Dynamic Environments: The Development of a New Tool to Assist Practitioners

Jemma M. Harris and Mark W. Wiggins
University of Western Sydney

Scott Taylor
Eastern Australia Airlines

Matthew J.W. Thomas
University of South Australia

Successful performance in complex operating environments requires that operators function with a level of proficiency in cognitive skills such as situation awareness, planning, problem-solving, and decision-making. However, assessing the application of these skills within the operational environment has been difficult, to the extent that their application is often simply inferred on the basis of a set of behavioural responses. The broad aim of the present study is to develop a tool to identify the application of cognitive skills within the applied environment. Twenty-one experienced airline pilots participated in a series of cognitive interviews in which they were asked to describe the tasks that they engage during a flight. Preliminary analyses suggest that different phases of flight are associated with different levels of cognitive complexity and can be used as a basis to test the validity of a cognitive assessment tool.

Introduction

Human error has long been recognized as a predominant factor in accidents within advanced domains. Indeed, accident data reveals that the operator is associated with at least 70–80 per cent of civil and military aviation mishaps (Hawkins, 1993; O'Hare *et al.*, 1994; Wiegmann and Shappell, 1999), 80 per cent of shipping misadventures (Lucas, 1997), and 58 per cent of medical accidents (Leape *et al.*, 1991). More importantly, while the rate of mechanical causes of accidents has declined markedly over the last 40 years, the rate of human error has declined at a much lesser rate (Wiegmann and Shappell, 2001). This suggests that there is a need to further develop the tools and strategies necessary to combat the preconditions of human error in advanced technology environments.

O'Hare *et al.* (1994) sought to examine the underlying causes of human error through an investigation of 284 aviation accidents that occurred in New Zealand between 1983 and 1989. Nagel's (1988) model of information processing (cognition) was used to classify those accidents involving human error. Specifically, Nagel (1988) argues that information processing involves three stages: Perception; decision; and action. The perceptual stage of information processing involves acquiring information from visual, auditory, olfactory and/or tactile indicators (cues) and interpreting this information in the context of a task. The decision stage consists of generating alternatives and planning a course of action, while the final, action stage of information processing involves executing the course of action that has been selected during the decision stage (Nagel, 1988).

On the basis of Nagel's (1988) information processing stages, 71 per cent of the 284 accidents studied were considered to involve human error, whereby 22 per cent were coded as information errors, 35 per cent as decision errors, and 43 per cent as action errors (O'Hare *et al.*, 1994). However, a different picture emerged when the mishaps were divided according to accident severity. Of the 34 accidents in which there was a fatality or serious injury, 62.5 per cent were attributable to decision errors, while only 25 per cent involved the action stage (O'Hare *et al.*, 1994). Of the 169 cases involving minor/non-injury occurrences, only 30.5 per cent were the result of decision errors, while 45.6 per cent were attributed to action errors. These results closely mirror those reported by Jensen and Benel (1977) in their examination of NTSB records between 1970 and 1974 in which the majority of non-fatal incidents appeared to involve perceptual-motor factors, whereas most fatal accidents were associated with decision-related factors (Jensen and Benel, 1977).

A similar picture emerged when Jentsch, Hitt and Bowers (2002) examined the 61 aviation training issues contained within Funk and Lyall's (1997) human-automation interaction database of accident and incident reports, interviews, research reviews, and questionnaire data. Jentsch, Hitt and Bowers (2002) used an information processing model similar to that proposed by Nagel (1988) to classify automation-related problems. The results indicated that 86.9 per cent of issues were associated with perception and decision-making, while only 13.1 per cent of issues were associated with the action stage of human performance. Similarly, in air traffic control, 91.1 per cent of problems emerging from the introduction of a new air traffic control facility were classified as perceptual or decision-related in nature, while action issues accounted for only 8.9 per cent of concerns (Jentsch, Hitt and Bowers, 2002).

Overall, the results of various taxonomic investigations indicate that the majority of adverse occurrences in complex operating environments tend to be associated with the perceptual and decision-related (cognitive) stages of information processing. Further, it appears clear that the severity of these occurrences tends to be greater than the severity associated with errors at the action stage of information processing. These results suggest that there is a need to consider the development of tools that enable the assessment of cognitive skills amongst operators as a means of identifying and responding to potential sources of error.

The management of errors in advanced environments: assessment and training considerations

Typically, the management of errors within advanced technology environments begins with the development of a training initiative as a means of either creating an awareness of the potential for errors amongst operators or, alternatively, developing a skill base to enable operators to recognize and avoid errors in the future. However, as training and education account for approximately 20 per cent of the operational costs in advanced environments, there is some concern as to whether this investment is justified in terms of the improvements in performance (Fister, 2000). Indeed, there are a number of deficiencies that have been observed in the assessment and training of operators, the most significant of which is a focus on the development of procedural skills to the exclusion of cognitive skills (Seamster, Redding and Kaempf, 1997).

The development of automated systems in advanced technology environments such as aviation has resulted in a change in the demands for cognitive skills. However, contemporary assessment and training is primarily orientated towards the acquisition of procedural skills. This approach emphasizes overt behaviour as the basis of performance and has been evident in the establishment of behavioural competencies for the assessment of skills (Seamster, Redding and Kaempf, 1997). The difficulty with this approach is that the appropriateness, or otherwise, of cognitive skills is unlikely to be accurately assessed through inferences that are made on the basis of overt behaviour.

Both for practitioners and for organizations, there is a need to assess directly the application of cognitive skills. At the level of the operator, it enables training to be delivered more efficiently to address a particular deficiency and facilitate an improvement in performance. For an organization, there are financial incentives associated with the prevention of errors and improvements in productivity. However, in each case, a tool is necessary to identify when and how cognitive skills should be used by operators. The notion of cognitive complexity provides a starting point for the development for this type of cognitive skill assessment tool.

Cognitive complexity: the missing piece of the puzzle?

Woods (1988) argues that it is the cognitive complexity associated with a task that will determine the level and the types of cognitive skills that are necessary for successful performance. Here, cognitive complexity is concerned with the objective state of the task relative to other tasks, rather than the perceived level of cognitive difficulty which is reflected in models of cognitive load (*e.g.*, Sweller, Merrienboer and Pass, 1998). The cognitive complexity of a task is determined, initially, by considering the task as a response to a change that has occurred within the system state.

A change in a system state is normally reflected in a series of indicators, and it is the nature of these indicators and the required response of the operator that determines the level of cognitive complexity of the task (Woods, 1988). For instance, the indicators may be dynamic such that events will occur at times that are indeterminate. This inability to predict the nature of changes requires a level of situational awareness and the capacity to monitor a series of indicators over a period of time (Woods, 1988). Similarly, when subsystems are highly interconnected, a failure reflected by one indicator may have multiple consequences, so that attention must be directed towards multiple sources of failure and the potential consequences of these failures.

In responding to indicators of system function, there is always an associated level of uncertainty as to whether the indicators are accurate and/or whether the pattern of indication is an accurate interpretation of the system state (Woods, 1988). This uncertainty demands a level of information acquisition and the generation and evaluation of potential options from memory (Woods, 1988). To implement a response to these changes, there is a requirement for risk assessment to establish whether the response is appropriate given the circumstances.

Woods (1988) argues that as the cognitive complexity of the environment increases, higher-order cognitive skills (*e.g.*, decision making) are likely to be engaged, while relatively lower levels of cognitive complexity will demand lower-order cognitive skills (*e.g.*, situation assessment) (Woods, 1988). Moreover, cognitive complexity can be measured by assessing dynamism, the interrelationships with subsystems, and the level of uncertainty and risk associated with the indicators that are used to manage a task (Woods, 1988). The primary objective of the present study was to construct a methodological technique to identify the cognitive complexity associated with various stages of flight.

Method

Participants

The participants consisted of 21 pilots from a regional airline. Of these pilots, six were first officers, six were captains, five were training captains and the remainder were check captains. The total flight experience ranged from 3,500 to 16,500 hours.

Stimuli/procedure

The participants were advised of a scenario in which they were the pilot flying on the first flight of the day from Sydney to Coffs Harbour. A semi-structured interview using the cognitive task analysis technique was used to acquire information pertaining to the tasks required for each stage of the flight. The stages included pre flight; taxi and take off; climb; cruise; descent; approach and landing; and post landing. The discussion of the stages was counterbalanced to control for possible order effects. For each of the tasks identified, the participants were asked four questions relating to different aspects of cognitive complexity. A fifth question pertaining to perceived task difficulty was included as a subjective estimate of the cognitive load. The questions and the scoring procedure are listed in Table 14.1. The responses from participants were tape recorded for later transcription.

Table 14.1 Cognitive task analysis interview questions

Questions	Scoring
How often can you rely on the information that you use to perform the task?	(1) Always (2) Sometimes (3) Never
To what extent does the information that you use to perform the task change over time?	(1) On a daily basis (2) On an hourly basis (3) On a minute by minute basis
How often do you get the task wrong?	(1) Almost never (2) Sometimes (3) Most of the time
How important is it to get the task right?	(1) Not very important (2) Moderately important (3) Extremely important
How difficult is the task?	(1) Not very difficult (2) Moderately difficult (3) Extremely difficult

Results

Number of tasks

The following results represent preliminary findings from the investigation. As evident in Table 14.2, the pre flight stage involved the greatest number of tasks, while the descent stage involved the fewest tasks.

Table 14.2 Mean number of tasks described as a function of stage of flight

	Mean	Standard deviation
Pre flight	35.40	26.49
Taxi and take off	22.20	15.83
Climb	11.00	8.63
Cruise	13.00	7.02
Descent	7.88	5.52
Approach and landing	13.07	4.81
Post landing	12.60	7.27

Cognitive complexity

For each participant, their mean scores on the four cognitive complexity items were summed to provide an overall mean score for the level of cognitive complexity for each stage of flight, with a possible range of 4 through to 12. As is evident in Table 14.3, the approach and landing stage was associated with highest mean cognitive complexity score relative to the other stages of flight, while the post landing stage was associated with the lowest mean cognitive complexity score. However, it should be noted that when the cognitive complexity scores for the seven stages of flight were compared using a one-way analysis of variance (ANOVA), using $\alpha = 0.05$, the result failed to reach statistical significance, $F_{6,141} = 1.75$, $\rho = 0.19$, $\eta^2 = 0.107$.

Table 14.3 Mean cognitive complexity scores as function of stage of flight

	Mean	Standard deviation
Pre flight	6.92	0.76
Taxi and take off	7.50	0.85
Climb	7.21	0.89
Cruise	6.86	0.86
Descent	7.21	1.12
Approach and landing	7.62	0.65
Post landing	6.77	1.01

Table 14.4 Mean perceived difficulty scores as function of stage of flight

	Mean	Standard deviation
Pre flight	1.08	0.28
Taxi and take off	1.14	0.36
Climb	1.50	0.52
Cruise	1.21	0.43
Descent	1.50	0.52
Approach and landing	1.69	0.48
Post landing	1.08	0.28

Perceived difficulty

The mean perceived difficulty scores for the seven stages of flight were analysed using a one-way ANOVA, with critical alpha set at 0.05. The ANOVA test assumptions were satisfactory, and the result was statistically significant, $F_{6,141} = 4.53$, $\rho = 0.00$, $\eta^2 = 0.24$. *Post hoc* comparisons using the Tukey HSD test revealed a statistically significant difference between the approach and landing stage of flight and the pre flight, taxi and take off, and post landing stages. Inspection of the means indicated that the higher perceived difficulty score was associated with the approach and landing stage. Descriptive statistics are provided in Table 14.4.

Discussion

The primary objective of the present investigation was to construct a methodological technique to identify the cognitive complexity associated various phases of flight. The process involved the application of a cognitive task analysis in which participants were asked to describe the key tasks that they perceived to be necessary to successfully complete each stage of a flight from Sydney to Coffs Harbour. For each of the tasks that they identified, the participants were asked to rate the level of cognitive complexity and their perception of the difficulty associated with the performance of the task.

The results revealed that the methodological technique was relatively successful in distinguishing the different levels of cognitive complexity for the key tasks required to successfully perform the various stages of flight. The approach and landing stage of flight appeared to be associated with the highest level of cognitive complexity, while the post-landing stage was associated with the lowest level of complexity. However, it should be noted that the results from a one-way ANOVA failed to establish a statistically significant difference between these assessments.

In relation to the perceived difficulty associated with the performance of the tasks, a statistically significant difference was evident between the stages of flight, whereby the approach and landing stage of flight was rated as significantly more difficult than the pre flight, post landing, and taxi and take off stages. In combination with the results associated with the assessments of cognitive complexity, this outcome suggests that there are differences between the cognitive demands associated with different stages of flight and that these differences may relate to the cognitive complexity of the tasks involved.

The outcomes of the present study are broadly consistent with previous investigations of the differences in the demands associated with different stages of flight. Indeed, in an investigation of accident occurrences in corporate flight operations within the USA over a one-year period, 47 per cent of accidents were cited as occurring during final descent and touchdown phases of flight (Trammell, 1980). Similarly, in an examination of the worldwide commercial jet fleet, Boeing reports that from 1995 through to 2004, the final approach and landing phases of flight generated 51 per cent of the total accidents (Boeing, 2005).

The present study is part of a larger investigation that involves the development of a tool to assist the identification of the application of cognitive skills within the applied environment. It draws on Woods' (1988) assertion that as the cognitive complexity of the environment increases, higher-order cognitive skills *(e.g.*, decision-making) are likely to be engaged, while relatively lower levels of cognitive complexity will demand lower-order cognitive skills (*e.g.*, situation assessment). Therefore, as part of this larger investigation, the relationship between the cognitive complexity of a task and the cognitive skills required will be further examined.

It is anticipated that the qualitative data generated during the present study will be examined with reference to the frequency with which higher and lower cognitive concepts are used by participants to describe the key tasks as a function of cognitive complexity and stage of flight. It is expected that tasks that are associated with a relatively greater level of cognitive complexity will be described using concepts that reflect the application of higher-order cognitive skills (*e.g.*, terms such as 'decide'). Conversely, tasks that are associated with a relatively lower level of cognitive complexity are expected to be described using concepts that reflect relatively lower-order cognitive skills (*e.g.*, terms such as 'look').

To validate the relationship between the cognitive complexity of a task and the cognitive skill required to successfully perform that task, a further study will examine the pattern of cue acquisition for tasks that are structured to embody a high or low level of cognitive complexity. Cues represent features of an environment that have some meaning or association for the receiver and may be visual, auditory, olfactory, or tactile in form (Ratcliff and McKoon, 1995). Previous research has identified that an operator's pattern of cue acquisition will vary as a function of the demands of the task (*e.g.*, Bellenkes, Wickens and Kramer, 1997; Wiggins *et al.*, 2002; Underwood, 2005).

Overall, the results of the present research provide the basis for a methodology that will be used subsequently to test the validity of a cognitive skill assessment tool. In particular, there is evidence to suggest that different stages of flight are associated with different levels of perceived cognitive complexity and perceived difficulty. Ultimately, it is anticipated that this research will provide the foundation for a more efficient approach to cognitive skills training whereby training can be directed to correct a particular deficiency. In the long-term, this approach is expected to yield positive consequences for the management of human error and system performance within advanced domains.

References

Bellenkes, A.H., Wickens, C.D. and Kramer, A.F. (1997), Visual Scanning and Pilot Expertise: The Role of Attentional Flexibility and Mental Model Development *Aviation Space and Environmental Medicine*, **68**, 569–579.

Boeing (2005), Statistical Summary of Commercial Jet Airplane Accidents: Worldwide Operations 1959-2004 from www.boeing.com/news/techissues/pdf/statsum.pdf, Accessed on 11 December 2005.

Fister, S. (2000), Reinventing Training at Rockwell Collins, *Training*, **54**, 64–70.

Funk, K. and Lyall, E. (1997), Flight Deck Automation Issues, from http://www.flightdeckautomation.com/fdai.aspx.

Hawkins, F.H. (1993), *Human Factors in Flight*, Aldershot: Ashgate.

Jensen, R.S. and Benel, R.A. (1977), Judgement Evaluation and Instruction in Civil Pilot Training, *Final Report FAA-RD-78-24*, Springfield, VA., USA: National Technical Information Service.

Jentsch, F., Hitt II, J.M. and Bowers, C., (2002), Identifying training areas of advanced automated aircraft: Application of an information processing model, *Advances in Human Performance and Cognitive Engineering Research*, 2, 123–137.

Leape, L.L., Brennan, T.A., Laird, N., Lawthers, A.G., Localio, A.R., Barnes, B.A., Hemert, L., Newhouse, J.P., Nagel, D., Weiler, P.C. and Hiatt, H. (1991), The Nature of Adverse Events in Hospitalized Patients *New England Journal of*

Medicine, 524, 377–384.

Lucas, D. (1997), The Causes of Human Error, in *Human Factors in Safety Critical Systems* Redmill, F. and Rajan, J eds., Oxford: Butterworth-Heinemann, 57–65.

Nagel, D. (1988), Human Error in Aviation Operations, in *Human Factors in Aviation*, Wiener E. L. and Nagel, D. eds., London: Academic Press, 263–303.

O'Hare, D., Wiggins, M., Batt, R. and Morrison, D. (1994), Cognitive Failure Analysis for Aircraft Accident Investigation *Ergonomics*, **11**, 1855–1869.

Ratcliff, R. and McKoon, G. (1995), Sequential Effects in Lexical Decision: Tests of Compound-Cue Retrieval Theory *Journal of Experimental Psychology: Learning, Memory, and Cognition*, **21**, 1380–1388. [PubMed 8744970] [DOI: 10.1037/0278-7393.21.5.1380]

Seamster, T.L., Redding, R.E. and Kaempf, G.L. (1997), *Applied Cognitive Task Analysis in Aviation*, Aldershot: Avebury Technical.

Sweller, J., van Merrienboer, J.G. and Pass, F.G. (1998), Cognitive Architecture and Instructional Design *Educational Psychology Review*, **10**, 251–279. [DOI: 10.1023/A%3A1022193728205]

Trammell, A. (1980), *Cause and Circumstance*, New York: Ziff: Davis.

Underwood, J. (2005), Novice and Expert Performance with a Dynamic Control Task: Scanpaths during a Computer Game, in *Cognitive Processes in Eye Guidance* edited by: Underwood, G., New York: Oxford University Press, pp. 303–320.

Wiegmann, D. and Shappell, S. (1999), Human Error and Crew Resource Management Failures in Naval Aviation Mishaps: A Review of U.S. Naval Safety Centre Data, 1990-96 *Aviation Space and Environmental Medicine*, **70**, 1147–1151.

Wiegmann, D. and Shappell, S. (2001), Human Error Analysis of Commercial Aviation Accidents: Application of the Human Factors Analysis and Classification System (HFACS) *Aviation Space and Environmental Medicine*, **72**, 1006–1016.

Wiggins, M.W., Stevens, C., Howard, A., Henley, I. and O'Hare, D. (2002), Expert, Intermediate and Novice Performance during Simulated Pre-Flight Decision-Making *Australian Journal of Psychology*, **54**, 162–167. [DOI: 10.1080/00049 530412331312744]

Chapter 15

Error Management Training: Identification of Core Cognitive and Interpersonal Skill Dimensions

Matthew J.W. Thomas and Renée M. Petrilli
University of South Australia

Human Error remains a significant causal factor in incidents and accidents across a range of high-risk industries. The development of specific error management training programs has recently been suggested as a key strategic response to the ongoing impact of error at the individual and team level. However, we still lack a sophisticated understanding of how error management training might best be deployed, and we lack a clear scientific definition of best practice in error management training. This chapter presents the findings of a preliminary interview-based study designed to explore and document the informal error management strategies used by expert Training Captains in the commercial aviation environment. The interview data provides a range of insights into the processes of error management employed by expert pilots.

Introduction

The safe actions and satisfactory performance of personnel are essential aspects of maintaining safety across a wide range of high-risk industries. Accepted models of accident trajectory typically include both active failures of personnel and systems, as well as latent conditions which may lie dormant in an organization's operational system for considerable time (Reason, 1990). Closely aligned to the concept of active failures and latent conditions are the terms error and threat respectively, concepts which have recently been the focus of considerable research in the commercial aviation setting.

Operational personnel act as the last line of defence in complex operational environments (Reason, 1997). Safety is often maintained through the actions of individuals 'at the coal-face' through their response to complex and sometimes ill-defined problems. Accordingly, the management of threat and error has been suggested to be a necessary focus of any organization's attempts to effectively maintain safety in high-risk operations (Klinect, Wilhelm and Helmreich, 1999). Furthermore, human error is now accepted as a natural part of everyday performance, and can occur both spontaneously or can be precipitated by a variety of environmental and personal factors such as individual proficiency, workload, fatigue, and team-dynamics.

As Helmreich (2000) suggests, given the ubiquity of human error, and the wide range of factors which promote error, a key to safety lies in effective error management by operational personnel. In response to the increasing sophistication in our understanding of the role of error management in enhancing operational performance and safety, error management training programs are becoming innovative new elements of many airlines' training systems (Phillips, 2000). However, recent research has highlighted a number of challenges facing effective error management training programs. Thomas (2003), in a study of Line Training in the commercial airline setting, highlights the difficulties associated with the effective detection of error events during training, and the lack of instructor debrief and analysis of errors as they occur. Current error management training practices lack large-scale empirical investigation, and existing evidence suggests that they may require considerable refinement and improvement. Indeed, the continued findings of low levels of error detection during audits of normal operations suggest that error management training is an area requiring urgent investigation.

This chapter presents the results of a project that sought to investigate effective strategies for error management training within commercial aviation. More specifically, the study set out to investigate the wide range of individual expertise and informal strategies used in the management of errors during normal flight operations, and the development of error management skills in the training environment.

Method

Participants

Participants were volunteer pilots who each had significant instructional experience, and who were currently working, or had recently worked, in Training and Checking roles within the commercial airline environment. Volunteers were recruited using an information sheet for the study, which was distributed through both formal and informal networks of airline Check and Training pilots. A total of 14 experienced aviators (instructors, check captains, and training captains) were recruited for the study.

Design and procedure

The study adopted an interview-based approach to the investigation of the core components of effective error management, and error management training within the commercial aviation environment. As an explicit objective of the study was to collect data with respect to the tacit knowledge of domain experts, the study was designed within an interpretive framework.

The study used an adapted form of the Critical Decision Method, which is a well-established cognitive interview technique (Klein, Calderwood and Macgregor, 1989). The Critical Decision Method involves a semi-structured interview of experienced personnel and is designed to efficiently gather data on the bases of the proficient performance of complex tasks. Specifically, the Critical Decision Method is used to identify aspects of expert performance from a cognitive perspective, and serves to highlight the frequently hidden components of expertise that underlie and drive the overt observable behaviours. Furthermore, this technique has been used in the process of training system specification, as a method for identifying and defining training requirements.

The Critical Decision Method, as adapted for this study, involves an interview of approximately one-hour duration for each participant. The interview proper was broken into three major sections, with the Critical Decision Method being used three times to explore different aspects of the management of error in normal operations as well as training. The three sections of the interview were broadly: 1) the management of slips and lapses; 2) the management of mistakes; and 3) error management from an instructional perspective.

Analysis

As the objective of the study was to explore and document the error management strategies used by expert pilots, in order to inform the syllabus and curriculum structure of an error management training package, the analysis of interview data was undertaken with explicit reference to the interpretive research design. Accordingly, the study did not seek to test pre-defined hypotheses or adopt any detailed quantitative analysis of data.

Results

The interview data provided a range of insights into the processes of error management employed by expert pilots. The scope of this chapter allows only for a brief discussion of the major categories of error management competencies identified by the expert pilots in the interview data. A more detailed exploration of each of these categories is provided in the full report (Thomas and Petrilli, 2004).

Error avoidance is the first stage in error management, and adopts the perspective that the minimization of error is a critical first step in enhancing safety in normal operations. While it is accepted that the task of eliminating human error is an impossible goal, there are certainly a wide range of techniques that can be employed to avoid, and thus reduce the occurrence, of error. Fourteen components of error avoidance were identified from the analysis of interview data, which have been grouped under four broad categories:

Situation Awareness

- Attention, vigilance and comprehension
- Pre-action attention
- Self-monitoring

Multi-Crew Functions

- Monitoring other crewmember(s)

- Communication
- Teamwork and support

Task Management
- Avoidance of error producing conditions
- Active dependence on standard operating procedures
- Planning and preparation
- Gates
- Deliberate and systematic decision-making
- Review and evaluation

Attitudinal Factors
- Conservatism
- Diligence

Error detection mechanisms were found to share much in common with the strategies identified above for error avoidance, as effective error detection depends to a large degree on maintaining situation awareness. However, aside from a focus on situation awareness, the interview data suggest that effective error detection also includes a range of multi-crew coordination factors, as well as unique attitudinal factors. A total of nine components of error detection were identified, under three broad categories.

Situation Awareness
- Maintaining a mental model of the flight
- Monitoring: scan and systematic check
- Detecting divergence
- Self-monitoring

Multi-Crew Functions
- Monitoring and cross checking of other crewmember(s)
- Familiarity with other crewmember(s)
- Communication

Attitudinal Factors
- Expectation of error
- Comfort levels and intuition

Error response is the third phase of error management, and involves the rectification of the error, or the resolution of any problem-state caused by an error. Once the error has been detected, this process should present few problems for the crew who has regained situation awareness. The following discussion highlights areas in which error response can be managed in order to create efficiencies in the error management process and enhance the safety of normal flight operations. From the interview data, a total of nine components of error response were identified under four broad categories:

Situation Awareness

- Information gathering and problem diagnosis
- Projection into the future and identification of alternatives

Task Management

- The maintenance of safety
- Deliberate decision-making process

Multi-Crew Functions

- Communication and information sharing
- Management of error response
- Workload management

Attitudinal Factors

- Acceptance of error
- Avoidance of rumination

Error management training refers to the structured development of error management skills in the training environment. A critical premise for error management training is that it should not form a separate element of a training curriculum, but rather elements of error management training should be integrated into ground, simulator and line training, as well as inform aspects of ongoing development of expertise in already experienced pilots. Data from the interviews highlighted eight core components of error management training grouped under three broad categories.

Experiential Factors

- Exposure to error producing conditions
- Exposure to errors and problem states
- Structured 'hands-on' training in error management and solutions

Attitudinal Factors

- Understanding cause and effect
- Development of confidence
- Development of self-analysis skills

Debriefing

- Choice of error-events to debrief
- Focus of error-event debrief

Discussion and conclusion

This study has provided a detailed account of error management, through the analysis of experts' understanding of error avoidance, detection and resolution. From the results of this study, it is evident that elements of error management share much in common with our current understanding of Crew Resource Management (CRM).

Aspects such as situation awareness, task management and communication are all common elements of CRM programs in modern airlines.

However, this study has provided specific new detail with respect to the actual cognitive processes of error management that we commonly group together under such categories as situation awareness. Moreover, the results of this study place significant emphasis on metacognitive processes that underlie the more evident cognitive, affective and interpersonal components of error management.

The findings of this study reinforce a model of error management that emphasizes the process of mismatch emergence as the driver of error detection, problem identification and error resolution (see Figure 15.1).

This model of error management, as first proposed by Rizzo, Ferrante and Bagnara (1995) and subsequently expended in this study, has several unique features that suggest it offers considerable explanatory power in relation to the detection of safety breaches in dynamic real-world contexts.

Firstly, the model makes explicit that the crucial mechanism involved in error detection is that of mismatch emergence, whereby conflict arises between the expected state of the work system and the actual observed state of the system. The results of this study emphasize the role of situation awareness, and in particular the construction of a dynamic mental model of the desired and expected state of the 'system': the aircraft status and configuration in space and time. This study has identified a range of cognitive and metacognitive processes that provide an adequately comprehensive framework for mismatch emergence to reliably occur.

Secondly, the model is consistent with naturalistic explanations of expert behaviour whereby action schema are frequently activated in response to environmental stimuli with little or no conscious processing on behalf of the operator. Indeed, the model allows for situations whereby safety breaches are contained through immediate response to the emergence of a mismatch in system state. Accordingly, error management can itself be seen as dynamic expert behaviour rather than a serial and rational process.

In everyday environments, corrective action is frequently applied before a clear understanding of the exact nature of the problem, or even recognition that a specific error has been committed. This error management pathway is illustrated through the bottom arrow in the figure above. Using an example of expert pilot performance, a significant glide-slope deviation is likely to be corrected through automatized actions of changes in attitude and thrust long before the problem state is clearly defined and the originating error, such as the speed-brake remaining extended, is actually detected. In this example, mismatch emergence leads directly to correction, then through a backwards flow of information through the model, the problem-state is diagnosed and the original error can then be detected. Using the language of Rasmussen's (1987) levels of performance, this form of error management could be termed skill-based error management with respect to the high levels of automatic cognitive processing at play.

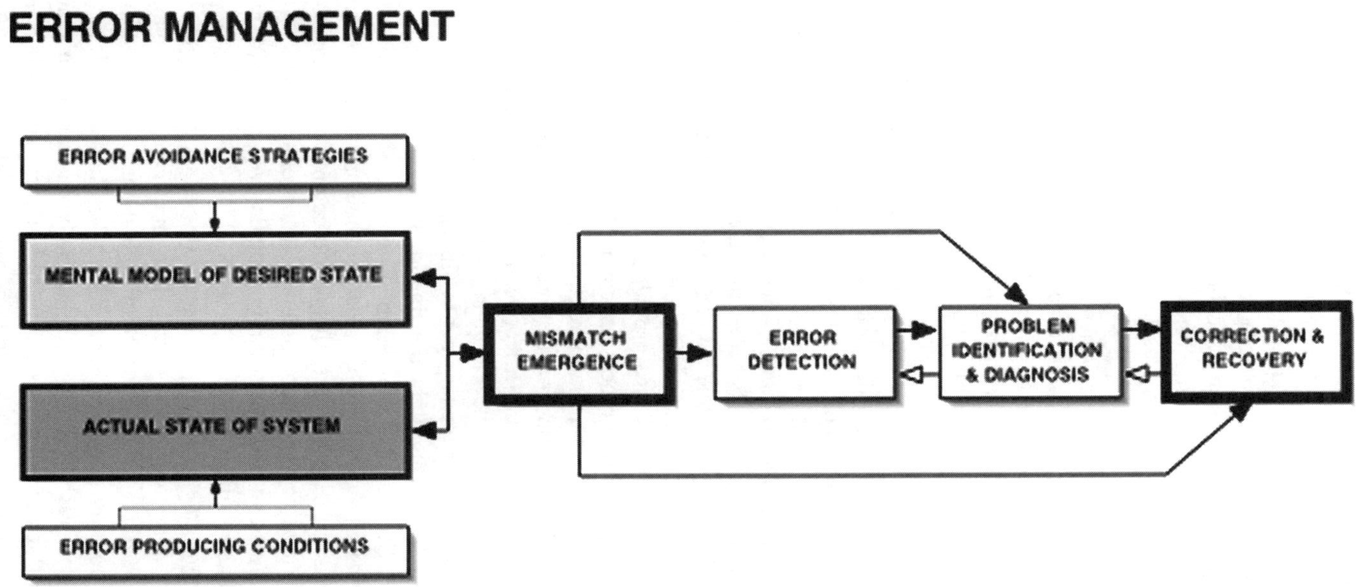

Figure 15.1 A model of error management

In situations where the mismatch emergence does not immediately activate any specific corrective schema, a process of problem identification might in turn activate a variety of possible corrective schema through the application of stored rules. For instance, the detection of adverse weather in close proximity to the aircraft might activate a variety of stored rules relating to the impact of wind strength and direction on the movement of weather cells, the requirements for aircraft configuration in weather, and the impact of the direction of deviations from track on total track miles and time to destination. Accordingly, this form of error management could be termed rule-based error management, as it involves the identification and activation of stored rules from the long-term memory.

In certain situations, where the mismatch emergence presents ill-defined or ambiguous system states, more conscious processing might be required to actively search for an error in order to accurately identify and diagnose the problem-state. Only once this conscious search for errors has been undertaken, and the error detected, can appropriate recovery action be taken. For instance, the detection of problems with the descent profile as governed by the automatic flight system might lead to a conscious search for an error in the programming of the flight management computer. It therefore follows that this form of error management can be termed knowledge-based error management, with respect to the high levels of conscious processing involved.

This study has provided a wide range of perspectives that in turn can inform a comprehensive curriculum structure for error management training. Much of what has been identified in this study is in line with existing human factors knowledge, and many parallels are to be found with existing Crew Resource Management training programs. However, this study has also provided a range of new perspectives on error management, and the types of processes that might contribute to effective error management training programs. Furthermore, the study highlights three important new developments for error management training: 1) the need to focus more on cognitive skill development and the affective domain; 2) the need to integrate technical and non-technical skill development; and 3) the need to increase the experiential components of error management training.

The results of this study highlight the important role of non-technical skills in error management. Such skills as situation awareness, construction of accurate mental models and mental simulation, anticipation and contingency planning, self-monitoring, and deviation detection are all generic cognitive processes that have been identified as critical components of error management. Accordingly, any successful error management training program should develop an explicit focus on the development of these skills in pilots.

This study has highlighted the complementary role of technical knowledge and skill alongside a wide range of non-technical skills in effective error management. Therefore, the integrated development of technical and non-technical skills through carefully designed training programs appears to be fundamental in the error management training process. Similarly, it is unlikely that any error management training program will achieve the greatest possible benefit unless it is integrated within existing forms of simulator-based training and aircraft-based training.

Finally, the results of this study have highlighted that error management training cannot be seen just as a 'classroom' activity. Rather, to explore and develop the wide range of competencies that underpin effective error management, specific experiential forms of training must be used. The error management competencies highlighted by the experienced pilots interviewed in this study are all context-driven, and likely to be affected themselves by a range of error producing conditions such as high-workload, stress and distraction. Accordingly, the task management elements of effective error management dictate a need to embed an error management training focus within existing experiential forms of training in commercial aviation.

Acknowledgements

The study was undertaken as the first of two studies funded through a grant from the Commonwealth through the Department of Transport and Regional Services on behalf of the Australian Transport Safety Bureau. The authors sincerely thank the Australian Commonwealth Government for the support provided to undertake this research.

The authors also wish to thank the volunteer Check and Training Captains who generously gave up their time to share their expertise. This study would not have been possibly without the generosity of the participants and their willingness to discuss openly their own strategies to deal with their errors, and the errors of their trainees within the commercial aviation environment.

References

Helmreich, R.L. (2000), On Error Management: Lessons from Aviation, *British Medical Journal*, **320**, 781–785. [PubMed 10720367] [DOI: 10.1136/bmj.320.7237.781]

Klein, G.A., Calderwood, R. and Macgregor, D. (1989), Critical Decision Method for Eliciting Knowledge, *IEEE Transactions on Systems, Man, and Cybernetics*, **19**, 462–472. [DOI: 10.1109/21.31053]

Klinect, J.R., Wilhelm, J.A. and Helmreich, R.L. (1999), Threat and Error Management: Data from Line Operations Safety Audits, in *Proceedings of the Tenth International Symposium on Aviation Psychology* edited by: Jensen, R. S., Columbus, OH: Ohio State University.

Phillips, E.H. (2000), Managing Error at Centre of Pilot Training Program, *Aviation Week and Space Technology*, **153**, 61–62.

Rasmussen, J. (1987), Cognitive Control and Human Error Mechanisms, in *New Technology and Human Error* eds Rasmussen, J., Duncan, K. and Leplat, J., Chichester, UK: John Wiley & Sons.

Reason, J. (1990), *Human Error*, Cambridge, UK: Cambridge University Press.

Reason, J. (1997), *Managing the Risks of Organizational Accidents*, Aldershot: Ashgate.

Rizzo, A., Ferrante, D. and Bagnara, S. (1995), Handling Human Error, in *Expertise and Technology: Cognition & Human-Computer Cooperation* edited by: Hoc,

J.M., Cacciabue, P.C. and Hollnagel, E., Hillsdale, NJ: Lawrence Erlbaum Associates.

Thomas, M.J.W. (2003), Improving Organizational Safety through the Integrated Evaluation of Operational and Training Performance: An Adaptation of the Line Operations Safety Audit (LOSA) Methodology, *Human Factors and Aerospace Safety*, **3**, 25–45.

Thomas, M.J.W. and Petrilli, R.M. (2004), Error Management Training: An Investigation of Expert Pilots' Error Management Strategies During Normal Line Operations and Training – Study One Report, Adelaide, SA: University of South Australia.

Chapter 16

Confidential Observations of Rail Safety (CORS): An Adaptation of Line Operations Safety Audit (LOSA)

Allison McDonald and Brett Garrigan
Queensland Rail

Lisette Kanse
University of Queensland

Line Operations Safety Audit (LOSA) is a proactive safety management tool, developed for and used in aviation to collect data on threats and errors occurring during everyday operations, and how these are managed. The present research demonstrates that LOSA is readily adaptable for use in the rail industry; however, the transfer required a number of modifications. An outline is given of some key aspects of the Confidential Observations of Rail Safety (CORS) development process, including: design of observation materials; marketing the program to unions and drivers; selection and training of observers; ethical considerations; and issues surrounding legal and safety accountability.

Introduction

Although a significant amount of human factors research has been conducted within the rail industry in recent years, still a far greater amount of such research is available from other transport industries such as aviation and road transport (Wilson and Norris, 2005). There are many similarities in the types of human factors challenges encountered across various transport operations, so there is likely to be much the rail industry can learn from the way in which other transport industries have applied human factors research and expertise. Prior research (Queensland Rail, 2003) highlighted a number of opportunities to adapt existing human error initiatives from the aviation domain to rail. One such initiative is the Line Operations Safety Audit (LOSA) program.

Line Operations Safety Audit (LOSA)

Line Operations Safety Audit (LOSA) was developed as a joint program involving the University of Texas and Continental Airlines, with funding from the Federal

Aviation Administration Human Factors Division (ICAO, 2002; Flight Safety Foundation, 2005). LOSA is a proactive safety management tool that uses trained observers to collect data about external threats and crew behaviour during normal flight operations (Klinect, Wilhelm and Helmreich, 1999). The observations are conducted under non-jeopardy conditions by pilots trained in the use of specially designed observation materials. Data gained through LOSA can be used to guide improvements to training and operations (Flight Safety Foundation, 2005; Helmreich, in press). LOSA is not limited to identifying safety issues or potential improvements; it also identifies examples of positive safety performance, and successful application of threat and error management countermeasures (Helmreich, Klinect and Wilhelm, 1999). LOSA uses the Threat and Error Management Model (Helmreich, Klinect and Wilhelm, 1999) as a framework for data collection and analysis.

Threat and Error Management Model

The Threat and Error Management Model (Helmreich, Klinect and Wilhelm, 1999) as used in aviation, defines threats as external situations that must be managed by the cockpit crew during normal, everyday flights. Threats increase the operational complexity of the flight and pose a safety risk to the flight at some level (ICAO, 2002). The operational definition of flight crew error on the other hand, is action or inaction that leads to deviation from crew and/or organizational intentions or expectations (Helmreich, Klinect and Wilhelm, 1999). Threats and errors are considered to be normal parts of everyday operations that must be managed. Threats and errors may sometimes go undetected, on other occasions they may be effectively managed, or may result in additional errors which require subsequent detection and response. By collecting and analysing data using this framework, it is possible to learn about more and less successful threat and error countermeasures being used during everyday operations. The Threat and Error Management Model appears to be relevant to many applications outside of aviation, including the rail operating environment.

Confidential Observations of Rail Safety (CORS)

The rail adaptation of LOSA was named Confidential Observations of Rail Safety (CORS). This title was generated in an attempt to highlight the confidential nature of the observations, and to differentiate these from other existing monitoring or audit programs in place within the organization. CORS differs from LOSA in the type of data collected, as the focus of CORS observations was limited to data related to threat and error management. Unlike aviation, evaluation of behavioural markers was excluded from the observation process, as there was concern that such judgements may be perceived by participants to be subjective and evaluative, and that this in turn may reduce initial acceptance of the program. The authors also considered CORS observation data to be an important input into the development of behavioural markers.

The results of CORS observations will be used together with outputs of the National Rail Resource Management (RRM) project to generate rail-specific behavioural markers for future use in training and assessment. The National RRM project, managed by the Department of Infrastructure Victoria, aims to adapt Crew Resource Management training for the rail industry (Public Transport Safety Victoria, 2005).

CORS will provide a snapshot of operational safety performance, and it is anticipated that the results of CORS may highlight future directions for training and awareness, and potential improvements to organizational systems and processes. CORS may also provide a baseline measure of drivers' threat and error management skills before implementing Threat and Error Management training as part of the existing Human Factors training program.

The University of Texas proposes 10 LOSA Operating Characteristics which are critical success factors in implementing LOSA (ICAO, 2002). CORS was developed in accordance with these characteristics as follows:

- Jump-seat observations during normal flight operations: CORS observations are conducted on normal timetabled revenue services, with participants fulfilling their normal rostered workings. There are no observations conducted during driver monitoring or assessment journeys.
- Joint management / pilot sponsorship: A Steering Committee was developed during the planning stage of CORS. This provided an opportunity for management and union involvement and endorsement of the program. The Steering Committee is involved in all stages of CORS, including development of observation materials, observer selection and training, scheduling of observations, review of progress, and interpretation and communication of results.
- Voluntary crew participation: CORS observations are only conducted with permission of the participating driver. When observers approach drivers, they are required to provide information about CORS, and obtain voluntary informed consent before conducting an observation. If a driver declines, the observer continues to approach other drivers until consent is obtained. Participants may withdraw from the program at any stage during an observation.
- De-identified, confidential and safety-minded data collection: CORS observation worksheets were designed to ensure that no identifying information is recorded, such as names, train numbers, or dates of observation.
- Targeted observation instrument: CORS observers use specially designed observation materials based upon the Threat and Error Management Model, and target external threats, crew errors, how these are detected and managed, and actual outcomes. Observers record and code events according to threat and error code lists specifically developed for this project by driver representatives and Human Factors experts.
- Trusted, trained and calibrated observers: To promote trust in CORS, observers are peer train drivers, selected from expressions of interest, and specially trained in the technical and non-technical skills required for conducting observations. Observers participate in regular calibration sessions to promote consistency and reliability of coding.
- Trusted data collection site: All observation worksheets and signed informed

consent forms are sent to the University of Queensland for data entry and storage. No one within the rail organization has access to individual observation paperwork. Feedback from drivers has shown that this discernable level of independence greatly contributed to their confidence in and acceptance of the CORS program.

- Data verification roundtables: Calibration sessions are conducted regularly to promote reliability and consistency of coding. These sessions have also provided observers with an opportunity to discuss techniques that worked well when approaching participants, or conducting observations, and to explore any barriers or difficulties experienced.
- Data-derived targets for enhancement: At the time of writing the chapter, CORS observations were not yet complete, however, the next stage of CORS involves analysing the data obtained in the first round of observations to determine trends or key issues, and to prioritize these for attention. An action plan will be developed to address key issues that have emerged.
- Feedback of results to the line pilots: Results of the data analysis will be communicated to drivers, and other key stakeholders within the organization via the Steering Committee. It will be important to link these results to planned improvement actions to demonstrate that the program has been used to generate change and safety improvement.

Design of observation materials

CORS observation materials follow a similar format to that of their LOSA equivalents; however, the content of threat and error code lists is specific to the types of threats and errors that could be encountered in the rail domain. Threat and error code lists were generated using information gained through task analysis information, accident/incident investigation data, and focus groups with driver trainers and senior train crew. These lists were further developed during a trial phase of observations conducted as part of observer training. Threat and error lists were limited to those items that may impact on operational and/or passenger safety, and excluded those that would primarily impact on service efficiency or comfort.

CORS Worksheets are designed to collect data about:

- Threats – when the threat occurred, who detected it, how it was detected, and how it was managed
- Errors – when the error occurred, who committed the error, whether it was associated with a threat, who detected it, how it was detected, how the crew managed or responded to the error, and the outcome of the error, and
- In addition, a separate page was designed to record demographic data such as day and time of observation, rollingstock type, length of driving experience, line of route, whether it was express running, and a general description of the context of the journey.

Ethical and legal considerations

Given the sensitivities surrounding collection of threat and error data, the CORS Steering Committee sought ethical clearance for the program through the University of Queensland's Behavioural and Social Sciences Ethical Review Committee. In obtaining ethical clearance, it was demonstrated to the review committee that adequate provision had been made to address key considerations such as:

- Train driver participation in observation sessions being entirely voluntary
- Lodgement of participants' informed consent
- Freedom for participants to withdraw from observations at any stage
- Protection of anonymity and confidentiality in the handling and storage of data
- CORS Observers signing a confidentiality agreement
- Assurances that CORS data would not be used for disciplinary purposes
- Limitations imposed on access to CORS data in the event of incident investigations.

It was acknowledged from the outset that the success of CORS hinged on safeguarding confidentiality and reasonable limits of accountability. In establishing these parameters, extensive consultation occurred with management, train crew unions, legal professionals and the University of Queensland.

To protect confidentiality and anonymity, CORS observation worksheets do not contain any personal identifying information. As an added measure, observers send signed informed consent forms and completed observation worksheets in separate sealed envelopes to the University of Queensland. There the consent forms are securely stored separately from the observation data, and only the notes from the observation worksheets are entered in a password protected data file before being securely stored as well.

Despite the assurances and safeguards to confidentiality, there remained concerns that threat and error data could still be used for disciplinary purposes if the participating driver was involved in a safety incident whilst being observed. To alleviate these concerns, it was necessary to articulate clear Limits of Accountability that preserved confidentiality and anonymity as far as practical, but without compromising legal and workplace health and safety obligations. These formed part of the Informed Consent process and were readily accepted by train crew once the limits to how the information could be used were clearly defined.

Marketing the program

Initially, CORS was marketed via a Steering Committee and existing communication channels such as newsletters and information boards within depots. A distinctive logo was developed to mark all CORS documentation for easy identification. There were two key components critical to successfully marketing CORS. The University of Queensland was engaged as a neutral third party, responsible for receiving and storing observation data off-site, which seemed to promote participants' trust in the confidentiality and anonymity of observation data. Observers were trained in promoting CORS, and in dealing with negative or difficult responses; however, observers had very few negative responses to the program. Observer feedback suggests that the program has been well received amongst drivers, and observers have reported very few rejections when drivers were approached to participate. The most significant challenge in marketing CORS was to differentiate it from other observations such as driver performance monitoring and assessment, and other observations collecting data for reviewing timetables.

Observer selection and training

Although both pilots and a small number of non-pilots may act as LOSA observers (ICAO, 2002), the decision was made to include only 'peer' train drivers in the team of CORS observers. The rationale behind this decision was to promote trust in the program, to maintain role clarity of senior train crew, and to differentiate CORS observations from other compliance-based monitoring or assessment. Observers were selected via expressions of interest, and undertook an interview and written work sample activity, designed to assess: their appreciation of confidentiality requirements, ethical behaviour, observation skills, written communication skills, accuracy of coding information, and willingness to follow rules and procedures. Observer selection results were endorsed by train crew union representatives via the CORS Steering Committee.

Observers undertook three days of training before commencement of the program. The training involved presentations, case studies, facilitated discussions, simulated practice, and observation practice on revenue services. The training provided a brief introduction to Human Factors and system safety, a background to LOSA and Threat and Error Management, key concepts in ethics and confidentiality, communicating and marketing the program to potential participants, and dealing with difficult situations. Observers were trained to use the CORS observation materials, and they provided feedback which was used to further modify the threat and error lists. Observation practice was conducted using videotaped driving scenarios and a train driving simulator, and before the final day of training, observers worked in pairs to conduct trial observations on normal revenue services. Feedback from these trial observations helped further refine the observation materials, and provided an opportunity for observer calibration before conducting observations that would form part of the CORS data set. It is interesting to note that observers found the experience of conducting observations using video and simulator scenarios more difficult than the trial observations on revenue services. This appears to be due to relative lack of sensory cues, other than visual and auditory, in the video and simulator scenarios, and the higher number of threats and errors occurring in these scenarios compared with 'normal' journeys.

Observer calibration

Calibration sessions were held regularly during the CORS data collection period. These sessions provided an opportunity to maintain high levels of consistency and reliability in coding and documenting observation results. The calibration sessions provided an opportunity to agree upon coding of unusual or unexpected events, and to discuss any issues arising from the data set, such as missing demographic data, or inconsistent coding of similar events.

As part of these sessions, observers also discussed methods they used in recruiting and observing participants, and were able to share strategies for dealing with unusual events or difficult situations. Some of the issues observers worked collaboratively to resolve were the tendency for participants to engage in conversations whilst being observed, rostering constraints, establishing links between threats and errors, and incorrectly classifying error outcomes *viz.* whether the error was inconsequential, led to an undesired state, or resulted in an additional error.

Preliminary results

Although threat and error occurrence rates observed through LOSA differ across airlines, initial CORS results reveal rates within a similar range to those recorded in LOSA. Table 16.1 compares LOSA data for three airlines (Klinect, Wilhelm and Helmreich, 1999) with preliminary CORS data.

Table 16.1 A comparison of LOSA data (Klinect, Wilhelm and Helmreich, 1999) and CORS data

	Airline A	Airline B	Airline C	CORS
Threats per segment/ journey	3.3	2.5	0.4	4.6
Errors per segment/ journey	0.86	1.9	2.5	1.6

Furthermore, LOSA data (Klinect, Wilhelm and Helmreich, 1999) reveals that threats occurred in 72 per cent of flight segments observed, compared with 100 per cent of journeys in CORS; and errors occurred in 64 per cent of flight segments in LOSA, compared with 77 per cent of journeys in CORS.

A promising finding of the CORS observations was that about 80 per cent of the observed threats were managed effectively. This percentage was somewhat lower for the observed errors. In nearly all cases where there was an identifiable consequence, this was typically an undesired state, with no additional error. As could be expected, not all threats and errors occurred with equal frequency, and certain clusters of recurring threats and errors are starting to emerge from the data. The lists have proven to be sufficiently complete: for only 5 per cent of the observed threats and 7 per cent of the observed errors so far none of the existing codes could be assigned.

Conclusion

This chapter demonstrates how the LOSA methodology has been successfully adapted to a rail equivalent, CORS, and highlights a number of issues that needed to be tackled in the process. Initial results are promising: the program appears to be well accepted among all stakeholder groups, observers have no difficulties in recruiting participants, and the collected data are starting to provide insights not only in which threats and errors occur but also how well these are managed. Future research will aim to improve the quality and quantity of threat and error management data collected through CORS, and may include the addition of behavioural markers to observations. The next generation of CORS observations will aim to measure any change in threat and error management skills after train crew have been trained in key threat and error management principles drawn from the first round of CORS observations.

Acknowledgements

We would like to thank James Klinect and Bob Helmreich from the LOSA Collaborative for their advice and assistance with the development of CORS. Thank you to the CORS Observers who have helped to develop and implement CORS, and to the drivers who volunteered to be observed. The authors appreciate the support of Queensland Rail and The University of Queensland.

References

Flight Safety Foundation (2005), Line Operations Safety Audit (LOSA) Provides Data on Threats and Errors, *Flight Safety Digest,* **24**, 4–18.

Helmreich, R.L. (1998), Error Management as an Organizational Strategy, *Proceedings of the IATA Human Factors Seminar, Bangkok*, Thailand, pp. 1–7.

Helmreich, R.L. (in press), Culture, Threat, and Error: Assessing System Safety, in *Safety in Aviation: The Management Commitment: Proceedings of a Conference*, London: Royal Aeronautical Society.

Helmreich, R.L., Klinect, J.R. and Wilhelm, J.A. (1999), Models of Threat, Error, and CRM in Flight Operations, *Proceedings of the Tenth International Symposium on Aviation Psychology*, Ohio State University, Columbus, OH, pp. 677–682.

International Civil Aviation Organization (ICAO) (2002), 'Line Operations Safety Audit (LOSA)', *ICAO Doc 9803 AN/761.*

Klinect, J.R., Wilhelm, J.A. and Helmreich, R.L. (1999), Threat and Error Management: Data from Line Operations Safety Audits, *Proceedings of the Tenth International Symposium on Aviation Psychology*, Ohio State University,

Columbus, OH, pp. 638–688.

Public Transport Safety Victoria (PTSV) (2005), *Rail Resource Management Project Bulletin*, August, 2005, Department of Infrastructure Victoria.

Queensland Rail PSG Human Factors Office (2003), *Human Error Research Project*, Queensland Rail Internal Document.

Wilson, J.R. and Norris, B.J. (2005), Rail Human Factors: Past, Present and Future, *Applied Ergonomics*, **36**, 649–660. [PubMed 16238999] [DOI: 10.1016/j.apergo.2005.07.001]

Chapter 17

Human Factors in Air Traffic Control: An Integrated Approach

Christine C. Boag-Hodgson
Airservices Australia

Human Factors (HF) are risks that exist within any environment that has a human operator. The effective management of HF risks requires identification, analysis, mitigation and review. The management of HF therefore requires the implementation of these four steps within the organization's operations. Airservices Australia uses an integrated approach to manage HF risks. The first component involves providing HF awareness through education and training. This ensures that employees have sufficient HF knowledge to recognize when HF may impact on operations and react accordingly. The second component is a change management system that incorporates HF assessments. The third component relates to the identification of HF risks during incident investigation, while the final component in the Airservices Australia integrated HF approach is the use of surveillance activities to monitor HF impacts on the system as a whole. The on-going implementation of this integrated approach to manage HF risks will be described in further detail.

Introduction

Human Factors is a body of knowledge about the human abilities and human limitations that are relevant to the design of tools and machines, tasks and environments, in addition to systems (Chapanis, 1996).

Human Factors continues to gain popularity within cognitively demanding domains, such as transportation. Within aviation, and indeed aviation psychology, HF is not a new concept. The International Civil Aviation Organization (ICAO) recognized the importance of HF in 1986 with the adoption of the Resolution A26-9 on Flight Safety and Human Factors. One of the means by which this resolution was to be met was through the publication of regular digest (ICAO Circulars) on HF and safety in aviation. In addition, in 1998, the first edition of the *Human Factors Training Manual* (ICAO) was published to assist in the development of training programmes for aviation organization personnel.

Assessing the human operator within his/her task, environmental and organizational systems has never been more important than in the present increasingly technological age. Air Traffic Control (ATC) is one domain where the importance of HF has been grasped with both hands.

Airservices Australia is the civil air traffic control provider for the Australian airspace, which covers approximately 11 per cent of the world's airspace. With the establishment of its subsidiary Airservices Pacific Incorporated in 2005, the organization now provides services at five international air traffic control towers, with additional international advancements already being progressed. Within Airservices Australia, the concept of Human Factors has been integrated into many different areas of the organization. HF experts have been working within the organization for many years to assist in the on-going management of HF risks. The roles of these experts are both strategic and tactical, requiring a variety of skills across a number of subject areas including data analysis and evaluation, equipment design, change management and incident/event investigation.

An integrated approach

The effective integration of HF within Airservices Australia has involved a multi-faceted approach. Simply providing HF training to the air traffic controllers, fire fighters and technicians within the organization would not be sufficient to ensure the effective monitoring of human abilities and limitations. In other words, it would not ensure the effective on-going assessment and mitigation of HF risks. As a result, HF assessments have been integrated within the organization using a variety of approaches that can be largely categorized into four areas: education, change management, investigation, and systemic monitoring. This integrated approach provides greater potential for the identification, intervention and on-going analysis of the relevant HF risks that exist within the various areas of the organization.

Education

The provision of adequate HF education, including training and awareness programs, is one component of the Airservices Australia integrated HF approach. It is impractical for every one of the almost 3,000 employees to be a qualified HF specialist. Providing employees with an awareness of HF concepts, however, gives them the knowledge to recognize when a formal HF assessment may be required, facilitating the likelihood that appropriate action will be undertaken.

HF concepts are introduced in the training syllabus for *ab initio* controllers in the Airservices Australia Training College. This early introduction of HF risks occurs for a variety of reasons. Common HF outcomes drive the justification for many ATC procedures and practices. For example, problems with correctly receiving and processing (hearing) aural information, such as instructions, has resulted in the read back/hear back procedure whereby a controller issues an instruction, the pilot reads back the content of the instruction, and the controller confirms the pilot's read back as correct. Understanding the possible human errors that can occur during the issuing of an instruction, such as problems with the phonological loop, allows trainee controllers to appreciate why the read back/hear back procedure has been incorporated into everyday operations.

Another reason for the early introduction of HF training is the underlying belief that awareness has the potential to reduce susceptibility to errors. That is, if the controller knows his/her limitations and constraints, then he/she is less likely to be susceptible to performance errors directly associated with these factors (Garland, Stein and Muller, 1999).

In addition to *ab initio* HF training, Airservices Australia also provides HF training specifically on Team Resource Management (a similar concept to Crew Resource Management in the flight deck). Other more general HF training is addressed at team development day training sessions, as well as safety related publications such as lessons learned, and information papers on specific human limitations. A recent example was an information paper that was distributed to all operational personnel on the expectation and confirmation biases (Safety Evaluation Unit, DSEA, 2005).

Airservices Australia technicians also undergo training in HF and the management of human error. This HF training is similar to that outlined in the ICAO *Human Factors Training Manual* (1998) and incorporates HF fundamentals such as conceptual models of error, cultural and organizational factors, human performance, communication, teamwork and environmental issues.

As part of the five and a half day Airservices Australia incident investigator training course, trainee investigators or investigators renewing their qualifications undertake HF and human performance training. In addition, Airservices Australia technicians undergo a five-day human error investigation course. The nature of these investigator training courses will be described in more detail in the Investigation section of this chapter.

It becomes apparent that HF training is being provided throughout the controllers' and technicians' careers – during initial training, throughout routine operations and during incident analysis. The method of distribution of these HF messages includes formal workshop training, team day briefings, safety briefing documentation including safety newsletters and lessons learned following incidents, monthly reports summarizing the contributory factors to safety related incidents and events, as well as safety videos on specific HF topics.

It is important to recall that the focus of these various education and awareness methods is not to make the individual an HF expert. Rather it is intended to give the individual the necessary information to identify when a formal HF assessment or intervention by an HF expert may be required. Additionally, individuals are left in a better position to recognize operational situations where more specifically applied behaviours are warranted.

Change management

As the workplace becomes more technological, the importance of proactively assessing the HF impacts of change becomes paramount (Benel and Benel, 1998). Any change to the workplace should assess the impact of the change on the human operators and the systems within which they operate. Optimally, this HF assessment should be done before implementation. This preliminary HF assessment is therefore an integral component of every project plan.

Within Airservices Australia, most changes must comply with the requirements of the Airservices Australia Safety Management System. This system requires that the safety related impacts of the change undergo a safety assessment as a minimum analysis. The impact of the change upon the individual must be analysed as part of this assessment.

One of the largest HF assessments that was undertaken by Airservices Australia occurred during the implementation of The Australian Advanced Air Traffic System (TAAATS). The HF assessment was undertaken over several years, during the design and implementation of TAAATS. The issues that were addressed were very diverse, ranging from the ergonomics of the new workplace and room layout, to the functionality of the displays (see Figure 17.1), to the colour palette to be used on the displays. The on-going enhancement of TAAATS has also required subsequent HF assessments. For example, the recent introduction of a graded system of audible and visual alerts and alarms underwent an HF assessment before implementation.

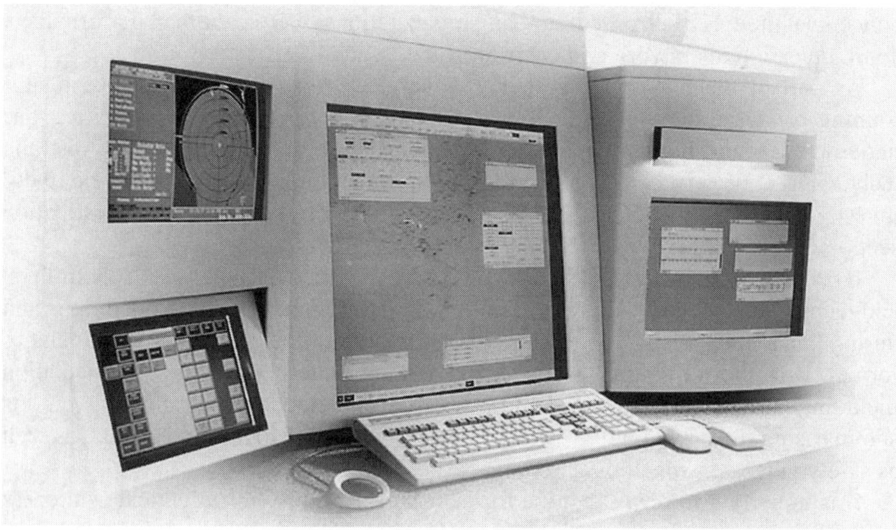

Figure 17.1 TAAATS console display

In a more recent project that involved the integration of user-preferred routes (UPR) in the airspace between Australia and New Zealand, the HF assessment actually resulted in a procedural change being postponed until additional technological advancements had been integrated into TAAATS. The introduction of UPR was being requested by industry, as it resulted in fuel savings by allowing the airlines to choose their own routes based on factors such as prevailing winds, rather than flying according to the currently designated flight routes. During the design phase, an HF assessment was undertaken of the impact of UPR on controller mental workload, situation awareness and performance. On the basis of the results of this HF assessment, the implementation of UPR was postponed until additional technology was available within TAAATS to mitigate against the HF risks that had been identified.

Many Airservices Australia project plans specifically include an HF assessment of the proposed changes as part of the project development phase. Consequently, an appropriately qualified HF specialist is assigned to the project to carry out the necessary assessments during the project's development rather than simply providing an assessment at implementation, which would then require workarounds to overcome any identified human capability limitations.

Investigation

The investigation of incidents and events is another area where HF principles are used extensively within Airservices Australia. Conceptual models are used to facilitate the investigation process and ensure that all contributory factors, including HF, are identified for appropriate mitigation. The incorporation of an HF checklist into the Airservices Australia Investigation Manual reminds investigators of the many different areas of HF that might contribute to incidents and events. The use of conceptual models such as the Reason Model (Reason, 1990) and the SHELL Model (Hawkins, 1987) ensure that the investigation examines how the human operator was interacting with the other systems (environmental and organizational) and how this may have contributed to the incident.

Airservices Australia incident investigators receive training to enhance their ability to recognize the contribution of HF to incidents. This training is tailored for the type of investigation being undertaken, be they air traffic services or technical. Air traffic service investigation training focuses on HF topics such as situation awareness, decision making, ergonomics and fatigue, in addition to training on human performance related topics including information processing, perception and attention.

In contrast, our technical investigations are centred on two primary considerations, (i) The aspects of complex systems that increase the probability of error, and (ii) those aspects that of a system that permit an error to have negative consequences. Investigators undergo Human Factors Event training and learn to identify different types and sources of error. This training addresses a variety of HF fundamentals including information processing, physiological variables, team issues, hardware, procedures, the environment, and investigative interviewing, as well as conceptual models. The course also incorporates practical training in the Tool for the Investigation of Maintenance Error (TIME) error and factor classification system.

In addition to HF training for investigators, the contribution of HF to air traffic services incidents is undertaken as part of the investigation process. The contribution of a comprehensive set of local factors, organizational factors and contributory factors is rated on a 1 to 5 Likert scale as a mandatory component of the investigation process. This allows for systemic HF issues to be readily identified and appropriate mitigation/management strategies to be implemented.

Systemic monitoring

Surveillance of HF risks is carried out through a variety of systemic monitoring activities. Annual, quarterly, monthly and weekly reports of incident and event data are a primary source of the HF posing risks to the air traffic system. These reports are able to identify on-going threats that may be low in magnitude, but persistent in nature, or threats that are becoming more substantial over time.

Once identified, the specific human factor risk will need to be analysed as well as managed and mitigated. Typically this will involve a more detailed assessment of the risk, and a systemic review or systemic investigation will be undertaken, involving a qualified HF expert, in addition to subject matter experts as required. Example reviews of this type include a Systemic Review of Breakdown of Separation Occurrences, a Review of Pilot Violations of Controlled Airspace, and ongoing reviews of both Runway Incursion Incidents and Pilot Operational Deviations.

Proactive indicators have recently been incorporated into these systemic monitoring activities. These predictive indicators include the development of a Safety Lead Indicator Model (SLIM) which assesses lead indicators as opposed to lag indicators to determine operational safety (Dumsa *et al.*, 2003).

Another proactive means of monitoring HF is the development of a Normal Operations Safety Survey (NOSS). This means of surveillance is similar to the Line Oriented Safety Audit (LOSA) developed by the University of Texas for identifying the threats, errors and undesired states within the flight deck (Helmreich *et al.*, 1999). NOSS is being used within the air traffic system to identify similar risks, including HF risks that are being created by the human operators within the system. Once identified, these risks can then be assessed and managed or mitigated as required.

Another way by which Airservices Australia is trying to proactively manage HF risks is by conducting an annual Safety Climate Survey of all (operational and non-operational) staff. The premise underlying this assertion is that an individual's perceptions of safety affect his/her safety related behaviour (Dobbie, 2003; Health and Safety Executive 2001).

All of these systemic monitoring and data collection activities are on-going to allow the analysis of HF risks over time. The combination of lead and lag indicators ensures that Airservices Australia is proactive in its review of HF risks. These proactive assessments facilitate the organization's capacity to identify, assess and manage HF risks through on-going reviews.

An example of the integrated safety approach

The four components of the Airservices Australia integrated HF approach are used interactively to ensure a comprehensive HF assessment is being routinely undertaken. In one project, an HF assessment of two major Australian radar tower air traffic operations was undertaken. The rationale for the project stemmed from the investigation data that had been collected during a two-year period. In addition, systemic monitoring had identified trends that warranted further analysis and review. The HF education that had previously been presented to controllers ensured that the subject matter experts chosen to participate in the review had the necessary background knowledge to assist the HF expert who was the project leader. The project examined all relevant sources of data and made recommendations for change. These changes underwent further HF assessment before implementation.

As described, each of the four components of the Airservices Australia integrated approach to managing HF risks do not occur in isolation. Neither does each component happen only once. Management of the HF risks is an on-going activity that requires corroboration and reliable data sources before changes will result.

Conclusion

The integrated approach that Airservices Australia uses for the ongoing management of HF maximizes the likelihood that the limitations of human capabilities are identified, analysed, mitigated and managed as well as reviewed. This includes the limitations of the individual human operator, in addition to any limitations within the interaction between the human operator and their various systems (task, environment and organization).

The multi-faceted integrated approach used within Airservices Australia encompasses HF awareness through a variety of educational mediums; the management of HF associated with change; the assessment of HF as contributory factors within incident and events; as well as the on-going review and analysis of HF at a systemic level. It is through this integrated approach that Airservices Australia seeks to proactively identify HF risks and mitigate or manage these risks effectively.

References

Benel, R.A. and Benel, D.C.R. (1998), A Systems View of Air Traffic Control, In, *Human Factors in Air Traffic Control* eds Smolensky, M. W. and Stein, E. S. San Francisco, CA: Academic Press.

Chapanis, A. (1996), *Human Factors in Systems Engineering*. New York: Wiley.

Dobbie, K.L. (2003), The Safety Climate of Airservices Australia, *The Proceedings of the Sixth Australian Aviation Psychology Association Symposium: Setting the Standards*. Sydney, AU: Australian Aviation Psychology Association.

Dumsa, A., Marrison, C., Boag, C.C., Hayes, J., Clancy, J., Casey, S. and Grey, E. (2003), Predicting Organizational Safety Performance: Lead Safety Indicators for Air Traffic Control, *The Proceedings of the Sixth Australian Aviation Psychology Association Symposium: Setting the Standards*. Sydney, AU: Australian Aviation Psychology Association.

Garland, D.J., Stein, E.S. and Muller, J.K. (1999), Air Traffic Controller Memory: Capabilities, Limitations and Volatility, in *Handbook of Aviation Human Factors: Human Factors in Transportation* eds Garland, D. J., Wise, J. A. and Hopkin, V. D. Mahwah, NJ: Lawrence Erlbaum Associates.

Hawkins, F.H. (1987), *Human Factors in Flight*. Aldershot: Ashgate.

Health and Safety Executive, (2001), *Reducing Risks, Protecting People*. Norwich:

Her Majesty's Stationery Office.

Helmreich, R.L., Klinect, J.R., Wilhelm, J.A. and Jones, S.G. (1999), The Line LOS Checklist, Version 6.0: A checklist for human factors skills assessment, a log for external threats, and a worksheet for flightcrew error management', *The University of Texas Team Research Project Technical Report 99-01.*

International Civil Aviation Organization (1998), *Human Factors Training Manual (Doc 9683-AN/950)*. Montréal: ICAO.

Reason, J. (1990), *Human Error*, Cambridge: Cambridge University Press.

Safety Evaluation Unit, DSEA (2005), *The Low-down on Biases.* Canberra, AU: Airservices Australia.

PART 4
The Modality of Human Factors: Exploring the Management of Human Error

Part 4 is a collection of human factors interventions from the different industry modes. It will introduce the concept that human error is indeed 'ubiquitous' and therefore, its management should be approached from multiple applications: training and development, leadership and supervision, situational awareness, distraction management etc. Novel elements in this part include research on driver distraction, medical team resource management, a new look at situational awareness, rail risk management and fatigue management in the medical field.

Chapter 18

Human Factors at Railway Level Crossings: Key Issues and Target Road User Groups

Jeremy Davey, Nadja Ibrahim and Angela Wallace
Queensland University of Technology

As the interface between roads and train lines, railway level crossings are the potential site for vehicle-train collisions. Due to the lack of financial viability in solely approaching this issue from an engineering perspective, there is increased interest in a human factors perspective. This chapter will outline the findings from preliminary research conducted as part of a three-year study of motorist behaviour at railway level crossings in Australia. In-depth interviews were undertaken with 122 participants from the at-risk road user groups of: heavy vehicles; older drivers and younger drivers. Risk behaviours and attitudes were found to differ between sub groups and by setting which has strong implications for the design of educational interventions to improve motorist safety at level crossings.

Introduction

This chapter discusses selected results of preliminary research that was conducted as part of the Level Crossing Risk Management Project. This three-year project is investigating motorist behaviour at railway level crossings and the role of educational interventions for different road user groups.

With the direction of the Australian National Road Safety Action Plan (Australian Transport Council, 2000) and an emerging lack of financial viability in continuing to approach level crossing risk management from an engineering perspective, there is increased interest in approaching this issue from a human factors and educational perspective. However, due to the small numbers of fatalities and the limited data available on the characteristics of all fatal vehicle-train crashes nationally, it is difficult to determine the key risk behaviours and the road user groups that are 'at risk' for involvement in a level crossing accident.

To establish road user groups that may warrant investigation, a consensus survey utilizing a modified Delphi technique was conducted nationally with experts in the fields of road and rail safety in 2004 and confirmed with current literature and research with Train Drivers as key informants. Older drivers, younger drivers and heavy vehicles drivers were indicated to be potential at-risk groups by the panel of experts, which was supported by road crash statistics and Train Driver accounts. Older drivers are overrepresented in level crossing fatalities with 26 per cent of the fatalities being in the 60 + age group compared to 10 per cent of drivers in other fatal road crashes for this age group (Australian Transport Safety Bureau, 2002). Younger drivers (17–25 years) are represented in 25 per cent of road deaths in 2003 and 25 per cent of persons seriously injured on the road in 2002 (Australian Transport Safety Bureau, 2004). The importance of heavy vehicles as a target group includes the potential derailments of trains and possible catastrophic consequences. It has been predicted that the Australian road freight task will double by 2015 (Bureau of Transport Economics, 1999), substantially increasing heavy vehicles on the roads.

A range of unsafe and risk taking behaviours were noted by the panel of experts and Train Drivers with differences noted between the groups, between urban or regional settings and by the types of protection systems the road user is exposed to.

Background

Railway level crossing safety has been an ongoing concern to road and rail authorities for many years. It has been included as a major action area in the 2003–2004 Australian National Road Safety Action Plan (2000) and a supporting measure in the 2005–2006 plan (Australian Transport Council, 2005). Collisions between vehicles and trains at railway crossings account for only a small per cent of all road casualties; however, they are amongst the most severe of road crashes. A review of the literature indicated that the rate of fatalities in vehicle/train crashes is somewhere between three (Afxentis, 1994) and 30 (National Highway Traffic Safety Administration, 2005) times higher than all other types of road crashes. It is approximated that there are around 100 crashes between a road vehicle and a train in Australia each year (Australian Transport Council, 2005). From 1997 to 2002, there were 74 deaths due to collisions between trains and motor vehicles at level crossings in Australia (Australian Transport Safety Bureau, 2003). A Commonwealth investigation of fatal crashes at railway level crossings has supported the notion that human fault is a high source of accidents (Australian Transport Safety Bureau, 2002). Based on an analysis of accidents spanning from 1988 to 1998, it was revealed that unlike other fatal road crashes, accidents at railway crossings were less likely to involve fatigue, speeding, drugs or alcohol (Australian Transport Safety Bureau, 2002). These accidents were more likely to be attributed to errors in driver behaviour (Australian Transport Safety Bureau, 2002). Unintended road user error was more common in level crossing crashes (46 per cent) than in other fatal road crashes (22 per cent) (Australian Transport Safety Bureau, 2002).

There are approximately 9,400 public railway level crossings in Australia, of which 30 per cent have active protection (flashing lights or flashing lights and boom barriers), 64 per cent have passive protection (static array of signs including STOP or GIVE WAY signs) and the remainder have other control or protection (Ford and Matthews, 2002). Accidents at crossings with some form of active protection are more common than accidents at passive crossings in Australia (Ford and Matthews, 2002).

Method

Following the identification of potentially at-risk road user groups, an examination of the psychosocial and environmental characteristics of the target groups in relation to driving at level crossings was conducted. The research was undertaken in 2005 in regional and urban settings in Queensland with the target groups of: older drivers, younger drivers and heavy vehicle drivers. Participants were recruited from community groups, Technical and Further Education (TAFE) institutes, universities, businesses and professional associations. The study was both exploratory and descriptive in nature and aimed to explore and describe motorists' knowledge, attitudes and behaviours in relation to railway level crossings. To date there has been no research conducted with specific road user groups in relation to distinct level crossing knowledge levels, attitudes and behaviours.

This study was specifically undertaken in order to guide and inform the development of railway level crossing educational strategies for specific road user groups. As it is beyond the scope of this chapter to report on the complete findings from each group, this chapter will provide an overview of the key behaviours and underlying beliefs of the target groups.

Participants

A total of 122 people participated in the study, 51 were female (41.8 per cent) and 71 were male (58.2 per cent). Overall, 70 participants (57.4 per cent) took part in the semi-structured interviews and 52 (42.6 per cent) in the focus group sessions. The mean age of participants was 42.67 years (SD = 23.96, Range = 17–89 years). Demographic details are presented in Table 18.1.

	Male	**Female**	**Mean age**	**Mean years licensed**	**Number of participants**
Older Drivers Urban Area	10 (43.5%)	13 (56.5%)	76.3 (61–89 years)	50.22 (17–75 years)	23
Older Drivers Regional Area	11 (55%)	9 (45%)	67.99 (61–85 years)	46.95 (30–69 years)	20
Younger Drivers Urban Area	14 (56%)	11 (44%)	20.4 (17–28 years)	2.74 (0.17–10 years)	25

	Male	Female	Mean age	Mean years licensed	Number of participants
Younger Drivers Regional Area	10 (35.7%)	18 (64.3%)	19.78 (17–24 years)	2.56 (0.08–8 years)	28
Truck Drivers All Areas	26 (100%)	—	46.31 (26–68 years)	29.19 (8–54 years)	26

All participants in the Truck Driver group were male, reflecting the gender profile of this workforce. Home address postcodes of the truck drivers indicated the participants were from a range of areas within the states of South Australia, Victoria, New South Wales and Queensland. The mean number of years driving trucks was 23.25 years (range 4–54 years).

Procedure

The data collection took place in a variety of Queensland metropolitan and regional centre settings. All participants completed a survey questionnaire plus a semi-structured interview or focus group. Both purposive and convenience sampling techniques were used to recruit participants. All participants were volunteers and received either: a gift of AU$10; catered refreshment; academic credit; or a travel mug as a gift for their time.

Older drivers (aged 60 years or over) were mainly recruited through existing social and charity organizations including bowls clubs and social groups. Younger drivers (aged 17–28 years) were recruited through the education facilities of TAFE colleges and universities. Selection criteria included age and possession of a Driver's Licence or Learner's Permit. Truck drivers were recruited through convenience sampling at a large 'Truckstop' south of Brisbane.

Results

Older drivers (regional area)

Knowledge—This group demonstrated the highest level of knowledge about the road rules and types of level crossings. There was however, a very low level of knowledge about the 'yellow hatching' road marking that is present on level crossings to denote a clear way ('Keep Clear'). Overall safety issues at level crossings were well known and there was a very high level of knowledge that accidents occur at level crossings with many participants able to recall two or more incidents.

Behaviours and influencing factors Emerging themes on the topic of behaviour included: the use of protective behaviours when driving at crossings; the unsafe community 'norm' behaviours and the influencing factor of 'dicey crossings'. Protective behaviours were felt to be 'special actions' to help ensure their safety when driving at level crossings such as: extra careful looking or slowing, taking longer and stretching to see, avoidance of bad crossings, listening, and ignoring pressure from other drivers. Pressure from other drivers was noted to include: beeping, yelling, overtaking at the crossing, driving out from behind stopped vehicles to go through the crossing and generally creating a perception of being pressured to go through the crossing in the participants. Dicey crossings were perceived to be crossings that are dangerous through their design, visibility, or environment. Two features of a dicey crossing were that: even when the driver is doing the 'right thing' they are at risk due to the actual crossing; and that at these crossings decisions have to be made without adequate 'information' leading to inadvertent risk taking.

Risk perception Those who perceived that they were not at risk gave a number of reasons for this belief including: they follow the rules; they use protective behaviours; they avoid dicey crossings; and that being familiar/knowledgeable/experienced with the crossings leads to 'taking more care'. Inattention and dicey crossings were factors thought to increase their risk. There was general agreement that all risk taking is unacceptable and risk taking behaviours were not reported. Three participants however recalled situations where they had been distracted or inattentive and had a close call at a dicey crossing. Most participants felt no confusion at level crossings; however a number of participants had experienced confusion at unfamiliar crossings and also at some of the dicey and complicated crossings within their local area. There was a general agreement that judging the distance of a train from a crossing was 'hard' and that is why you should use 'extra caution'.

Perceived risk factors After expressing initial disbelief that their age group could be involved in crashes due to their strong safety attitude and behaviour, a number of age related factors were offered by the participants themselves. The theme of 'balance' was developed to reflect the age-related risk factors that were put forward by the groups which were then strongly interconnected with a range of compensatory behaviours. The concept of balance included: balancing those admitted age-related risk factors with perceived compensatory behaviours; balancing aspects of risk at level crossings, like dicey crossings with protective behaviours; and balancing limitations or perceived risks with confidence based on knowledge and experience.

Perceived age-related risk factors included: decreased vision, night vision problems, and difficulties with glare, slower reflexes, reduced awareness/alertness, poor hearing and economic issues (car problems). Compensatory behaviours/attitudes included: increased time looking, wearing glasses, avoidance of night driving or unfamiliar places, alert/cautious driving, slowing on approach, awareness of limits, and anticipation of hazards.

Older drivers (urban area)

Knowledge This group did not have the same level of knowledge about the different types of level crossings as their regional counterparts. While active crossings with flashing lights and boom gates were known, only a small number knew of passive STOP signed crossings. GIVE WAY signed or 'lights only' crossings were not well known. There was very low exposure to passive crossings and although some participants had encountered them before it was often many years ago, leading to an uncertainty of whether these types of crossings are still in existence.

There was a very high level of knowledge of the road rules in this group even for the types of crossings that were unknown or rarely encountered. Community norm behaviours, such as going through the flashing lights before and after the boom gates, were noted. Similar to the regional older drivers, the theme of pressure from other drivers was also present in the urban older driver group. Knowledge of the meaning of the yellow, cross-hatched, road markings at crossings was generally poor with comments that they are to 'show that the crossing is there' and they are, 'new … to highlight the crossings' (b).

There was a low level of awareness generally of safety issues at level crossings. A few people felt that it was not really an issue at all: 'unless you're going over them, you forget they're there' (c). All of the participants knew of accidents that had happened at level crossings and there was a general perception that these accidents are 'not rare' but that most happen in 'country areas' where there are no boom gates. It was very interesting to note that the specific accidents recalled by each group were those that involved drivers from their peer group.

Behaviours and influencing factors In terms of driving behaviours at level crossings this group also emphasized 'special actions' to help ensure their safety or protective behaviours. These included: following the rules, slowing on approach, stopping, looking, listening, ignoring pressure from other drivers, being alert and 'not being in a hurry'. As with the regional group no risk taking behaviours were acceptable or reported.

Risk perception While the majority of the participants felt that level crossings were areas of risk, they believed that their protective behaviours reduced their risk of involvement in a level crossing accident, as did having low exposure to crossings in general. Others believed that there was always some risk with risk factors of: malfunctioning equipment; stalling; failing brakes; and high exposure given. In general their safety attitudes and behaviours led to a low perception of risk and high perception of protection.

Perceived risk factors The perceived risk factors for their age group were very similar to those of the regional group. Confusion and distraction were discussed, with comments that unfamiliar crossings can be 'confusing', roads with many turning lanes were 'confusing' and signs, billboards and other 'things' on the road were distracting.

Similarly to the regional group the theme of balance emerged clearly, as in difficult driving situations, protective or compensatory behaviours were reported, including, being 'diligent' and alert, avoiding unfamiliar areas, or getting a relative to drive. Their perceived age-related risk factors included: decreased vision to see flashing lights or the actual crossing; slower reflexes; slower reaction time; confusion at unfamiliar crossings; panic (if stalled or trapped); and distraction. The compensatory behaviours/attitudes cited included: careful looking; avoidance of night driving or driving to unfamiliar places; alert/cautious driving; slowing on approach; awareness of limits; and being 'safe drivers'.

Younger drivers (regional area)

Knowledge '(A railway crossing is) … a really annoying thing that you have to stop for and the train takes ages to cross.' (Participant A)

The younger drivers from the regional areas were familiar with all types of level crossings and most could recount the correct rules. Some incomplete rules, with no mention of looking for or giving way to trains, were noted. Community norms were indicated, with a number of participants stating the correct rules, then immediately commenting 'you're supposed to stop, but nobody does' (in relation to passive crossings). Pressure by other drivers to behave within the community norm was also noted, with some participants expressing concern that the cars behind them 'beep' if they wait for the lights to stop flashing.

This group displayed a high awareness of safety issues related to level crossings. Reflecting their exposure to passive crossings, many comments were made on 'bad' crossings along the same themes as discussed by the older regional drivers group (dicey crossings). Unlike the younger driver urban group, almost half of the regional participants displayed knowledge about level crossing accidents and there were some cases of personal knowledge of accidents and near misses at level crossings.

There was a perception that enforcement by police was very low and the there was a low likelihood of being caught for disobeying the road rules at level crossings. There also appeared to be a low level of knowledge of the severe consequences of involvement in a vehicle – train collision with a number of participants actually failing to mention 'living' or 'not dying' as a benefit of driving safely at level crossings.

Influencing factors, risk perception and behaviours Police presence influenced risk taking with comments of, 'stop at lights … but if cops aren't there … go through'. Other influencing factors for acting safely included past experiences of near misses, having passengers in the car and wanting to avoid accidents. Negative factors associated with driving safely at level crossings included frustration with waiting for the train, and pressure from other drivers.

Being familiar with crossings was thought to increase safety, although others admitted that they drove more complacently, less cautiously and might be 'driving on autopilot'. With unfamiliar crossings some felt that they 'might not see' the crossing at all or even be less cautious with the expectation that remote crossings are rarely used.

There was a low risk perception of being involved in a level crossing crash. Using protective behaviours, familiarity with trains and crossings, and the low number of trains at crossings were given as reasons for this perception. The lack of experienced repercussions could also be seen to influence risk perception with comments including, 'Its okay … you know … I'm still here' (about results of risk taking). Those who perceived some degree of risk focused mainly on feelings of fear, acknowledging risk taking or inattention, and factors that were beyond their control including: visibility, stalling and being a passenger. Attempts were made to justify risk taking behaviour with community norms of 'everybody does it' and being able to 'see there is no train'. These were strengthened by past experiences of low enforcement or no repercussions.

The behaviours reported by this group could be classified as perceived: 'low risk', 'medium risk' and 'high risk' behaviours; with 'low risk' behaviours being considerably more acceptable than 'high risk'. It is extremely important to note the extreme differences between the participants and the researcher assigned level of risk. The most alarming difference is the acceptance and performing of actual 'highest risk' behaviours by participants with the perception that these behaviours are 'low risk'. This disparity is much more prevalent in the younger regional drivers than the younger urban drivers. Some of the most concerning behaviours reported were of an extremely high risk level but were perceived as low risk and acceptable behaviours by the participants. These included not stopping at passive crossings, driving through flashing lights at flashing lights only crossings and 'cutting the train off' (by driving along-side the train and turning in front of it). There was a very high level of risk taking acknowledged although some participants stated their behaviours could be 'risky' but that it 'doesn't feel like a risk' because of community norms ('everyone does it').

Education on level crossing safety was generally very limited with only a small number of participants taught about level crossings by their parents or driving instructors when learning to drive. Many said that they never received any information about level crossings except what was in the learner's booklet or learner's test. These participants stated that they simply learnt by driving when they got their licence.

Younger drivers (urban area)

Knowledge Alarmingly there was a very low level of knowledge of the types of crossings found outside the metropolitan area, which obviously raises issues of safety with this group if and when they drive in a regional area. There was a high level of knowledge of boom gated active crossings, but no knowledge of the existence of 'flashing lights only' crossings. A very low level of knowledge of passive crossings was demonstrated.

In terms of road rules, only a small number stated incorrect rules such as, stating that flashing lights meant the same as 'amber lights at traffic lights'. Incomplete rules were common as most participants did not specifically mention looking, listening or giving way to the train at passive crossings. Similarly to the older drivers there was a low level of knowledge of the meaning of road markings at level crossings.

Community norms were indicated with a number of participants stating the correct road rules, then immediately commenting 'that nobody does that though' (in relation to waiting until the lights have stopped flashing before driving at active crossings).

In contrast to the younger regional drivers, level crossing accidents were perceived to be 'very rare' events and awareness of safety issues was quite low. Enforcement was perceived to be low at level crossings. It was thought that it was 'very unlikely' that you would be caught for disobeying the rules at a level crossing.

Influencing factors, risk perception and behaviours Police presence was felt to influence behaviour with comments that if there were no police present, 'then you take more risks' and not taking risks was influenced by not wanting to 'get into trouble'. Pressure by other drivers to behave within the community norm was also noted with some participants expressing concern that the cars behind them would 'beep at you' if you waited for the lights to stop flashing. Associated negative factors included waiting (especially when in a hurry) and the reactions and pressure from their peers as passengers.

There was a distinct perception with this group that there was a very low risk of involvement in a level crossing crash, 'a slight to zero risk'. This perception was underpinned by the reasoning that the majority of participants 'followed the rules'. Some participants however, attempted to justify risk taking with beliefs that they can ensure that they are 'safe', for example: not coming to a complete stop at passive crossings by reasoning that it was 'okay' if they 'could see that no trains were coming'.

As discussed with the younger driver regional group, self reported behaviours could be categorized into three main levels of risk (of a crash) – as perceived by the participants. 'Low risk' behaviours were considerably more acceptable than 'high risk'. Most self reported that they took part in 'low risk' behaviours such as driving through flashing lights before the boom gates come down, 'because everyone does it'. Behaviours that were perceived to be of a very high risk level and were not self reported, with comments that, 'Trying to beat the train is risky … going around the booms is just plain stupid'. Again, there were extreme differences between the participants perceived risk levels and the researcher assigned level of risk. The most alarming difference was the acceptance and willingness to perform actual 'highest risk' behaviours by participants with the perception that these behaviours are only 'low risk' behaviours; for example, there was a disturbing willingness to take risks at passive crossings reported. Low education levels on level crossings safety when learning to drive was evidenced as with the regional group.

Truck drivers (all areas)

Knowledge This group displayed a very high level of knowledge of the road rules and types of level crossings. Most participants could report exactly how many and what type of crossings they go over on their route. In terms of awareness of safety issues, many drivers knew of accidents involving trucks and trains. Personal definitions of driving safely at level crossings included obeying the rules and the use of protective behaviours such as slowing, looking and scanning. Being careful and alert, avoiding queuing and generally using 'common sense' were also felt to increase safety.

Overall it was felt that their ability to drive safely at level crossings was limited by engineering issues of crossing design and visibility. The identified problems included: the angle of approach; poor visibility, lighting, and road surfaces at crossings; short stacking problems; inadequate warning signs and stopping distances. Other problems identified included issues with slowing or stopping in heavy vehicles and the time taken for re-acceleration. Level crossings were generally felt to be an important issue but were not ranked as highly as the main 'other issues' facing the industry: fatigue, speeding and drugs.

Influencing factors, risk perception and behaviours There was perception of low enforcement of level crossings. The majority of drivers felt that a police or camera presence at level crossings would deter people from breaching the rules. Fines for breaching the rules were also noted to be a strong deterrent.

More than half of the participants believed that there was some risk for a crash due to engineering issues, truck cabin visibility and stopping issues, equipment malfunctions and human error. Errors were noted to include: distraction, judgement errors, fatigue, being unfamiliar with the crossing, complacent or driving unsafely. Comments included, '… all it takes is a little lapse of judgment and you could be in danger' and, 'complacency … go over the same crossing 400 times and never see a train'. Those who felt there was a 'low risk' believed that risks are reduced due to the safety actions they undertake.

The majority of participants stated that they 'always' drive safely at level crossings and attributed this safe behaviour to their knowledge of dangers associated with crossings and train crashes. A very small number of drivers indicated that they did not always drive safely at level crossings by: not stopping at passive crossings if no train was seen on approach; or trying to beat the train. Situations recounted where the participants nearly had an accident at a level crossing (close calls) often involved some level of distraction or inattention and had multiple factors involved including fatigue, sun glare, poor visibility or short stacking.

The majority of participants had never experienced any form of level crossing safety education and comments indicated that both engineering and education needs should be addressed for truck driver safety:

- Fix physical road issues first and follow with education – education is pointless if you cannot perform it and
- Make people safe. You can make the crossing as safe as you like but it's people that do the unsafe things; make the people safe.

A number of risk taking behaviours were reported as the behaviours of other truck drivers. The most common of these behaviours was 'trying to beat the train'. The suggested reasons given for this behaviour provided an insight into a number of concerns, including time pressures. Time pressure comments included:

- Express loads trucks try and race the train because of the time pressure. Coming from 100 to a stop and then back up to 100 again takes about five minutes and that takes five minutes off their travel time (5b)
- Complacency due to familiarity, risk taking 'if it's safe' or there is good visibility, and driving onto the wrong side of the road to avoid bad road surfaces were also mentioned. Other behaviours included: overhanging the crossing, not stopping at passive crossings and poor scanning. Reflecting the general safety attitude displayed by the participants interviewed, comments were made that these behaviours were unacceptable to the majority of drivers who strive to drive safely.

Conclusion

This chapter discussed some of the emerging issues from preliminary research with three distinct road user groups by setting. Risk behaviours and attitudes differed between the groups and by urban or regional setting. While older drivers displayed high levels of knowledge and low risk taking behaviours, age related factors were acknowledged by the group to be risk factors at level crossings. Compensatory and protective factors employed by older drivers were believed to reduce or control their risk. The younger drivers demonstrated a low perceived risk of consequences in relation to level crossing behaviour and subsequently reported the highest levels of risk taking of all the sub groups. The regional group in particular reported a high acceptance of a range of risk taking behaviours. As a group the truck drivers indicated a high level of knowledge of safety and engineering issues with level crossings and had direct experience of close calls at crossings. Risk taking was acknowledged as minority behaviour in the industry, with time pressures, complacency, inattention and fatigue noted to be risk factors. The findings of this study will inform the design and implementation of specific education strategies to improve level crossing safety behaviour in motorists.

Acknowledgements

Funding by the Cooperative Research Centre for Railway Engineering and Technologies is gratefully acknowledged. The researchers would like to thank and acknowledge all the participants from industry, government and the community for their time and participation.

References

Afxentis, D. (1994), *Urban Railway Level Crossings: Civil Engineering Working Paper*. Melbourne: Monash University.

Australian Transport Council (2000), *National Road Safety Action Plan, 2003 and 2004*. Canberra: Australian Transport Council.

Australian Transport Council (2005), *National Road Safety Action Plan for 2005 and 2006*. Canberra: Australian Transport Council.

Australian Transport Safety Bureau (2002), *Level Crossing Accidents: Monograph 10*. Canberra: Commonwealth Department of Transport and Regional Services.

Australian Transport Safety Bureau (2003), *Level Crossing Accident Fatalities*. Canberra: Australian Transport Safety Bureau.

Australian Transport Safety Bureau (2004), *Young People and Road Crashes*. Canberra: Australian Transport Safety Bureau.

Bureau of Transport Economics (1999), *Trends in Trucks and Traffic*. Canberra: Commonwealth of Australia.

Ford, G. and Matthews, A. (2002), *Analysis of Australian Grade Crossing Accident Statistics*, Paper Presented at the 7th International Symposium on Rail-Road Highway Grade Crossing Research and Safety, Melbourne: Monash University.

National Highway Traffic Safety Administration. (2005), Railroad Safety: U.S. Department of Transportation.

Chapter 19

Attitudes to Safety and Teamwork in the Operating Theatre, and the Effects of a Program of Simulation-Based Team Training

Brendan Flanagan, Michele Joseph and Michael Bujor
Southern Health Simulation and Skills Centre, Victoria

Stuart Marshall
The Alfred Hospital

The aim of this chapter is to determine if the safety and teamwork culture of an operating suite changes with a structured simulation-based team training program. The attitudes of the staff working within the complex social environment of the operating theatre are determinants of the outcome factors of safety, efficiency and job satisfaction. High reliability organizations have recognized that an initial evaluation of attitudes to safety is an important part of managing the changes of the environment, and the assessment of 'Crew Resource Management' (CRM) training in improving decision making and team-building. The composition of this training must be devised for each institution in turn.

A Safety Attitudes Questionnaire was administered to operating theatre staff before and after a simulation- based team training program, to assess the effectiveness of the program in changing the safety and teamwork culture. At present, insufficient data are currently available to allow statistically valid conclusions to be drawn.

Introduction

Team training has not been a traditional mode of preparation for professionals working within the complex social environment of an operating theatre; nevertheless, surgery is increasingly provided by a variety of craft groups, including semi-skilled technicians, and several specialties of medical, and nursing personnel, all of whom have differing training, experience, knowledge, skill and value systems (Kohn, 1999; Murray and Foster, 2001). The interactions between individuals from these various groups have been shown to be determinants of the outcome factors of safety, efficiency and job satisfaction (Yule, 2003, Firth-Cozens 2001). These outcomes may suffer if the organizational culture does not support a climate of teamwork and safety (Helmreich and Davies, 1996; Davies, 2005) and previous studies have demonstrated this culture is often lacking (Fox, 1994; Opie, 1997; Lingard *et al.*, 2002; Flin *et al.*, 2003).

The collective attitudes of the staff in the workplace define the climate of the organization, which partly explains the performance of the institution. Climate and culture are often used interchangeably; however, climate refers to day-to-day perceptions of work, whereas culture refers to shared beliefs, values and attitudes (Yule, 2003). Measurement of these attitudes to safety and teamwork correlate in other industries with indices of error (Sexton and Klinect, 2001), and data from incidents related to anaesthesia appear to support this (Williamson *et al.*, 1993). The aviation industry has recognized that an initial evaluation of attitudes to safety is an important part of managing the changes of the environment, and the assessment of 'Crew Resource Management' (CRM) training in improving decision making and team-building (Gregorich and Wilhelm, 1993), and this concept is gradually being accepted in healthcare (Risser *et al.*, 1999; Davies, 2001; Murray and Foster, 2001). In the aviation setting, CRM training has been shown to reduce absenteeism and staff turnover rates, and improve reporting of potential hazards in the workplace (Sexton and Klinect, 2001). This initial assessment of attitudes is termed the 'Safety Attitudes Questionnaire' (SAQ). The SAQ has been validated in aviation for 20 years, and has shown to be sensitive to predict performance of the organization (Schaefer and Helmreich, 1993; Helmreich, Sexton and Merrit, 1997; Sexton and Klinect, 2001). The questionnaire has been adapted for use in the operating theatre and intensive care units and has been validated in over 200 healthcare institutions worldwide. The survey and others like it have mainly been used in longitudinal surveys of these centres to specifically assess the culture of safety (Sexton, Thomas and Helmreich, 2000; Sexton *et al.*, 2004). However only a few studies have examined its use in assessing the impact of interventions on safety (Helmreich and Davies, 1996; Pronovost *et al.*, 2003).

We describe the use of the survey to assess the impact of a program of simulation-based team training in the operating suite of a large university hospital in Melbourne, Australia. Based on the results of the initial SAQ the program was designed to specifically target aspects of teamwork and safety. The SAQ was then readministered after the program. The hospital itself is an adult university-affiliated teaching hospital, which provides all types of specialized surgical and anaesthetic services including neurosurgery, cardiothoracic surgery (heart and lung transplantations) and trauma care.

Method

After institutional ethics approval and permission from the University of Texas to use the SAQ, a covering note and the anonymous postal questionnaire were sent to all staff identified as having a role in the functioning of the operating theatre suite. The survey was administered to the usual members of staff present in the operating theatre (surgeons, anaesthetists, nurses and technicians), plus preoperative-assessment and day-of-admissions nursing and administrative staff. Wherever possible the study was explained to the participants with a short presentation before administration of the questionnaire.

After assessment of the preliminary data, a course was developed addressing the main themes of the questionnaire responses, namely teamwork and communication. As described by Beaubien and Baker, the most effective way of delivering team training to existing teams under stressful or ambiguous conditions is a high-psychological fidelity simulation scenario approach (Beaubien and Baker, 2004; Salas *et al.*, 2005). Unfortunately, this excluded participation by the relatively small numbers of administrative and ward staff. As this was an exercise in team training, it was important to ensure that the constitution of the team reflected that which would normally be present in a routine operating theatre session. Hence the scenarios were designed to incorporate all the usual members of the operating theatre team: surgeon, surgical assistant, surgical nurse, scout nurse, anaesthetist, anaesthetic nurse, and operating theatre technician. The session itself consisted of a brief introductory lecture, a familiarization with the mannequin, a practice debriefing of an aviation video, and two scenarios with debriefings assisted by the facilitators. This format is an extension of previous CRM-style training in healthcare (Howard *et al.*, 1992; Gaba *et al.*, 1998), in which the debriefing explores both medical/technical as well as non-technical aspects of the scenario. An advantage of this approach to training is the discussion between participants, and exchange of points of view and mental models as part of the debriefing process, as this is often quoted as one of the failings of teams (Flin and Maran, 2004).

The sessions were conducted by certified healthcare CRM instructors, all of whom were experienced members of the operating theatre team. As such they had insight into the unique problems encountered in the operating theatre environment.

To ensure a realistic environment for the participants, the training took place on-site in a spare operating theatre within the participants' normal working environment. This was considered important for logistics reasons to ensure all of the members of the team would be available to participate.

At the completion of the program, each participant was given a separate evaluation form specifically related to the half-day program and asked to complete it after several days of reflection on the experience. The SAQ was then administered to the same staff groups for a second time three months after the course, whether they had taken part in the training or not. Analysis of the results was undertaken in accordance with the scales published by the University of Texas (Sexton and Thomas, 2003; Sexton *et al.*, 2004).

The power of the study to detect a 50 per cent improvement in the number of respondents reporting positive teamwork and safety from the median value of the index institutions was calculated to be 0.77 and 0.99 respectively ($\alpha = 0.05$) assuming a 50 per cent response rate to both the pre- and post-intervention questionnaires.

Results

In total, 328 initial questionnaires were administered to all members of the nursing, surgical, anaesthetic and support staff, with 176 returns; a response rate of 54 per cent. The responses were analysed in accordance with the University of Texas guidelines, to examine the six domains of workplace climate (Sexton *et al.*, 2004). The six domains are: job satisfaction, teamwork climate, safety climate, perceptions of management, stress recognition, and working conditions. All of the domains scored below average in the distribution of 50 index healthcare organizations used for validation. Only 22 per cent of respondents (n = 38) reported a positive teamwork climate in the operating theatres, and 32 per cent of respondents (n = 56) reported a positive safety climate.

During the simulation-based team training, eight half-day sessions were run for a total of 59 participants. The sessions involved surgical and anaesthetic specialist nursing staff (n = 21 and n = 12 respectively), anaesthetic specialists and trainees (n = 8 and n = 6), theatre technicians (n = 8) and surgical trainees (n = 4). A separate evaluation of the half-day program was provided by 18 of the 59 participants (response rate of 31 per cent); a summary is provided in Table 19.1.

Table 19.1 Post course postal evaluation (N = 18)

Question	Number (%) of participants agreeing with statement
The lecture/video introducing CRM was organized and comprehensive	17 (94%)
Knowledge gained about CRM will help my practice	17 (94%)
This course will help me work more safely	14 (78%)
Clarified importance of team work	17 (94%)

Three months after the training 346 safety attitude questionnaires were administered. At the time of writing there have been 41 returns (response rate of 12 per cent). Insufficient data are currently available to allow statistically valid conclusions to be drawn. Interim analysis suggests no significant change in the climates of safety or teamwork. Preliminary data are displayed in Table 19.2.

Table 19.2 Preliminary data on the effect of the CRM training program on teamwork and safety climates

	Number (%) of respondents reporting positive climate		
	Before training period (n = 176)	*Three months after training period* (n = 51)	
Teamwork climate	38 (22)	9 (18)	p = 0.540
Safety climate	56 (32)	13 (25)	p = 0.387

Qualitative data

The free text responses on the evaluation forms for the half-day program were analysed. These responses support previous thematic analyses of course evaluation forms and structured interviews of participants of similar CRM-based courses (Wilson *et al.*, 2005). The five themes identified were:

Realism:
- (Positive) 'Surprisingly realistic'
- (Positive) 'Much more realistic in your own OR than at [the simulation centre]. Good to do it with your usual staff and assistants. Would be good to have surgeons participate'
- (Negative) 'A more life-like mannequin. That is frustrating – being unsure whether a response/lack of response is "real" or failed technology'

Team:
- (Positive) 'Open discussion and team approach to scenarios. Non judgemental discussions by each team member'
- (Positive) 'I think participating in this course with people you regularly work with makes it more worthwhile as you are aware of levels of experience and areas of specialist knowledge – who to support/guide/protect: who can function well under pressure: who needs direction and who doesn't'

Self-belief:
- (Positive) 'I feel much more comfortable with the prospect of a crisis situation arising in theatre, after reviewing and observing its handling in this scenario'
- (Negative) 'I am now unsure as to whether I wish to continue with peri-operative nursing if the simulation experience represents 'real life'

Learning:
- (Positive) 'Overall very valuable experience which we would all benefit doing every year'

Process:
- (Positive) 'This was an excellent challenging but non threatening environment. Excellent work by the instructors'

The SAQ has a free text response area for recommendations to improve safety in the clinical area. A total of 124 suggestions from the pre-program questionnaire were made, these were classified into themes as illustrated in Table 19.3.

Table 19.3 Most frequent recommendations for improved safety in the OR

Response	Frequency (n = 124)
Improved communication	26
Improved training/supervision of trainees	15
Improved/new equipment	11
Workload distribution	10

Response	Frequency (n = 124)
More safety guidelines and briefings	9
More staff	9
Improved teamwork	9
Miscellaneous	35>

Discussion

With the data currently available, this study is unable to demonstrate the effect or lack of an effect of exposure to a CRM program with respect to the culture of safety and teamwork in a medical setting. We are continuing to collect data from the second, post-intervention, survey.

The initial questionnaire return rate of 54 per cent is comparable with other surveys and is not unusual when surveying professional groups (Flin *et al.*, 2003). Healthcare consultants are generally reticent to respond in postal surveys, and are dismissive of interview-based qualitative data (Kellerman and Herold, 2001). Future investigators using questionnaires with these groups should be prepared to follow up respondents both personally and by repeated reminders. Incentives for completing the questionnaire were not used in this study to prevent the possibility of introducing bias into the results.

The Safety Attitudes Questionnaire is a well-validated survey, and recent reports on the survey claim over 200,000 medically related questionnaires in circulation (Sexton *et al.*, 2004). Given cost constraints and the number of staff involved, the SAQ was the method thought to be most efficient in reaching all staff within the area, and most likely to produce a valid, and meaningful quantitative result. Other surveys have recently been validated for use in the healthcare setting, including the Safety Climate Survey (SCSu) and Safety Culture Scale (SCSc) (Pronovost *et al.*, 2003; Kho *et al.*, 2005). These surveys show a high internal consistency for safety culture; however, they do not measure teamwork, a priority aim of this study. Despite the SAQ having a high internal consistency for both safety and teamwork climates, there is currently little evidence that it reflects observable practice in the operating theatre workplace. Research is continuing on the construct validity of these surveys (Pronovost and Sexton, 2005a).

This particular study may be perceived to be limited by virtue of it being a single centre study, and may also be criticized by the lack of a true control group. However, the measurement unit is the organization and not the individuals, or subgroups, and hence a control group within the organization is inappropriate. The climate of an organization is also variable both over time and between organizational subunits, and therefore comparisons of climate scores should be undertaken with caution (Coyle, Sleeman and Adams, 1995; Yule, 2003).

This type of training is challenging to conduct. The authors experienced a range of difficulties in the development and delivery of the course. These included the practicalities of achieving adequate participation rates from all groups, and designing scenarios with adequate fidelity for all members of the team. The practicalities of running a simulation course within a working operating theatre suite are considerable. The safety of the real patients in the operating suite remains the utmost priority, and isolation of simulation equipment must be meticulous. Some of the participants felt that the normal work occurring within the operating suite was a distraction from the learning opportunity, although the majority appreciated the fidelity of using their own equipment and environment.

Biomedical engineering support should be immediately available to troubleshoot unexpected problems in the unfamiliar environment of the remote simulation lab, and facilitators and simulation coordinators who are familiar with the equipment and scenarios are essential. Moreover, the cost of such a program may well be prohibitive if individual participants are expected to pay without incentive or reimbursement. This is less likely to be an issue if the aim of the session is enhancing teamwork and the costs are borne by the institution.

The development of any program to improve patient safety in a hospital setting should target a range of issues. These include the production and adherence to protocols reflecting best practice (Leape, Berwick and Bates, 2002) and appreciation of human factors with education of the staff regarding the limitations of the individual, the team and the system in which they work (Maurino *et al.*, 1995b). The front-line staff should also be engaged in any improvements as greater ownership and involvement by the staff enhances motivation and safety performance (Firth-Cozens, 2001; Clark, 2002). Embracing a safety culture within an organization consists of management of factors at both local and organizational levels (Maurino *et al.*, 1995a). Such a program has been developed by Pronovost *et al.* (Pronovost *et al.*, 2005b), and has proven to have an impact on the culture of safety using the Safety Climate Survey, and to reduce length of stay and staff turnover in the intensive care unit. This program uses intervention at a local level to produce cultural change in safety via formation of work units created to address specific problems. Our study took a different approach and aimed to engage all members of the multidisciplinary team in practical examples of both safety management and teamwork, to improve both domains.

The number of participants in the course ($n = 59$) may have been too small to influence the climate of the entire workspace to any great degree. Participation was entirely voluntary, and participation by senior surgeons was particularly low. Participation rates by surgeons should be a major focus in the design of future studies. Previous team training courses in the operating room setting experienced similar problems. Helmreich and Schaefer developed 'Team Orientated Medical Simulation' (TOMS) in the early 1990s. This program was not sustained due to the inability to engage the surgical team in scenarios with relevance to them (Helmreich and Schaefer, 1994).

Initial power calculations suggest that our choice of a 50 per cent improvement in teamwork and safety climates from the median values would require a sample size of 150 respondents before and after the training ($\rho = 0.05$) to accept the null hypothesis that there is no effect with simulation-based team training. The clinical significance of a 50 per cent improvement in these measures is unknown; furthermore, as no previous studies have examined the influence of simulation-based training on climate or safety and teamwork, the likely amplitude of effect was also unknown at the time of the design of the study. Due to the single-centre nature of this study, the number of participants (and hence its power) is limited by the number of staff working in the institution. A multi-centre study design with pooled data from other institutions would assume similar climates of safety and teamwork which may not hold true.

CRM in aviation has evolved over 25 years into a mandated, regulated discipline, and cultural change has been slow to occur (Musson and Helmreich, 2004). CRM in healthcare is currently still in its infancy with little or no regulation as to its content or providers. Adaptation of CRM training from other industries into health may not work, and input into this type of training must come from specialists with experience from within the health sector and be individualized to each institution (Musson and Helmreich, 2004). Further studies into the effects of CRM in healthcare are urgently needed to determine the needs and effects of this type of training.

Acknowledgements

We would like to thank all the staff of the Southern Health Simulation and Skills Centre, and Biomedical engineers who assisted in the development and delivery of the program, and the operating theatre staff of the Alfred hospital for their participation.

References

Beaubien, J.M. and Baker, D.P. (2004), The use of simulation for training teamwork skills in health care: How low can you go?, *Quality and Safety in Health Care*, **13**, i51–i56. [PubMed 15465956] [DOI: 10.1136/qshc.2004.009845]

Clark, G. (2002), Organisational culture and safety: an interdependent relationship, *Australian Health Review*, **25**, 181–189. [PubMed 12536878]

Coyle, I.R., Sleeman, S.D. and Adams, N. (1995), Safety Climate, *Journal of Safety Research*, **26**, 247–254. [DOI: 10.1016/0022-4375%2895%2900020-Q]

Davies, J.M. (2001), Medical applications of crew resource management. *In Improving Teamwork in Organizations* eds Salas, E., Bowers, C. and Edens, E., Mahwah, NJ: Erlbaum, pp. 265–281.

Davies, J.M. (2005), Team communication in the operating room, *Acta Anaesthesiology Scandinavica*, 2005, No. 49, 898–901.

Firth-Cozens, J., (2001), Cultures for improving patient safety through learning: The role of teamwork, *Quality and Safety in Health Care*, **10**, 70–75. [PubMed 11389314] [DOI: 10.1136/qhc.10.2.70]

Flin, R. and Maran, N. (2004), Identifying and training non-technical skills for teams in acute medicine, *Quality and Safety in Health Care*, **13**, i80–i84. [PubMed 15465960] [DOI: 10.1136/qshc.2004.009993]

Flin, R., Fletcher, G., Mc George, P., Sutherland, A. and Patey, R. (2003), Anaesthetists' attitudes to teamwork and safety, *Anaesthesia*, **58**, 233–242. [DOI: 10.1046/j.1365-2044.2003.03039.x]

Fox, N.J. (1994), Anaesthetists, the discourse on patient fitness and the organisation of surgery, *Sociology of Health and Illness*, **16**, 1–18. [DOI: 10.1111/1467-9566.ep11346987]

Gaba, D., Howard, S., Flanagan, B., Smith, B., Fish, K. and Botney, R. (1998), Assessment of clinical performance during simulated crises using both technical and behavioural ratings, *Anesthesiology*, **89**, 8–18. [PubMed 9667288] [DOI: 10.1097/00000542-199807000-00005]

Gregorich, S.E. and Wilhelm, J.A. (1993), Crew Resource Management Training Assessment, *in Cockpit Resource Management* eds Wiener, E. L., K., B. G. and Helmreich, R. L., San Diego: Academic Press, pp. 173–198.

Helmreich, R. and Schaefer, H.G. (1994), Team performance in the operating room, *in Human Error in Medicine* Ed, Bogner, M., Hillsdale, NJ: Erlbaum, pp. 224–253.

Helmreich, R.L. and Davies, J.M. (1996), Human factors in the operating room: interpersonal determinants of safety, efficiency and morale, *Balliere's Clinical Anesthesiology*, **10**, 277–294.

Helmreich, R.L., Sexton, B. and Merrit, A. (1997), *University of Texas Aerospace Crew Research Project Technical Report 97-6 The*, Austin: University of Texas.

Howard, S.K., Gaba, D.M., Fish, K.J., Yang, G. and Sarnquist, F.H. (1992), Anesthesia Crisis Resource Management Training: Teaching Anesthesiologists to Handle Critical Incidents, *Aviation Space and Environmental Medicine*, **63**, 763–770.

Kellerman, S.E. and Herold, J. (2001), Physician response to surveys. A review of the literature, *American Journal of Preventative Medicine*, **20**, 61–67.

Kho, M.E., Carbone, J.M., Lucas, J. and Cook, D.J. (2005), Safety Climate Survey: reliability of results from a multicenter ICU survey, *Quality and Safety in Health Care*, **14**, 273–278. [PubMed 16076792] [DOI: 10.1136/qshc.2005.014316]

Kohn, L.T., C., J.M. and Donaldson, M.S. (1999), *To Err is human: Building a Safer Health System*, Washington, DC: National Academy Publishing.

Leape, L.L., Berwick, D.M. and Bates, D.W. (2002), What practices will most improve safety? Evidence based medicine meets patient safety, *Journal of the American Medical Association*, **288**, 501–507. [PubMed 12132984] [DOI: 10.1001/jama.288.4.501]

Lingard, L., Reznick, R., Espin, S., Regehr, G. and DeVito, I. (2002), Team communications in the operating room: talk patterns, sites of tension, and implications for novices, *Academic Medicine*, **77**, 232–237. [PubMed 11891163] [DOI: 10.1097/00001888-200203000-00013]

Maurino, D.E., Reason, J., Johnston, N. and Lee, R.B. (1995a), *Beyond Aviation Human Factors*, Aldershot, UK: Ashgate, pp. 138–165.

Maurino, D.E., Reason, J., Johnston, N. and Lee, R.B. (1995b), *Beyond Aviation Human Factors*, Aldershot, UK: Ashgate, pp. 1–29.

Murray, W.B. and Foster, P.A. (2001), Crisis Resource Management Among Strangers: Principles of Organizing a Multidisciplinary Group for Crisis Resource Management, *Journal of Clinical Anesthesia*, **12**, 633–638. [DOI: 10.1016/S0952-8180%2800%2900223-3]

Musson, D.M. and Helmreich, R.L. (2004), Team training and resource management in health care: Current issues and future directions, *Harvard Health Policy Review*,

5, 25–35.

Opie, A. (1997), Thinking teams thinking clients: issues of discourse and representation in the work of health care teams, *Sociology of Health and Illness*, **19**, 259. [DOI: 10.1111/1467-9566.00051]

Pronovost, P. and Sexton, J.B. (2005a), Assessing safety culture: guidelines and recommendations, *Quality and Safety in Health Care*, **14**, 231–233. [DOI: 10.1136/qshc.2005.015180]

Pronovost, P., Weast, B., Holzmueller, C.G., Rosenstein, B.J., Kidwell, R.P., Haller, K.B., Feroli, E.R., Sexton, J.B. and Rubin, H.R. (2003), Evaluation of the culture of safety: survey of clinicians and managers in an academic medical center, *Quality and Safety in Health Care*, **12**, 405–410. [PubMed 14645754] [DOI: 10.1136/qhc.12.6.405]

Pronovost, P., Weast, B., Rosenstein, B., Sexton, J.B., Holzmueller, C.G., Paine, L., Davis, R. and Rubin, H.R. (2005b), *Journal of Patient Safety*, **1**, 33–40.

Risser, T., Rice, M., Salisbury, M., Simon, R., Jay, G.D. and Berns, S.D. (1999), *Annals of Emergency Medicine*, **34**, 373–383. [PubMed 10459096] [DOI: 10.1016/S0196-0644%2899%2970134-4]

Salas, E., Wilson, K.A., Burke, C.S. and Priest, H.A. (2005), Using Simulation-Based Training to Improve Patient Safety: What does it take? *Journal on Quality and Patient Safety*, **31**, 363–371.

Schaefer, H. and Helmreich, R.L. (1993), The Operating Room Management Attitudes Questionnaire (ORMAQ), *NASA/University of Texas FAA Technical Report University of Texas Austin*.

Sexton, B., Thomas, E. and Helmreich, R. (2000), Error, stress and teamwork in medicine and aviation: cross sectional surveys, *British Medical Journal*, **320**, 745–749. [PubMed 10720356] [DOI: 10.1136/bmj.320.7237.745]

Sexton, B., Thomas, E., Helmreich, R.L., Neilands, T.B., Rowan, K., Vella, K., Boyden, J. and Roberts, P.R. (2004), in Technical Report 04-01 The University of Texas Center of Excellence for Patient Safety and Practice.

Sexton, J.B. and Klinect, J.R. (2001), In Proceedings of the Eleventh International, *Symposium on*, Columbus, OH: Aviation Psychology The Ohio State University.

Sexton, J.B. and Thomas, E.J. (2003), The University of Texas Center of Excellence for Patient Safety Research and Practice.

Williamson, J., Webb, R., Sellen, A., Runciman, W.B. and Van der Walt, J.H. (1993), Human failure: Analysis of 2000 incident reports, *Anaesthesia and Intensive Care*, **21**, 678–683. [PubMed 8273898]

Wilson, V., Lacey, O., Kelly, L. and Flanagan, B. (2005), *SimTect 2005*, Brisbane: Healthcare Simulation Conference Simulation Industry Association of Australia.

Yule, S. (2003), Senior Management Influence on safety performance in the UK and US energy sectors, University of Aberdeen, Scotland.

Chapter 20

Human Factors and the QFI: Developing Tools through Experience

Boyd Falconer
University of New South Wales

There are several negative human factor issues that create significant challenges to improving pilot training in both civil and military aviation. In Australia, military aviation operations are undertaken predominantly by the Royal Australian Air Force (RAAF), and the challenges to improving pilot training include an organizational aversion to human factor issues, an audience effect created by the existence of 'authority' within the training environment and a low level of confidence amongst many trainees. The chapter identifies several tools for improving the quality of pilot training 'output' in Australia, based on the author's experiences as an RAAF Qualified Flying Instructor (QFI) and aspects of data gathered during a recent review of existing RAAF-related research.

Introduction

To the casual observer, operational aviation and the pilot training that supports it is shrouded in mystery. Indeed, the sheer spectacle of aviation never ceases to amaze many, both within aviation and outside of it. This alluring mix of mystery and spectacle is perhaps greatest in the form of military aviation where stringent selection standards and years of training are required to operate some of the fastest and most deadly transport platforms in the world. In Australia, the military aviation function is undertaken predominantly by the RAAF. This chapter represents an attempt to not only demystify the training that pilot candidates must complete before being awarded their 'Wings', but also provide an insight into the human factor difficulties experienced by instructional pilots – or 'QFIs' in RAAF parlance. These difficulties provide the springboard for tools that may assist training personnel in other human–machine systems where developing sound decision-making and behavioural habits is crucial to successful operations.

The chapter consists of three main parts. Firstly, the chapter provides an overview of RAAF pilot training; paying particular attention to the differences between civilian and RAAF pilot training, and the reasons for candidates' suspension from RAAF pilot training. Following this overview, the chapter describes the challenges to improving the quality of pilot graduates in Australia. Thirdly, the chapter investigates and describes several tools that can be used by both pilot candidates and instructor-pilots to overcome the aforementioned challenges.

RAAF pilot training – an overview

> 'The course is rigorous and the standards are high, but so too are the demands of today's military flying'.
> (Wing Commander Gareth Neilsen, Commanding Officer of the RAAF Advanced Flight Training School[1])

The first step for a civilian wanting to become an RAAF pilot is to successfully complete an extensive battery of aptitude, medical and psychological tests set by various specialist agencies within the Australian Defence Force (ADF). These tests enable the selection of personnel that are likely to hold the skills that will enable them to train as an RAAF pilot, and include such assessments as vigilance, physical coordination, spatial orientation and electrocardiographs. The candidates will then undergo 'flight screening' at the ADF Basic Flying Training School (BFTS) in Tamworth, approximately five hours drive north-west of Sydney. Successful candidates then complete approximately six months of officer training at RAAF Base Williams, just west of Melbourne. The pilot candidates then return to BFTS for basic flying training on the piston-engine CT-4B single-propeller aircraft. After successful completion of the BFTS phase, the candidates will enter the RAAF's 'Number 2 (advanced) Flying Training School' (2FTS) for the final and most in-depth phase of RAAF pilot training. Here they fly the advanced, fully aerobatic and ejection seat equipped PC-9 aircraft. This phase typically lasts 40 weeks, and frequently culls between 30 and 50 per cent of the remaining pilot candidates. This often translates into a graduation of pilots in magnitudes of only single-figures, after having commenced training with perhaps 50 other capable trainees selected from many hundreds.

Civilian and military training differences

In this light, it is worthwhile briefly exploring the differences or, more specifically, additional requirements that an RAAF pilot candidate must complete before being deemed competent for employment. These training items are listed in Table 20.1.

The table, while a relatively simple summary of RAAF pilot training, is extremely effective in demonstrating that significant additional requirements exist in RAAF pilot training that do not exist in civilian pilot training. Indeed, it is noteworthy that civilian training covers only one-third of the competency requirements that RAAF pilot training encompasses.

1 Interview by Private John Wellfare, 2005.

Table 20.1 Civilian and RAAF pilot training comparison

Item	Civilian training	RAAF training
Psychological test	No	Yes
Coordination test	No	Yes
Medical test	Yes	Yes
Personality assessment	No	Yes
Flight screening	No	Yes
Officer/leadership training	No	Yes
Basic flying (*e.g.* turns, stalls)	Yes	Yes
Instrument flying	Yes	Yes
Close-formation flying	No	Yes
Combat-formation flying	No	Yes
Night flying	Yes	Yes
Low-altitude navigation	No	Yes
Medium-altitude navigation	Yes	Yes
High-altitude navigation	No	Yes
Advanced aircraft handling (*e.g.* high-performance evasion manoeuvre, 'on-the-buffet')	No	Yes
Percentage of items completed	**33%**	**100%**

Why are pilot candidates suspended from pilot training?

Pilot candidates are generally suspended from pilot training due to insufficient progress being made at the required rate, usually due to one or more of the factors listed in Figure 20.1. Whilst the factors are listed in a linear fashion, in the author's experience the 'source' of a student's performance deficiencies is never neatly linear or singular.

Often reasons for suspension from training overlap, or are inter-related. The more common deficiencies are listed in Figure 20.1 and explained below.

Weak flying skills
(physical co-ordination)
⇓
Easily task-overloaded while airborne
(physical-mental co-ordination, task saturation)
⇓
Does not make decisions when required, or makes poor decisions
(cognition, prioritisation)

Figure 20.1 Typical deficiencies among RAAF student pilots

The first deficiency, weak flying skills, is rare. Cases of students simply not being able to manipulate the controls to stay flying are uncommon. Flight screening has generally weeded out the candidates that simply 'cannot fly'.

The second case is more common. This occurs where the student has a good attitude towards flying and applies much effort to their training, but cannot make the grade. Alternatively, the student's performance has deteriorated steadily as the flying syllabus became harder. The student may have been taking things easy and then by the time they realized that they needed to improve it was too late.

In the third case, a lack of or poor decision-making can have disastrous results. This performance trait is the most common for RAAF pilot candidates that struggle through pilot training (or ultimately fail). Pilots make many decisions each flight and hence if a student cannot make decisions, their performance will be poor. It is almost expected that a trainee pilot will make some poor decisions during the early phases of their flying career, as it is simply part of the learning process to develop captaincy. However, student pilots must be able to make a decision and not ignore a problem, or delay a decision to the point where it is no longer appropriate. Kern (1997, 1999) has also noted this trait in pilots, and pilot candidates, within the United States Air Force.

So what challenges are presented in needing to minimize these traits and maximize the quality of pilot graduates?

Challenges

There are several negative human factor issues that create significant challenges to improving pilot training in both civil and military aviation. The challenges include an organizational aversion to human factor issues, an audience effect created by the existence of 'authority' within the training environment and a low level of confidence amongst many trainees.

Organizational aversion to HF issues

There is widespread recognition that human performance factors remain the most significant issue in air safety (Lee, 1999; McCormack, 1999). Nevertheless, several operational ADF aviation personnel (*e.g.*, Falconer, 2002, 2005) and ADF psychologists (*e.g.*, Fellows, 1998) have suggested the level of understanding of human factor issues in ADF aviation is somewhat underdeveloped. More specifically, it has been argued that mission-focus has long created a degree of intolerance within the ADF towards human performance limitations that might hinder the mission. Indeed, the official history of Australia's involvement in the Second World War noted an instance where an A.I.F. battalion commander who handed over command to his second-in-charge while he napped in an effort to overcome his exhaustion during the retreat to Singapore was sacked for displaying 'weakness' (Falconer and Murphy, 2005).

Authority and rank

It can be difficult for some student pilots to fly well whilst a QFI is watching their every single move. It is especially difficult if the QFI is an intimidating instructor and the student pilot feels that the QFI is always 'on their back'. This idea can be likened to the Hawthorn or 'audience effect' whereby an individual's performance is influenced with the knowledge that they are being observed or examined. These are of course legitimate feelings, but they can be detrimental to a student pilot's performance. Nobody can fly well and have a high level of situational awareness, as well as worrying about what the instructor thinks of them. This interpersonal gradient is based on a profession-based construct (qualified pilot: student pilot).

An additional and perhaps even more pervasive gradient effect is based on rank, as shown in a survey of attitudes conducted amongst RAAF aircraft aircrew (Braithwaite, 2001). Respondents were asked what they would do if faced with a new rule that they considered to be unsafe. The question was designed to explore the effects of hierarchy upon employee attitudes towards safety context, and is reproduced in Figure 20.2.

Question: A senior officer introduces a new operating rule you consider to be unsafe, which if the following statements best describes your actions?

1. I would simply ignore the new rule; it's my life
2. I would complain about the rule to my colleagues
3. I would complain about the rule to my Wing Commander
4. I would complain directly to the officer responsible for the rule
5. I would obey the rule, as it is my job to obey the rules

Figure 20.2 Braithwaite's (2001) survey item designed to measure the influence of hierarchy among aircrew

Responses were almost divided between just two response options (see Figure 20.3). Approximately half of the respondents report that they 'would complain directly to the officer responsible for the rule', and are therefore arguably not 'influenced' by rank. In contrast, a sizeable proportion of respondents seemed to be strongly influenced by rank, or perhaps military culture more generally. These respondents reported that they 'would obey the new rule' despite considering the new rule to be unsafe. Interestingly, a small number (four per cent) responded that they 'would simply ignore the new rule'.

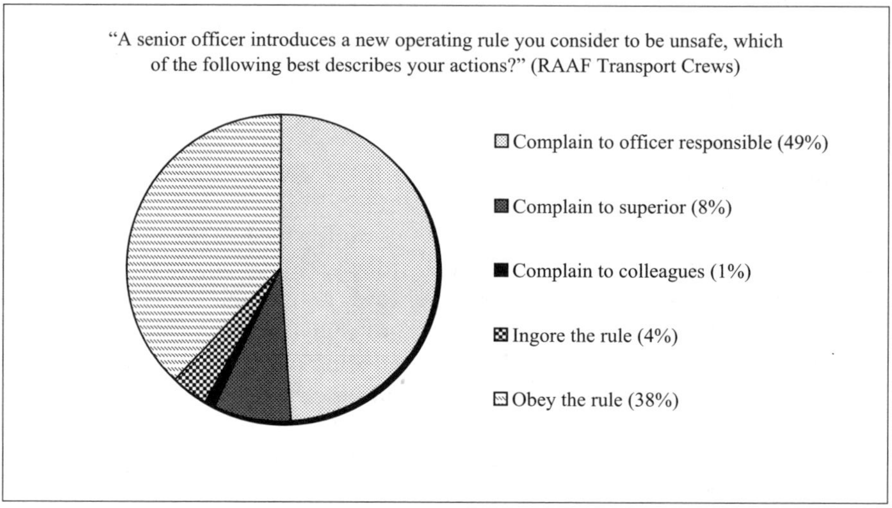

Figure 20.3 Response from RAAF aircrew to Braithwaite's survey item

These results raise important and fundamental questions. Does the existence of 'rank' influence aviation professionals to the extent that the safety could be inadvertently threatened? Does the reported willingness to obey unsafe rules suggest ingrained obedience/strong discipline, lack of initiative/autonomy or mission-focus? Conversely, does the small percentage of respondents reporting a willingness to ignore a new rule from a superior officer reflect a preference to avoid conflict?

Trainee confidence and the 'can-do' attitude

The much talked about 'can-do' attitude that is so crucial in military organizations has the potential to also undermine safety. At the heart of the 'can-do' attitude is a perennial tension. This conflict is common in safety-critical systems where people are constantly trying to reconcile what are often irreconcilable goals: staying safe and achieving the mission. This emphasis on mission achievement, irrespective of the associated human performance constraints and costs, has been termed the 'can-do' attitude. In the military, working successfully under pressures and resource constraints is a source of professional pride. The ability to achieve a task or mission on time and no matter the resource or human challenges is generally regarded as a sign of professionalism, an indication of expertise and competence. However, such conditions tend to generate trade-offs, and in safety-critical domains these compromises may eventually come at great cost in the form of accidents.

Dekker (2005) and Falconer (2005a) argue that 'can-do' is about informal work systems compensating for an organization's inability to provide the basic resources (time, tools, funding) needed for mission accomplishment. When things go wrong, it is often the individual who has tried to work on under resource constraints that gets blamed. The organization or system therefore changes very little, if at all, because no organizational learning has occurred. Organizations continue to find individuals, fuelled by the can-do attitude, who are willing to work harder, and take short cuts and risks to comply with professional obligations and expectations.

Tools for improvement

Self-analysis

Self-analysis is arguably the most important attribute for a pilot to possess. If an individual can analyse everything that they do whilst flying, then go away afterwards and reflect on how to improve it next time, then they will always become a better pilot. Self-criticism is an awareness of where an individual could have performed better, as well as the areas in which they performed well. Self-criticism does not constantly involve finding fault in one's actions, as this often results in disappointment if a task or skill is not executed perfectly. To be constantly down on oneself does damage confidence and self-esteem. A good knowledge of one's own strengths and weaknesses can only be gained through self-analysis. The mature individual will accept these weaknesses and work towards improving them.

Previous flying experience

On each Pilot's course there can be a wide variety of previous flying experience. It is worth spending some time thinking about the value of previous flying experience. In the past it has been a contributing factor in some students' suspension from the Pilot's course, while other students have found it to be extremely beneficial. There are students who have many hours of previous flying; others may have none at all. Occasionally, there is a student who does not even know how to drive a car. This previous experience can be an advantage; it can also be a disadvantage. If an individual does have previous flying experience, then they should carefully consider what that could mean in their present environment.

So how can pre-existing aviation knowledge and flying experience be a disadvantage? Military flying has different requirements to civilian flying. Each pilot needs to have similar basic skills but a different approach to flying. The military pilot must complete a military mission in some challenging conditions. Military pilots work with very different pressures to civilian pilots; they have a variety of missions to perform in extremely hazardous situations (see Merket, Bergondy and Salas, 1999; Soeters and Boer, 2000). As a direct result, military pilots have a different perception of flying than their civilian counterparts. Each of them is the best in their own environment. To handle the mission, military pilots must be very good at the basics of flying.

The fundamental item that determines if an individual's previous flying experience is beneficial or detrimental is their attitude toward both learning and flying. 'Attitude', in the context of military pilot training, is the way in which one perceives themselves and their role in relation to what they are doing. This concept does not involve one's personality or character but simply attempts to highlight the importance and value of a good attitude toward pilot training. The person who learns best is still the person with an appropriate attitude.

Learning flying skills

If we consider riding a bicycle through a road intersection, there is a lot to think about. You need to think about cars, so you do not get run over. There is the distance and speed of each car to estimate, an assessment of your own speed and if you have lived up until this point, then you have also become fairly good at planning a course of action should something go wrong. You can probably think of a few other things that the brain must process before riding through an intersection. Notice that I have not mentioned the thought required to steer and balance the bicycle. That is because we do not really think about it consciously; it is an acquired skill. So how did we go from training wheels to this complicated scenario without even thinking too much about it? Simply by using a building block approach of practicing our skill while gaining experience of riding through an intersection.

Study the techniques with attention to detail

⇓

Memorise techniques methodically

⇓

Understand what you are doing

⇓

During flight, let the experience 'soak in'

⇓

Learn from your debrief

⇓

Reinforce learning post-flight: controls, visual cues

⇓

Visualise the techniques regularly

Figure 20.4 Tools for maximizing learning development

Military flying training uses a very similar building block approach, but the learning pace is quicker and requires some specialized knowledge. To achieve good results the student pilot must have an understanding of how they are learning to fly. Flying an aircraft initially requires a lot of conscious thought about inputs *versus* aircraft movement. Each sequence learnt in the early phase of training develops into a skill through practice. It is important to use a consistent and correct technique in the early phases so that one learns the right skill.

So how can one maximize learning to develop these skills? There are a few key points to consider, and these are listed in Figure 20.4.

Visualization techniques

When it comes to visualization it seems almost everybody can describe it, but demonstrating it is another matter! This is because visualization is an imagination skill. It is popular in flying training as it very closely simulates actually flying the aircraft; your brain thinks it does anyhow. Visualization is a remarkably valuable tool as it assists individuals to learn at a much faster rate. There are different levels of visualization one can use. The easiest and most used is the simple daydream, where you imagine that you are flying.

When you are visualizing, try to see in your mind as much detail as possible, instruments, the places you will be flying over and how the aircraft will feel. Be as realistic as you can about control inputs and the aircraft's response. It should be just like a dream where it is all happening in real time with as many of the real pressures and time constraints as possible. The benefit of this method is that it really builds up your situational awareness once you are airborne. One of the hardest things to do whilst learning to fly is to deal with a situation airborne that you had never thought of before. If you have already lived it through using visualization, your awareness of the factors involved will be much higher. This could be the difference between handling a simulated emergency well or making a poor decision and handling the emergency very poorly.

An advantage of this method is that it will be obvious to you if you have gaps in your knowledge. If you have to stop the daydream to think of what an emergency checklist is, then you do not know that checklist well enough. If you have to try to remember how to fly a particular sequence and you cannot continue your visualization of being airborne at the same time, then you do not know that sequence well enough. If you go on your sortie and you experience a simulated emergency that you do not know well enough, your flying of the aircraft will be poor. Your brain will have to devote a lot its power into remembering the checklist. As a result you will get behind the aircraft as you begin to react to the errors in your flying accuracy. Situational awareness of what is happening around you will reduce as you spend more time trying to fly the aircraft. This scenario is the classic setup to make some bad mistakes or poor decisions. The simple daydream visualization before flight can help you handle these situations.

Early in your training a different type of visualization can help to learn flying skills. A daydream is not really going to achieve the aim of developing your flying skills. To learn flying skills you need to study the techniques from the relevant training manual, then you need to memorize the technique and sequence of events. In each sequence a student must fully understand what they are aiming to achieve. You should fully understand the sequence, including control inputs, timing, what you will see out the front and how it all fits together: now you are ready to visualize.

Close your eyes and imagine flying the manoeuvre from the beginning to the end. At first you will find it difficult to do; it will feel very cumbersome and disjointed and you will probably forget parts of what you are doing and perhaps 'get stuck' halfway through. Without visualization this is exactly what would happen to you airborne. Have a few goes and then have a break. In your break identify areas that need improvement. For example, did I forget to move the control column, or did I forget to look at the airspeed on entry?

When you are feeling tired be disciplined and force yourself to imagine everything in detail; do not drift off into daydream visualization or you will fall asleep. This type of visualization is useful post-flight to consolidate the sequences that you have learnt. The difference in learning rate between somebody who visualizes well and somebody who does not do any is quite remarkable, and easily observed amongst the student pilots at 2FTS.

Mission-type visualization of this kind is mentally draining; if you are learning it for the first time, limit these sessions to just one manoeuvre at a time. Take short regular breaks of up to a minute and then attempt another. Each session of visualization should not go for more than half an hour at a time. You will then need to take a longer break to let your mind relax. See Table 20.2.

Table 20.2 A taxonomy of visualization in flying training

Visualization type	Flying phase	Advantages
'Daydream'	Basic skills	• Low-stress • Creates an expectation or 'benchmark' for future flying
'Mission'	Advanced skills	• Technique-focus • Simplification of complex flying sequences

Conclusion

There are many negative human factor issues that create significant human factor challenges to improving pilot training in both civil and military aviation. Within Australian military aviation, the predominant challenges to improving pilot training include overcoming an organizational aversion to human factor issues, a Hawthorn effect created by the existence of 'authority' within the training environment and a low level of confidence among many trainees. This chapter has identified and provided an insight into the human factor difficulties experienced by instructional pilots – 'QFIs'. The chapter has also shown that these difficulties provide the springboard for tools that may assist training personnel in other human-machine systems where developing sound decision-making and behavioural habits is crucial to successful operations. These tools include unlocking the value of previous flying experience, adapting learning skills to a dynamic airborne environment, and using visualization. These tools could – and should – be used by both pilot candidates and instructor-pilots to improve the quality of pilot training 'output' in Australia.

Acknowledgements

The author wishes to acknowledge the support of numerous former colleagues and student pilots at 2FTS who provided many useful insights and advice relating to flying instruction. Specifically, Flight Lieutenants Dean 'Bags' McCluskey, Grant 'Fish' Fichera, Liam 'Pully' Pulford, Steve 'Axel' Ackerman and Scott 'Maddog' Tully are acknowledged, in addition to former Commanding Officer of 2FTS, Group Captain Peter 'Norf' Norford. The author also wishes to acknowledge the assistance of Alexandra Troutman at the University of New South Wales for her editorial support.

References

Braithwaite, G.R. (2001), Attitude or Latitude?, *Australian Aviation Safety*, Aldershot, UK: Ashgate.

Dekker, S.W.A. (2005), *Ten Questions about Human Error: A new view of human factors and system safety*, London: Lawrence Erlbaum Associates.

Falconer, B.T. (2002), Human Error or Human Endeavour? *ADF Flying Safety Spotlight*, **02**/2002, 2–4. Canberra, Australia: DFS-ADF.

Falconer, B.T. (2005), Cultural Challenges in Australian Military Aviation: Soft Issues at the Sharp End, *Human Factors and Aerospace Safety*, **5**(1), 61–79.

Falconer, B.T. (2005a), Human Factors and the 'Can-Do' Attitude, *Australian Defence Force Journal*, 169.

Falconer, B.T. and Murphy, P. (2005), The Can-Do Attitude: Strength or Weakness?, *Focus – Special Human Factors Edition*, Canberra, Australia: Department of Defence.

Fellows, R., (1998), 'An Analysis of Pilot Training in the ADF', Unpublished manuscript.

Kern, T. (1997), *Redefining Airmanship*, New York: McGraw-Hill.

Kern, T. (1999), *Darker Shades of Blue: A Case Study of Failed Leadership*, New York: McGraw-Hill.

Lee, R.B. (1999), DFS-ADF Flying Safety Special: Military Pilots and General Aviation Aircraft, in *DFS-ADF* edited by: Canberra, A.

McCormack, Air Marshal E., (1999), *Flying Safety within the Australian Defence Force*, Paper presented at the Australian Society of Air Safety Investigators Conference, Surfer's Paradise, Australia, June 5.

Merket, D., Bergondy, M. and Salas, E. (1999), Making Sense out of Team Performance Errors in military Aviation Environments, Transportation, *Human Factors*, **1**, 231–242.

Soeters, J.L. and Boer, P.C. (2000), Culture and Flight Safety in Military Aviation,

International Journal of Aviation Psychology, **10**(2), 111–133. [DOI: 10.1207/S15327108IJAP1002_1]

Wellfare, J. (2005), Pilots Course is one Step in a Long Journey, in *Air Force News* May 19, p. 11.

Chapter 21

Advanced Driver Assistance Systems and Road Safety: The Human Factor

Michael A. Regan and Kristie L. Young
Monash University Accident Research Centre

A wide range of entertainment, information and communication and advanced driver assistance systems are finding their way into the car cockpit. Whilst these can greatly enhance the safety, enjoyment, and amenity of driving, the potential benefits to be derived from them could be compromised if they are used inappropriately and poorly designed. The critical human factors issues that will underpin the effectiveness of these systems are discussed, with a particular focus on advanced driver assistance systems. Some research undertaken at the Monash University Accident Research Centre (MUARC) that bears on these issues is presented, including recent research on driver distraction.

Introduction

In 2004, around 1,600 people were killed and over 22,000 seriously injured in crashes on Australian roads. Since records began, over 171,000 people have died on Australia's roads (Australian Transport Safety Bureau, 2005). It is predicted that Intelligent Transport Systems (ITS), particularly those for vehicles, have great potential to enhance road safety (Rumar *et al.*, 1999; Regan *et al.*, 2001) and to lead to a significant reduction in the road toll. The term 'Intelligent Transport System' refers to the application of advanced electronics, information and communication and computing technologies to the transport sector to improve safety and efficiency, reduce congestion and minimize harm to the environment (Regan, 2004a).

ITS systems are at various stages of development. Those that have been developed mainly to enhance vehicle safety are referred to as Advanced Driver Assistance Systems (ADAS; see three and four for reviews of these technologies). ADAS have potential to maximize safety at one or more stages of a vehicle crash. Some are designed to prevent crashes. These include systems that warn drivers if they are exceeding the speed limit, are following a lead vehicle too closely, are about to crash, are about to fall asleep, and so on. Satellite Navigation Systems and Adaptive Cruise Control similarly have potential to prevent crashes.

Other ADAS technologies are designed to minimize trauma to vehicle occupants in a crash. These include Intelligent Speed Adaptation (which has the effect of reducing vehicle speed, and hence impact forces in a crash) and Seat Belt Reminder

Systems (which signal if any vehicle occupant is unbelted). Finally, ADAS can be designed to further reduce trauma to vehicle occupants after a vehicle has crashed. Automatic Crash Notification, for instance, is a system that uses global positioning and wireless communication to dial for an ambulance, even if vehicle occupants are unconscious.

Whilst these systems have great safety potential, at least in theory, the actual benefits to be derived from them are largely unknown. This is because ITS technologies in vehicles are yet to be deployed on a large enough scale over a long enough period of time in traffic to yield crash numbers that can be used to evaluate changes in safety (Rumar *et al.*, 1999). Very little is also known about how drivers will adapt to emerging systems over time and how this is affected by the design of systems, their level of training and other human factors issues.

This chapter examines some of the key human factors issues that will underpin the effectiveness of ADAS technologies, and describes some research bearing on these issues that has been undertaken by MUARC. Driver distraction is probably the first significant human factors issue to emerge as a by-product of the vehicle cockpit revolution, fuelled more by the proliferation of entertainment and information and communication systems than by ADAS technologies. Some current MUARC projects on this topic are also described.

Human factors issues

Large-scale research programs are being undertaken in Europe, North America and Japan to explore human factors and ergonomic issues relevant to the design, deployment and evaluation of ADAS technologies. Relatively little research on this topic has been conducted in Australia, although this about to change. In late 2005 the Australian Research Council (ARC) funded the establishment of a Cooperative Research Centre for Advanced Automotive Technology (known as the AutoCRC), which will have representatives from universities, industry and government. A core component of the 7-year program is research on the design and evaluation of the human–machine interface (HMI) for ADAS technologies in next generation road vehicles, to be led by Holden and MUARC. The human factors issues that underpin the safety effectiveness of these technologies are complex and varied. Some of the more important ones are discussed below (Regan, 2004b, 2004c).

HMI design issues

The Human–Machine Interface (HMI) through which the driver interacts with an ADAS must be designed in accordance with best ergonomic practice. While international ergonomics standards and guidelines are emerging to guide this process (*e.g.*, Rupp, 2004), further research is needed to support the design of the HMI in a manner that ensures that the systems do not distract, confuse and/or overload the driver. This must include consideration of both ADAS systems built into the vehicle as dedicated units and ADAS-equipped portable devices, such as mobile phones, brought into the vehicle.

Behavioural adaptation

Drivers can adapt positively or negatively to ADAS technologies. Negative behavioural adaptation, if it occurs, may take two forms. If vehicles become increasingly automated, and drivers have less to do, they may engage more in secondary, non-driving, activities (*e.g.*, talking on a mobile phone), which can compromise safety. This has been referred to as Task Difficulty Homeostasis (Fuller, 2002). The other form of negative behavioural adaptation is system-specific. People may go faster around a corner when they have an Intelligent Speed Adaptation system (which automatically warns them when they are exceeding the speed limit) because they wait for the system to warn them that they are driving over the speed limit rather than, as previously, driving to the prevailing conditions.

Awareness of system limitations

Drivers must understand the technological limitations of emerging systems. If their expectations of system performance are too high, they may expect the system to alert them to critical situations that the system is not capable of detecting.

Over-reliance

As drivers adapt to ADAS, they may over time become over-reliant on them. Driving with a Forward Collision Warning System, for example, may encourage drivers to scan less actively for vehicles in front of them, assuming that the system is doing this for them. Over-reliance can also pose problems when transitioning from automatic to manual control of a vehicle. In one study it was shown that, when an Adaptive Cruise Control system was made to fail unexpectedly in a simulator (these systems automatically maintain fixed following distances), at least one-quarter of the drivers tested failed to react appropriately and crashed (Stanton, Young and McCaulder, 1997).

Exposure to change

Little is known about the impact of ADAS technologies on travel exposure. It is known, for example, that satellite navigation systems can reduce mental workload and reduce unnecessary travel in unfamiliar locations. They may, however, encourage people to take more trips to locations they may previously not have travelled to, thus increasing their exposure to traffic risks.

Effects of automation

Increased automation of the vehicle can have several consequences. Drivers' danger avoidance skills and mechanisms, derived from the control of current generation vehicles, may be degraded if the vehicle itself becomes increasingly capable of scanning for, perceiving and responding automatically to traffic hazards. Automation of some driving tasks may also reduce drivers' attentiveness (Summala, 1997) and reduce their situational awareness of traffic events. As noted previously, automation may further encourage drivers to use their spare capacity to engage more in non-driving activities.

Driver acceptance

ADAS technologies are products. However, very little market research has been done to understand what features people want in these systems and what barriers there might be to their uptake and use. If user preferences are not well understood, the systems may be unacceptable to consumers and there will be little or no demand for them. If so, the potential safety benefits they offer will not be realized.

In summary, there are several human factors issues that will intervene to determine whether an ADAS will be effective in achieving its intended safety benefit. These issues must be addressed in designing, developing, deploying and evaluating ADAS technologies. For a more detailed coverage of these and other relevant issues, the reader is referred to Noy (1997); Barfield and Dingus (1998) and Regan *et al.* (2001).

Research at MUARC

The Monash University Accident Research Centre (MUARC) is one of the world's leading injury prevention research centres. It was established in 1987, has nearly 100 research staff and undertakes applied, multidisciplinary, research for industry and governments. The Centre is known, locally and internationally, for its research into the impact of ADAS technologies on driver performance, behaviour and safety. Some key projects that underpin the MUARC research program, and which address a number of the human factors issues discussed above, are described below.

- intelligent vehicle safety research
- intelligent transport systems: safety and human factors issues

In 2001, MUARC prepared for the Royal Automobile Club of Victoria (RACV) a report that reviewed a large body of literature on the safety and human factors issues relevant to the design, deployment and evaluation of ITS technologies (Regan *et al.*, 2001). The report reviewed autonomous, infrastructure-based and cooperative ITS that have the potential to prevent crashes, reduce crash trauma, and enhance post-crash trauma management.

Community acceptance of in-vehicle ITS technologies

As noted, little is currently known about community views about ADAS technologies. In 2002, MUARC conducted research for the RACV to assess community views about seven ITS technologies: Forward Collision Warning; Intelligent Speed Adaptation; Automatic Crash Notification; Electronic Licence; Alcohol Interlock; Driver Drowsiness Warning; and Lane Departure Warning (Regan *et al.*, 2002). Drivers identified several things that would prevent them from purchasing the systems (*e.g.*, a high false or nuisance alarm rate; systems not perceived as being useful; systems that cannot be manually overridden; systems that take too much vehicle control from them). The Fatigue Monitoring and Intelligent Speed Adaptation systems were deemed most acceptable, while the Electronic Licence was deemed least acceptable. Paradoxically, the systems that were perceived as least acceptable were estimated to have the greatest potential safety benefit. In descending order of benefit, these were the Alcohol Interlock, Electronic Licence and Intelligent Speed Adaptation systems. Drivers appear willing to pay surprisingly little for ADAS technologies (on average, around 200 Australian Dollars). Similar research, involving young novice drivers and some additional systems (Young, Regan and Mitsopoulos, 2004) was conducted by MUARC in 2003 for the Motor Accidents Authority of NSW (MAA).

Intelligent speed adaptation and heavy vehicles

Many heavy vehicles regularly exceed the speed limit. The aim of this 3-stage project, funded by Austroads, is to evaluate the effectiveness of Intelligent Speed Adaptation (ISA) in reducing heavy vehicle speeding. In Stage 1, MUARC reviewed the international literature on intelligent speed adaptation for both light and heavy vehicles (Regan *et al.*, 2003). In Stage 2, MUARC developed a research and logistics plan for conducting Phase 3 of the study (Regan *et al.*, 2003). Phase 3, if it goes ahead, will involve a field study to evaluate ISA in a fleet of heavy vehicles.

TAC SafeCar project

MUARC's centerpiece ADAS research activity has been the TAC SafeCar project. It began in June 1999 and involved as key partners the Victorian Transport Accident Commission (TAC), MUARC and Ford Australia. The study, which concluded in 2005, evaluated the technical operation, impact on driving performance and acceptability to drivers of three ADAS technologies: Intelligent Speed Adaptation (ISA); Following Distance Warning (FDW); and the Seatbelt Reminder (SBR).

Fifteen vehicles equipped with these technologies, called 'SafeCars', were sub-leased to nine public and private companies in and around Melbourne and 23 drivers each drove one of the vehicles for a distance of at least 16,500 km.

Drivers received advisory warnings from the ISA system if they were exceeding the speed limit by 2 km/hour or more. These consisted of visual, auditory and 'haptic' (upward accelerator pressure) warnings. The FDW system, which employed frontal radar, issued graded visual and audio warnings when the driver was travelling two seconds or less from the car in front. The SBR system provided visual and auditory warnings if any occupant in the vehicle was unrestrained. These became more aggressive as travel speed increased. Careful attention was given to the ergonomic design of the HMI for these systems, to ensure that the warnings issued were acceptable to drivers and did not confuse, distract or overload them. Training needs analysis was also conducted to define driver training requirements.

The ISA, FDW, and SBR systems had a positive effect on driving performance. Use of the ISA system resulted in a significant reduction in average and peak travel speeds and in the percentage of time travelling above the speed limit. The ISA system was most effective when there was no surrounding traffic and in 60 km/hour zones. There was no increase in travel times when using the system. The speed reduction effects of the ISA system were more pronounced when it operated in conjunction with the FDW system.

When the FDW system was active, drivers left a greater time gap between the SafeCar and the car in front and spent less time travelling at distances of less than one second. The system also increased the minimum distance reached between the SafeCar and vehicle in front on each trip.

Driver interaction with the SBR system led to large decreases in the percentage of trips driven where an occupant was unbelted for any part of the trip; in the percentage of total driving time spent unbelted; in the time taken to fasten a seat belt in response to the SBR warnings; in the average and peak speeds reached before buckling up, and in the time spent unbuckled while travelling at speeds of 40 km/hour and above.

Interestingly, the ISA and FDW systems were effective only while turned on; when they were turned off, drivers quickly reverted to their usual driving behaviours. The ISA and FDW systems appeared to be equally effective at night and during the day and for younger (aged less than 45 years) and older drivers (aged 45 years and over).

There was little evidence of any 'negative behavioural adaptation' to the systems. That is, there was little evidence that drivers compensated for the added safety benefits deriving from the systems by engaging in increased risk taking.

The ISA system is estimated to reduce fatal and serious injury crashes by up to 9 and 7 per cent, respectively. For the FDW system, the percentage of driving distance spent in rear-end collision mode (that is, where the vehicle would collide with the lead vehicle if it braked suddenly) is expected to reduce by up to 34 per cent when the system is active. Finally, use of the SBR system is expected to save the Australian community approximately 335 million Australian Dollars *per annum* in injury costs. The benefits might have been greater if the study drivers had been less conservative and if the study had not taken place at a time of stricter Police enforcement of speed laws.

Driver acceptance of the SafeCar systems was high. They were generally rated as being useful, effective, user-friendly and socially acceptable. This was the case both before, and after, experience with the systems. No system reportedly increased mental workload.

Some potential barriers to driver acceptance of the systems were identified. Participants reported that they would lose trust in systems that gave unreliable warnings. The cost of the systems was a potential barrier to acceptance, particularly maintenance and service costs. Some participants found the auditory warnings annoying, but suggested ways in which they could be made more acceptable. It is encouraging that participants found the systems acceptable in terms of their level of control, did not feel that they would rely too strongly on the systems at the expense of their own judgement, and did not think the systems would distract them from driving.

The results of the study, which is well known internationally, suggest that the ISA and SBR systems should be deployed on a wider scale in Victoria, but that the deployment of FDW should be delayed until there is a reduction in the unit cost of the system, until nuisance and false warnings issued by it can be minimized, and until the ergonomic design of the warnings issued is optimized.

Driver distraction research

The proliferation of entertainment, information and communication and ADAS technologies within the vehicle cockpit has increased community awareness of the role of distraction as a causal factor in road crashes. Findings from various studies suggest that anywhere between about 10 and 40 per cent of road crashes involve distraction as a contributing factor. Given that ADAS technologies are specifically designed to support drivers and enhance their safety, significant effort has gone into designing them in a manner which minimizes distraction. There is, however, converging evidence that entertainment and information and communication technologies constitute a major source of in-vehicle distraction (see Young and Regan, 2006, for a recent review). Distraction is arguably the first significant human factors issue to emerge as a by-product of the vehicle cockpit revolution. Some current MUARC projects in this area are described briefly below.

HMI and driver distraction – AutoCRC

As noted previously, MUARC and Holden are leading a large research program concerned with the design and evaluation of the HMI for ADAS technologies in next generation road vehicles, as part of the activities of the ARC-funded Cooperative Research Centre for Advanced Automotive Technology. Work will commence in early 2006.

ISO TC 22 SC13

The senior author of this chapter is the Australian representative on International Organization for Standardization (ISO) Committee TC 22, Sub-Committee SC13 (Ergonomics Applicable to Road Vehicles). Working Group 8 of this Sub-Committee develops human factors and ergonomic standards for the design and evaluation of the HMI for ADAS technologies for road vehicles, including standards for the measurement of driver distraction.

Bus driver distraction

This project, funded by the State Transit Authority of New South Wales (STA NSW), has been progressed in three phases: an analysis of the functions and tasks currently undertaken by STA bus drivers; the identification of actual and potential sources of bus driver distraction; and the conduct of a risk assessment of bus driver distraction. These phases have involved the conduct of subject matter expert (*SME*) interviews, observational studies, focus groups, a review of STA NSW and road transport policy rules and regulations, the development of a hierarchical task analysis of bus operations, an ergonomic assessment of buses and the conduct of a human error analysis for a bus operation task. So far, it has been found that STA NSW bus drivers undertake a wide range of tasks, both driving and non-driving-related, while driving. Further, in undertaking these tasks, drivers are exposed to a range of distractions including technology-related distractions, operational distractions, passenger-related distractions, environmental distractions and personal distractions. Recommendations for mitigating the effects of distraction for bus drivers have been made (Salmon, Young, Regan and Salmon, 2005).

Fleet driver distraction audit

In 2005 MUARC prepared, for Orica Consumer Products, a position paper outlining options for enhancing the safety of Orica employees who converse on hands-free mobile phones while driving for work purposes. The project involved several main tasks: a review of the literature on mobile phone use and its impact on driving performance and crash involvement; a review of Orica's current mobile phone policies; two focus groups with Orica employees; and an ergonomic assessment of some currently available mobile phones. It was found that Orica employees frequently use mobile phones while driving and that the nature of their work requires them to do this. Options for enhancing the safety of employees who use hands-free mobile phones whilst driving were recommended (Young, Regan and Salmon, 2005).

Simulator evaluation of in-vehicle distraction and road environment complexity

This study examined the effects of distraction on driving performance for drivers in three age groups. Two in-car distracter tasks were examined: operating the car entertainment system and conducting a simulated hands-free mobile phone conversation. The effect of visual clutter was also examined by requiring participants to drive in simple and complex road environments. Overall measures of driving performance were collected, together with responses to roadway hazards and subjective measures of driver perceived workload. The results revealed that the two in-car distraction tasks degraded speed control, degraded responses to hazards and increased subjective mental workload. These performance decrements were observed in both the simple and complex highway environments and for drivers in different age groups. It was concluded that both in-vehicle tasks impair several aspects of driving performance, with the entertainment system distracter having the greatest negative impact on performance (Horberry *et al.*, 2006).

Reviews of the driver distraction literature

In 2003, MUARC published a comprehensive review of the driver distraction literature (Young and Regan, 2003). The report reviewed a range of technologies and everyday activities that have the potential through distraction to degrade their driving performance and safety. In late 2005 MUARC updated the earlier literature reviewed in preparing a Submission to the Victorian Parliamentary Road Safety Committee Inquiry into Driver Distraction. This Submission also provided an integrated set of recommendations for managing the driver distraction problem in Australia.

Text messaging and young novice drivers

Using the advanced driving simulator located at MUARC this study examined, for young novice drivers aged between 18 and 21 years, the effects on driving performance of retrieving and sending text messages (Hosking, Young and Regan, 2005). It is the first to have done so. Twenty drivers participated and each drove on a simulated roadway containing a number of events, including a pedestrian emerging from behind parked cars, traffic lights, cars turning right in front of the driver, a car-following episode and a lane change task.

Retrieving and, in particular, sending text messages had a detrimental effect on a number of safety-critical driving measures. When sending and retrieving text messages, drivers' ability to maintain their lateral position on the road and to detect and respond appropriately to traffic signs was significantly reduced. In addition, drivers spent up to 400 per cent more time with their eyes off the road when text messaging than when not text messaging. While there was some evidence that drivers attempted to compensate for being distracted by increasing their following distance behind lead vehicles, they did not reduce their speed while distracted.

Despite the observed degradation in driving performance and legislation banning the use of hand-held phones while driving, a large proportion of the participants in this study reported that they regularly use hand-held phones while driving for talking and text messaging. The findings highlight the need for mobile phone safety awareness campaigns to target the young driver population in particular, in order to minimize the use of these devices among this age group.

Conclusion

The car cockpit is evolving rapidly. Drivers are interacting with an increasing range of entertainment, information and communication, and Advanced Driver Assistance Systems. The human factors issues that intervene to determine the extent to which these technologies are used safely by drivers are many and varied. This chapter has identified some of the more critical issues and has outlined a small sub-set of MUARC projects that have yielded new knowledge in the area.

References

Australian Transport Safety Bureau (2005), Road Deaths Australia: 2004 Statistical Summary, *Canberra, Australia*: ATSB.

Barfield, W. and Dingus, T. (1998), *Human Factors in Intelligent Transportation Systems.* Mahwah, NJ: Lawrence Erlbaum.

Fuller, R. (2002), Psychology and the Highway Engineer, in *Human Factors for Highway Engineers* eds Fuller, R. and Santos, J. A., England, UK: Elsevier Science Ltd.

Horberry, T., Anderson, J., Regan, M.A., Triggs, T.J. and Brown, J. (2006), Driver Distraction: The Effects of Concurrent In-Vehicle Tasks, Road Environment Complexity and Age on Driving Performance. *Accident Analysis and Prevention*, 38, 185–91.

Hosking, S., Young, K. and Regan, M. (2005), *The Effects of Text Messaging on Young Novice Driver Performance, Report Prepared for the NRMA by* Monash University Accident Research Centre, Clayton, Victoria.

Noy, I., ed. (1997), *Ergonomics and Safety of Intelligent Driver Interfaces*, New Jersey: Lawrence Erlbaum.

Regan, M.A. (2004a), A Sign of the Future I: Intelligent Transport Systems, in *The Human Factors of Transport Signs* eds Castro, C. and Horberry, T., London: CRC Press, pp 213–224).

Regan, M.A. (2004b), 'A sign of the future II: Human Factors', in Castro, C. and Horberry, T. (eds) *The Human Factors of Transport*, London: CRC Press, pp 225–238).

Regan, M.A. (2004c), New technologies in cars: Human factors and safety issues *Ergonomics Australia*, **18**, (3), pp 6–15.

Regan, M.A., Connelly, K., Young, K., Haworth, N., Mitsopoulos, E., Tomasevic, N., Triggs, T. and Hjalmdahl, M. (2003), *'Intelligent Speed Adaptation for heavy vehicles. Stage 2: Evaluation framework and action plan,' Austroads Report*, Sydney, Australia: Austroads (Under review).

Regan, M.A., Mitsopoulos, E., Haworth, N. and Young, K. (2002), Acceptability of Vehicle Intelligent Transport Systems to Victorian Car Drivers, *RACV Public Policy Report 02/02, November 2002*, Melbourne, Australia: Royal Automobile Club of Victoria.

Regan, M.A., Mitsopoulos, E., Triggs, T., J., Tomasevic, N., Healy, D., Tierney, P. and Connelly, K. (2003), Multiple In-Vehicle Intelligent Transport Systems: Update on the Australian TAC SafeCar Project, in *The Proceedings of the 10th World Congress on Intelligent Transport Systems*, Spain: Madrid, 16–19 November.

Regan, M.A., Oxley, J.A., Godley, S.T. and Tingvall, C. (2001), Intelligent Transport Systems: Safety and Human Factors Issues, in *01/01. RACV: Melbourne, Australia* edited by: Royal Automobile Club of Victoria Research Report.

Regan, M.A., Young, K. and Haworth, N. (2003), Intelligent Speed Adaptation for Light and Heavy Vehicles, *Austroads Report AP-R237*, Sydney, Australia: Austroads.

Rumar, K., Fleury, D., Kildebogaard, J., Lind, G., Mauro, V., Carsten, O., Heijer, T., Kulmala, R., Macheta, K. and Zackor, I.H. (1999), *Intelligent Transportation*

Systems and Road Safety, Brussels, Belgium: European Transportation Safety Council.

Rupp, G. (2004), Ergonomics in the Driving Seat, *ISO Focus*: *the magazine of the International Organization for Standardization,* Vol 1, No. 3, March, 11–15.

Salmon, P., Young, K. and Regan, M.A. (2005), Bus Driver Distraction: Analysis of Risk for State Transit Authority New South Wales Bus Drivers, *Report Prepared for State Transit Authority New South Wales by the*, Clayton: Monash University Accident Research Centre.

Stanton, N.A., Young, M. and McCaulder, B. (1997), Drive-by-wire: the Case of Driver Workload and Reclaiming Control with Adaptive Cruise Control, *Safety Science*, Vol. **27**, No. 2-3, 149–159. Summala, H. (1997), Ergonomics of Road Transport, *IATSS Research*, **21**, No. 2, 49–57.

Young, K. and Regan, M.A. (2006), Driver Distraction: Review of the Literature and Suggestions for Countermeasure Development, *The Proceedings of the Swinburne University Multi-Modal Symposium on Safety Management and Human Factors, 9-10 February 2006, Melbourne Australia.* (In Press).

Young, K. and Regan, M.A. and Hammer, M., (2003), 'Driver Distraction: A Review of the Literature', Monash University Accident Research Centre. Report 206. Melbourne, Australia.

Young, K., Regan, M.A. and Mitsopoulos, E. (2004), Acceptability of In-Vehicle Intelligent Transport Systems to Young Drivers, *Road and Transport Research*, **13**, No. 2, 6–16.

Young, K., Regan, M.A. and Salmon, P. (2005), Mobile Phone Use while Driving: Options Paper to Inform Orica Policy, *Report Prepared for Orica by the*, Clayton, Victoria: Monash University Accident Research Centre.

Young, K and Regan, M.A. (2006), 'Driver Distraction: Review of the Literature and Suggestions for Countermeasuare Development' (Chapter ?) in, J. Anca (ed) *Multimodal Safety Management and Human Factors: Crossing the Borders of Medical, Aviation, Road and Rail Industries*, pp 270–290, Ashgate: Aldershot.

Chapter 22

Recovery of Situational Awareness Using Operant Conditioning Techniques

Peter N. Rosenweg
Psyfactors Pty Ltd

The variable successes of conventional methods in dealing with the potential for human error have prompted a refocus on the pivotal need for the maintenance of Situational Awareness (SA) by operators. Practical operant conditioning techniques resulting in automatically triggered and cued recovery of alertness and cognitive ability under mental load were previously used in practice based interventions with chemical plant operators, pilots, police candidates and tanker drivers. In a preliminary trial with groups the method was shown to both generalize across contexts whilst avoiding adding another procedure to the safe execution of a role. Post training evaluation of a small sample comprising two dissimilar and contrasting groups indicated a relatively high degree of acceptance with approximately 80 per cent retention and subsequent usage of the method. The efficacy of the approach to invoke situational awareness as required is discussed.

Introduction

The attractiveness of systems and engineering approaches in reducing safety incidents has unfortunately not eliminated individual behaviour as the most frequently reported cause. An analysis by Endsley (1999), that most human error incidents resulted from a loss of situational awareness (SA) rather than judgement or skill based decisions, emphasized the importance of attention recovery mechanisms for safety critical roles. Many papers have been published dealing with attention recovery and fatigue countermeasures; to date, however, none have been found that include a mechanism to enable instantaneous and cued recovery at the moment of demand. This chapter formalizes a practice based approach to recovery of SA delivered in over 20 years of counselling and coaching performance with a diversity of clients.

Developmental criteria

Analysis of human errors in occupational settings suggests several psychological and functional criteria necessary for a practical 'person focused' solution to enhancing or recovering SA or Situational Safety Awareness (SSA):

- the need for a mechanism for quick recovery to an externally focused mental state;
- the recovery mechanism had to be cue-able or automatically invoked;
- the method must be applicable to both static and dynamic contexts;
- the resultant effect must generalize and benefit diverse scenarios rather than be context specific; and
- acquisition of the enhanced SA skill must be independent of prior knowledge, other skills and specific abilities.

This chapter considers one solution based on the criteria. The method benefits from the application of theory and known mechanisms to achieve targeted change in mental state with applicability to a diversity of individuals and contexts.

Mental state and unsafe behaviour

A person's mental state emanating from their fundamental coping style appears to directly influence their competency. The work on coping styles by Lazarus (1976), suggested that superior performance was more consistent with an externally focused direct action style than an internal (intrapsychic) style. A person's coping style and mental state could thus be seen to be a strong indicator of the ability to respond adequately to situations. Conversely, poorer performance under equal conditions suggested a failure in perception and processing at the moment of demand.

The behaviour based assessment of human error potential achieved by Broadbent *et al.* (1982), in the development of the 'Cognitive Failure Questionnaire' (CFQ), effectively translated mental state into a checklist of observable and self report mistakes, errors and risk prone behaviours. The CFQ items were synonymous with a degraded mental state, loss of external perspective and by implication a loss of SA.

A converging relationship between cognitive failure, unsafe behaviours and Endsley's (1995) loss of SA, an ability of the individual to see, understand and project outcomes is posited in Figure 22.1.

Mental state, performance and triggers

Achieving a positive mental state has historically been central to performance in combative and sports activities. Zen devotees refer to the pursuit of 'Mindfulness' (Walley, 1995), a parallel form of SA where perception and action occurs with precision when unfettered by introspective preoccupation prevalent in emotion based coping styles. More recently, work stemming from the findings by Lazarus (1976) and Lazarus and Folkman (1984), of the relationship between an external and goal directed mindset with improved performance, has been successfully applied in assessment of astronaut candidates and to mental training in sports. Sports people frequently refer to the fascination of achieving the 'Flow,' the 'Zone' or the 'Peak' experience Csikszentmihalyi (1990); Marr (2001) as the mental state required for ultimate performance.

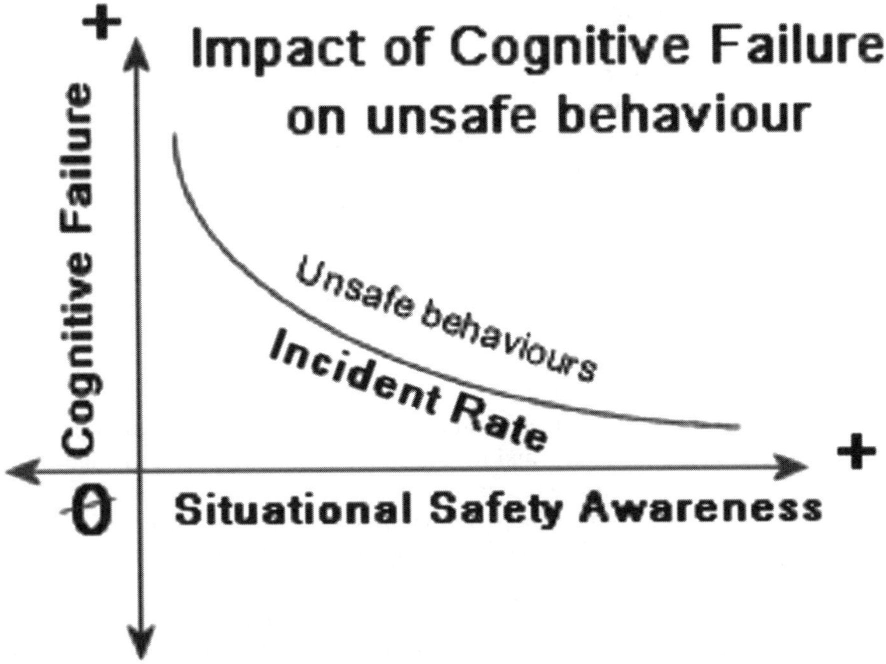

Figure 22.1 Cognitive failure and unsafe behaviour

Operational safety, however, requires a quicker way of achieving the necessary mental state than that provided by Cognitive Behavioural Therapy (CBT) based visualization-affirmation or somatic (meditation) methods now popular in sports coaching (Landers and Boutcher, 1993; Nideffer, 1993, Bull, Albinson and Shambrook, 1996, Doulliard, 2001; Behncke, 2004). Diaphragmatic breathing (DB) has been extensively used to achieve a rapid switch in mental state and is widely referenced as a clinical option for altering anxiety and panic disorder types of reactions (Peper and Tibbetts, 1997; Seaward, 2002). The DB method was assessed to be speedier in effect than the CBT or somatic methods, have greater simplicity, provide ease and uniformity of application to diverse groups, provide the response reinforcement needed for sustainability as well as have acceptability as a training option.

Sustaining peak vigilance or attention for long periods is difficult for most individuals and unrealistic in most work situations. Similarly, using the breathing technique (DB) to achieve a mental switch at the moment of demand would likely fail because of poor recall under stress, or it was displaced by more urgent priorities, or as is commonly the case when 'suddenly demanded' the person reverts to often unproductive or habitual responses. A technique of association or linking was required where one or another relevant cue would automatically invoke the switch in mental state without having to remember to do the DB breathing at that very moment.

Invoking an external mechanism or trigger as an initiator or reminder to 'switch on,' would obviate the need for exhausting and sustained attention. It had been found that image and semantic-based cueing and triggering as activating mechanisms linked to target associations, variously improved task performance, vigilance, search speed, reliable exercise of intentions, learning and judgement tasks with less error and greater retention than other methods (Compton *et al.* 2004, Hitchcock, Warm, Matthews, Dember, Shear, Tripp, Mayleben, Rosa and Parasuraman, R., 2003; Janelle Champenoy, Coombes and Mousseau. 2003; Guynn, McDaniel and Einstein, 1998; Morris, 1992; Pashler, *et al.* 1999, 2001; Janelle *et al.*, 2003; Warm *et al.*, 2003; Compton *et al.*, 2004). As an applied example, some sports coaches have instructed their tennis players to conserve their capacity for attention by using cues such as a ball bounce preparatory to serving to their opponent, to switch on focus; later, another cue such as turning their back on the net, to switch off.

Operant or 'response-reinforced conditioning' (Skinner, 1968) forms a positive association between a behaviour and a consequence. To help cues and triggers to work more effectively, aspects of the target outcome serve as the reinforcer to help prevent habituation and diminishing responsiveness. In the applied recovery of SA model, the behaviour is the triggered change or switch in mental state and the target consequence the enhanced attention. The secondary reinforcement of relief from fatigue or overload tension and the feeling of operating at closer to desired performance (often with elevation of self esteem) provides for immediate positive and continuous reinforcement; similar to the prescription of a 'Soon-Certain-Positive' reward for embedding a behaviour in the Behavioural Safety Training model (Krause, 1991).

The recovery of Situational Safety Awareness model

The action sequences in the SSA model are depicted in Figure 22.2. In the diagram the situation or scene initiates a trigger, which in turn cues the mental switch re-orientating perspective, allowing for a selection and retrieval of suitable skilled responses that usefully address the scenario.

Method

Objective

The objective of this preliminary study was to compare and evaluate the results of identical instruction in the cued recovery of situational awareness skill with two motivationally dissimilar groups operating in different settings.

Type of study

The study is of the differential type conducted on two dissimilar groups, with a qualitative assessment of uptake and use of the Recovery of SA method using operant techniques, assessed by observations, feedback and survey result methods.

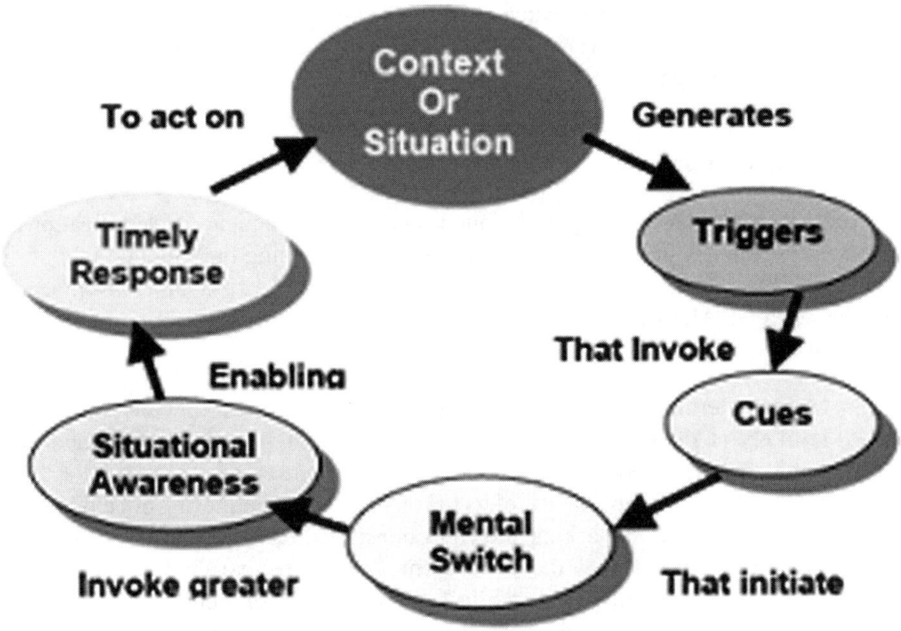

Figure 22.2 The situational safety awareness model

Participants

The sample size used for this report was a total of 39 participants in two dissimilar groups. The chemical plant group was assigned to the training by their manager whilst the police candidates were voluntary participants.

Participant characteristics

Group 1 participants comprised 19 male and two female married chemical plant personnel in the 35–55 age range involved in the management and maintenance of a continuous chemical process plant operation. The plant had many near misses and minor incidents but no serious injury resulting in lost time for nearly 10 years. The tenure of experience of personnel in the location was two to 10 years. All but two persons were qualified with some form of technical or trade training and similar work experience before joining the organization. All but four of the participants were assigned to a three-shift cycle. Organizational objectives were to maintain a high level of safety and safe behaviour during operations and extensive plant maintenance and refurbishment known as shutdowns.

Group 2 participants were candidates preparing for the rigours of the assessment process for police entry. They comprised three subgroups, 18 candidates in total, assembled in three sessions six at a time. Gender mix of the group was 14 males and four females. Candidates were in the 20–35 age range, of single status; educational standard was at the VCE level and four had some post high school technical or trades training. Participant attendance was voluntary; all candidates reported at commencement that they were uncertain and in some anxious about their likely performance in the panel interview process.

Instructional context

Instruction in the SA recovery technique proceeded subsequent to identification of the need for the SA skill for both groups and assessed using the 'Situational Safety Awareness' test (SSA test). Results were discussed and explored with emphasis on the 'Mental Alertness,' 'Safety Habits' and 'Coping Style' scales of the test (Rosenweg, 2001).

- Mental Alertness: Scores on this scale indicate the extent to which the person is likely to maintain their mental functioning and alertness to respond in a timely and appropriately way to hazardous circumstances.
- Safety Habits: Safety Habits refers to consistency in monitoring developments, anticipating actions, asking the right questions, *etc.*, generally, maintaining 'presence of mind' such that most events seem to be expected.
- Coping Skills: Involves the stability and orientation of mood and affect of the person as it impacts safety-orientated behaviour by way of their diligence, alertness and situational awareness, energy and responsiveness in addition to the adequacy of interaction with others.

Members of the first group (chemical plant operators) (G1) participating in the one-day session were provided with general instruction, topical exercises and self-appraisals regarding human factors risks and performance issues in the workplace, over a period of one seven-hour day. Participants were instructed in the attention and recovery technique at the beginning and end of each instructional topic, comprising a total of five practice events throughout the day. Projected images of the plant and critical equipment components served as objective 'triggers' and were included in the final four exercises.

Members of the second group (police candidates) (G2) were provided with general information about police selection procedures, self-appraisals of personality, coping and communication styles. Instruction was also provided in the behavioural interview process and question content of police selection interviews. Instruction in the SA recovery technique was provided at the beginning and end of each instructional topic, comprising a total of five practice events throughout the day. No specific images were used to invoke recovery. Instead 'subjective' cues such as perspiration, dry mouth, brow tension, hand wringing, foot shuffling, voice constriction, gaze avoidance and memory blocks were explored as subjective triggers to invoke or cue the recovery technique.

SA recovery procedure

Both groups received identical instruction and practice in Stage 1 and 2 of the DB procedure. Anticipation of the different contexts in which each group would finally apply the recovery technique, required a variation of the final stage.

Stage 1

Participants were instructed in the diaphragm breathing (DB) technique, involving body position, inspiration using stomach rather than thoracic movement and progressive or slow exhalation. The exercise was repeated five times until all participants had succeeded with the process and reported one or several of the predicted changes involving a perception of reduction of brow tension, change in tone of voice, relaxation of shoulders, some feeling of calmness, an evident mental stillness and an elevation of mood.

Stage 2

Both groups of participants were instructed to establish an external cue with an action of their choice after completing the DB, such as brushing the nose and a sharp intake of air through the nostrils. Each subsequent repetition of the DB exercise was associated with the use of the selected cue.

Stage 3

To achieve the link between triggers and the mental switch cue, participants in the chemical plant group (G1) repeated the selected cue, but not the actual DB when viewing a series of 3-second images of the plant, interspersed by a set of distracter images and an exercise of adding up numbers appearing on a screen.

The police candidates group (G2) were given no further instructions beyond identifying their own psycho-physiological triggers and preparing answers to five typical interview questions. Each candidate was then interviewed with participants taking turns as raters and members of the panel.

Assessment and follow-up

The panel rated candidates on their ability to maintain focus or recover a positive mental state in response to challenge questions. Observations were made of their fluency and completeness of answers, ability to maintain a positive demeanour, signs of nervous fidgeting and loss of eye contact when distracted by changes in question pace and tone of delivery by interviewers. At the conclusion participants discussed their perceptions and completed a brief survey regarding the process as summarized in Table 22.2. A phone follow-up with a sample of attendees from each session two weeks later confirmed that the recovery technique continued to be practised by those individuals.The chemical plant participants (G1) were observed closely in two follow-up exercises two weeks after the one-day seminar, for their usage of the cued response to the displayed 'trigger' images. A group discussion of usage and a brief survey regarding perceptions of the process provided feedback on the use of the method in other contexts as summarized in Table 22.2.

Results

Initial results of assessment of the two groups with the SSA Test depicted in Table 22.1 indicated differences on the 'Mental Alertness' and 'Coping Skills' dimensions and against the criterion score established by incident history free individuals in the SSA test validation sample. Overall results confirmed the expectation of a trend in differences between the two groups on the two scales of special interest. A descriptive and qualitative rather than statistical assessment of the results was made due to the small collective sample size.

Results on the Mental Alertness scale for the chemical plant operators (Table 22.1) indicated a mean score (71.5) only slightly below criterion for the group of 21 persons. Two thirds of the group achieved at or greater than criterion. By contrast, two-thirds of the police candidate group (Table 22.1) scored below the criterion (65) on this scale as expected, highlighting the differences in characteristics of the two groups.

Results on the Coping Skills scale for the chemical plant group (Table 22.1) indicated less than a quarter of the sample (23.8 per cent) scored at the criterion. Most of the group scored within 9 per cent of the criterion with a relatively small spread (SD = 7.3). By contrast, fewer still of the police candidate group scored at the criterion (16.7 per cent) showing nearly double the spread (SD = 13.2) within the group. Results of police candidates on this scale were in the expected direction when compared to the chemical plant operators.

Results on the Safety Habits scale reflecting anticipatory behaviour (as a notional reinforcement of coping skills) were opposite to the expected differences and the trend between the two groups on the other two scales. More than 70 per cent of the chemical plant group scored below criteria and more than 58 per cent of the police candidates scored at or greater than criteria.

Observation of the DB recovery skills practice showed that although a few participants were familiar with some aspects of the method none had applied it to any specific need. At the first practice session most participants in both groups took three to four attempts to achieve a moderate proficiency with the DB and generally required a reminder to include the cue for association. These had mostly become automatic by the fourth and fifth practice session of the day when visual triggers were introduced to the chemical plant group; and similarly the subjective triggers were introduced to the police candidates.

Table 22.1 **SSA test scale results for Group 1 and Group 2 participants**

Scale title	Scale criterion	Chemical plant operators			Police candidates		
		% Scoring => crit	Mean n = 21	SD	% Scoring => crit	Mean n = 18	SD
Coping skills	**85**	**23.81%**	**77.8**	**7.31**	**16.67%**	**68.2**	**13.23**
General hazard awareness	67	90.48%	75.2	8.63	66.67%	63.5	16.75
Mental alertness	**65**	**66.67%**	**71.5**	**11.59**	**33.33%**	**60.2**	**12.81**
Perception and comprehension	66	71.43%	70.5	10.45	58.33%	65.0	11.90
Responsible for safety	69	66.67%	76.0	9.96	50.00%	67.0	14.64
Risk avoidance	82	85.71%	88.7	10.84	41.67%	76.1	14.39
Safety conscientiousness	67	85.71%	81.3	10.79	75.00%	74.5	18.07

Table 22.2 Report of use and retention of the SA recovery skill

Post course evaluation	**Chemical plant operators (G1)**			**Police candidates (G2)**		
	High	Mod	Low	High	Mod	Low
Acceptance of method	90%			90%		
Ease of acquisition of recovery skill		76%		90%		
Ease of replication on triggered cues	90%				75%	
Retention of recovery skill	80%			80%		
Effectiveness of the skill	90%			90%		
Main context of reported usage	Alertness triggered on entry to plant			Maintaining calmness and focus during rigorous interviewing		
Other contexts of reported usage	Alertness recovery whilst driving home at end of 12 hours shift at 7 am and in managing interpersonal tension			Maintaining positive mood and memory recall during rapid questioning		

Discussion

The efficacy of the uptake and use of the SA recovery method was assessed in two groups dissimilar in most respects and in particular on their motivation. Overall, the post instruction outcomes (Table 22.2) for both groups suggested an experience of positive effect in switching their mental state reinforcing an acceptance of the method and a willingness to apply the technique in other contexts.

Differences between the two groups were evident in their results on the SSA test. The Mental Alertness and Coping Skills scales trended in the logical direction reflecting greater participant maturity, training and experience. The puzzling reverse of expected scores between the two groups with respect to the safety habits (anticipatory behaviours), could tentatively be explained by the extent to which the chemical plant group may have become complacent resulting from their highly automated and reliable environment, or had become selective (economic) with their attention and procedures, because of their job experience and relative personal maturity.

Post training evaluation in the follow up session for the chemical plant group indicated a high degree of acceptance of the method (Table 22.2). Participants were drawn to the greater relatedness of the skill to their needs and the immediacy of a positive effect, to that of other safety training. Similarly, the reported use of the method to maintain alertness whilst driving home, reportedly fatigued after the night shift, provided some evidence for the generalizing effect of the SA recovery process. Although each group used different triggers in the training to activate the essential cue and mental switch, the chemical plant group reported responding to their personal or subjective triggers in addition to external objective triggers.

Results for the police candidate group were more visible and immediate because of being in the spotlight with an audience as well as the panel interviewers during the interview session. Although, with this group, the responses to subjective triggers were reactive rather than anticipatory, the mechanism had the desired positive effect on mental state, helping to decrease tension and hindrance to memory recall when challenged. Seeing others benefit from the recovery process may have also reinforced the use of the method by successive interviewees.

The results of this preliminary investigation accorded with historical perspectives and recent sport coaching interventions of the critical nature and immediacy of mental state changes on performance. The person focused approach must be seen to be timely given the magnitude and frequency of human error attributed to incidents. The economy and utility of an intervention that operates in a 'one-to-many' relationship such as the recovery of SA process (one technique that addresses many types of situations), is an attractive prospect to that of the present 'many-to-many' (a specific procedure to deal with each situation) approach in safety training.

More investigation conducted with a control group to measure external performance criteria such as reduction in near miss incidents and in reinforcement methods to ensure greater effectiveness and sustainability seems warranted for the process.

References

Baxter, G.D. and Bass, E.D. (2003), Human Error Revisited: Some Lessons for Situation Awareness, in, *Proceedings of the Fourth Annual Symposium on Human Interaction with Complex Systems. Los*, Alamitos, CA: IEEE/Computer Society Publishing, pp. 81–87.

Behncke, L. (2004), Mental Skills Training for Sports, *Athletic Insight*, **6**, 1.

Broadbent, D.E., Cooper, P.F., Fitzgerald, P. and Parkes, K.R. (1982), Cognitive Failures Questionnaire (CFQ) and its Correlates, *British Journal of Clinical Psychology*, **21**, 1–16. [PubMed 7126941]

Bull, S.J., Albinson, J.G. and Shambrook, C.J. (1996), *The mental game plan, Getting psyched for sport.* Eastbourne: Sports Dynamics, Cheltenham: UK.

Byrne, E.A. and Parasuraman, R. (1996), Psychophysiology and Adaptive Automation, *Biological Psychology*, **42**, 249–268. [PubMed 8652747] [DOI: 10.1016/0301-0511%2895%2905161-9]

Compton, R., Wirtz, D., Pajoumand, G., Claus, E. and Heller, W. (2004), Association Between Positive Affect and Attentional Shifting, *Cognitive Therapy and Research*, **28**, No. 6, December, 2004, pp. 733–744(12): Springer. [DOI: 10.1007/s10608-004-0663-6]

Csikszentmihalyi, M. (1990), *Flow the Psychology of Optimal Experience*, Harper & Row.

Doulliard, J. (2001), *Body, Mind, and Sport: The Mind-Body Guide to Lifelong Health, Fitness, and Your Personal Best*, New York: Three Rivers Publishing.

Endsley, M.R. (1995), Measurement of Situation Awareness in Dynamic Systems, *Human Factors*, **37**, 65–84.

Endsley, M.R. (1999), Situation Awareness and Human Error: Designing to Support human Performance, *Proceedings of the High Consequence Systems Surety Conference*. Albuquerque NM 1999.

Fletcher, P.C. and Dolan, R.J. (1996), Brain Activity during Memory Retrieval, The Influence of Imagery and Semantic Cueing, *Brain*, October, **119**, No. 5, 1587–1596.

Guynn, M.J., McDaniel, M.A. and Einstein, G.O. (1998), Prospective Memory: When Reminders Fail, *Memory and Cognition*, **26**, 287–298.

Janelle, C.M., Champenoy, J.D., Coombes, S.A. and Mousseau, M.B. (2003), Mechanisms of Attentional Cueing during Observational Learning to Facilitate Motor Skill Acquisition, *Journal of Sports Sciences*, **21**, No. 10, 825–38.

Krause, T.R. (1991), The Behaviour Based Approach to Safety: A System for Continuous Improvement, *The Safety and Health Practitioner*, **9**, No. 8, 30–32.

Landers, D.M. and Boutcher, S.H. (1993), Energy-performance Relationships, in *Applied Sport Psychology: Personal Growth to Peak Performance*. edited by: Williams, J. M. *Mountain View, CA: Mayfield.*

Lazarus, R.S. (1976), *Patterns of Adjustment*, third edn, McGraw-Hill/Kogakusha Ltd.

Lazarus, R.S. and Folkman, S. (1984), *Stress, Appraisal and Coping*, New York: Springer.

Marr, A.J. (2001), the Zone – 'A Biobehavioral Theory of the Flow Experience',

Athletic Insight **3**, 1.

Morris, P.E. (1992), Prospective Memory: Remembering to Do Things, in *Aspects of Memory. Vol The Practical Aspects* eds Gruneberg, M. M. and Morris, P., second edn, London: Routledge.

Nideffer, R.M. (1993), Concentration and Attention Control Training, in *Applied Sport Psychology: Personal Growth to Peak Performance* ed Williams, J. M., *Mountain View, CA: Mayfield.*

Pashler, H., Johnston, J. and Ruthruff, E. (2001), Attention and Performance, *Annual Review of Psychology*, **2001**, No. 52, 629–651.

Pashler, H., Shiu, L-P. (1999), Do Images Involuntarily Trigger Search? A Test of Pillsbury's Hypothesis, *Psychonomic Bulletin and Review*, 1999, **6**, 3, 445–448.

Peper, E. and Tibbetts, V. (1997), Effortless Diaphragmatic Breathing, *Electromyography*, Biofeedback Foundation Europe.

Reason, J. (2000), Human Error: Models and Management, *British Medical Journal*, 2000, No. 320, 768–770. [DOI: 10.1136/bmj.320.7237.768]

Rosenweg, P. (2001), *Development of a test for Situational Safety Awareness*, http://www.psyfactors.com/validations.htm, accessed on 6 June 2006.

Seaward, B.L. (2002), *Managing Stress: Principles and Strategies for Health and Well Being*, third edn, Boston: Jones & Bartlett.

Skinner, B.F. (1968), *The Technology of Teaching*, New York: Appleton-Century-Crofts.

Walley, M. *(1995), The Attainment of Mental Health* – Contributions from Buddhist Psychology, in *Promotion of Mental health. Aldershot: Ashgate* eds Trent, D. and Reed, C.

Chapter 23

Effects of Flight Duty and Sleep on the Decision-Making of Commercial Airline Pilots

Renée M. Petrilli, Matthew J.W. Thomas, Nicole Lamon, Drew Dawson and Gregory D. Roach
University of South Australia

Decision errors have been identified as a key factor in numerous aviation incidents and accidents. Whilst previous research indicates that fatigue may impair decision-making, the specific effects of fatigue on crew decision-making remains unclear. Consequently, this study seeks to investigate the effects of fatigue on flight crew decision-making. Volunteer flight crew from an Australian commercial airline participated in a simulator-based study. Crew were tested in an operationally realistic scenario that required the crew to make a complex decision. Preliminary analyses indicate fatigued crews adopt different approaches to decision-making compared to non-fatigued crew, which highlights several implications for aviation safety.

Introduction

Pilot fatigue is increasingly being recognized as a significant risk in aviation operations. This is largely due to both domestic and international pilots having to deal with sleep irregularities, long duty days, early starts and night flying (Graeber, 1988; Caldwell, 2005). Furthermore, international pilots must also deal with multiple time zone changes, which can lead to jetlag (Caldwell, 2005). Jetlag is broadly defined as the desynchrony between circadian rhythms and local time cues caused by travel across time-zones (Arendt, Stone and Skene, 2000), and is associated with impaired performance and increased sleepiness (Caldwell and Caldwell, 2003).

Importantly, reports from the National Transportation Safety Board (NTSB) have recognized fatigue as a contributing factor in several aviation incidents, and accidents. For example, fatigue was determined as a major factor in the *Collision with Trees on Final Approach accident*, where a Boeing 727 crashed short of the runway (NTSB, 2002). Similarly, the Swiss Aircraft Accident Investigation Bureau (Beaa, 2001) found that in the Flight CRX 3,597 accident, the captain's 'ability to concentrate and take appropriate decisions as well as his ability to analyse complex processes were adversely affected by fatigue' (p. 11). In Australia, fatigue was implicated in a B737 ground proximity caution alert near Canberra, when a pilot entered incorrect information in the flight management computer (ATSB, 2004).

Previous laboratory studies have shown that fatigue associated with restricted sleep greatly impairs performance. Specifically, research has shown that human performance can be impaired after six or seven hours of sleep restriction, and sleep restriction of four hours or less can lead to microsleeps (Caldwell and Caldwell, 2003). Notably, data from aviation studies have found evidence of pilots experiencing microsleeps, which are more likely to occur during international in-bound trips (Samel, Wegmann and Vejvoda, 1997).

Despite sleep factors playing a significant role in the safety of flight operations, flight and duty limitations around the world do not take this into account. Flight and duty limitations are typically based on prescriptive rules that are inflexible and are not built on any scientific evidence. Furthermore, they do not account for individual differences in fatigue-related risk, and fail to distinguish between work-related fatigue and fatigue related to other factors (*e.g.*, sleep/wake) (see Dawson and McCulloch, 2005 for a review). Specifically, flight and duty limitations are designed to ensure that pilots are not exposed to unacceptable levels of fatigue through maximum shift lengths and minimum rest breaks. As previous research suggests, however, a pilot's experience of fatigue appears not only to be based on their work/rest history but on the amount of sleep they have been able to obtain between and during duty periods.

To address these issues, the current study was designed to systematically examine the effects of fatigue associated with work/rest and sleep/wake factors on the operational performance of commercial airline flight crews. As data collection was only recently completed, this chapter will present the preliminary analyses of the first 50 per cent of crews.

Method

Study design

The study employed a randomized factorial design with one between-subjects factor (duty history) with two conditions (rested, non-rested). Each condition consisted of a 3-hour simulator session in Sydney, Australia. Rested crews attended the simulator session at 14:00h following at least four consecutive local nights rest, whilst crews in the non-rested condition attended a simulator session after returning to Sydney from an international pattern. Return international trips were Australia–Europe, and Australia–US.

Participants

The preliminary analyses include 36 crews (72 pilots; mean age = 46 years, SD = 8.1) from a commercial airline. Mean total flying hours was 13,187 hours (SD = 4,738 hours), mean total years pilots had flown long-haul aircraft was 18.5 years (SD = 8), and mean total hours pilots had flown B747-400 aircraft was 3,411 hours (SD = 2,795). Each crew consisted of one Captain and one First Officer. Given that all pilots from the commercial airline are subject to twice-yearly physical examinations as part of their employment, all pilots from the commercial airline were assumed fit to participate in the study.

Materials and measures

Subjective fatigue questionnaire Subjective fatigue was assessed using the Samn-Perelli fatigue scale. The Samn-Perelli is a seven-point Likert scale with 1 = 'Fully alert, wide awake;' 2 = 'Very lively, responsive, but not at peak'; 3 = 'Okay, somewhat fresh;' 4 = 'A little tired, less than fresh;' 5 = 'Moderately tired, let down;' 6 = 'Extremely tired, very difficult to concentrate;' 7 = 'Completely exhausted, unable to function effectively'. The dependent measure derived from the Samn-Perelli fatigue scale was subjective fatigue.

Psychomotor vigilance task Pilots' neurobehavioural performance was assessed using the psychomotor vigilance task (PVT) (Palm-Pilot version; Walter Reed Army Institute of Research). The PVT requires participants to respond to a display for the duration of the test (five minutes), and to press the left or right response key (depending on their dominant hand) as quickly as possible after the appearance of a visual timer stimulus (*i.e.*, a number counting up from zero, in increments of 1 every millisecond). The participant's response stops the timer stimulus, and the response speed is displayed for a period of 500 m. The interval between stimuli presentations vary randomly from 2,000 to 10,000 m. Before commencing each PVT, pilots were instructed to respond to the stimuli as quickly as possible without making false starts. The PVT is widely used in field research because of its sensitivity to fatigue and its portability (Lamond, Dawson and Roach, 2005). The dependent measure derived from the PVT was the total mean reciprocal response time (RT) (in ms).

Sleep and work diaries Pilots kept a self-recorded sleep and duty diary for the entire study period. For the sleep diary, pilots recorded information for every sleep period. This included the start and end time of each sleep period, their subjective fatigue before and after each sleep period using the Samn-Perelli fatigue scale, and perceived sleep quality using a five-point Likert scale. Similarly, pilots recorded information for every work period in the duty diary. This included the on-blocks and off-blocks time (*i.e.*, start and end) of each flight, the origin and destination ports, and subjective fatigue before and after each flight using the Samn-Perelli fatigue scale. Time was recorded as universal time code (UTC). The independent measure derived from the work diary was whether pilots had flown an international pattern or not (*i.e.*, non-rested, rested). The independent measure derived from the sleep diary was taken as actual sleep in the prior 24 hours.

Actigraphs In support of the sleep/duty diaries, objective sleep/wake schedules were assessed using activity monitors (Mini Mitter, Sunriver, Oregon), which were worn by each pilot. Activity monitors are devices worn like a wristwatch on the wrist that allow for 24-hour recordings of activity (see Ancoli-Israeil, 2002). The activity monitor is a light-weight device that contains a piezo-electric accelerometer which measures the timing and quantity of body movement. The independent measure derived from the activity monitors was taken as actual sleep in the prior 24 hour (in mins).

Flight simulator

Crew performance was assessed in a Level D Boeing 747-400 flight simulator (CAE Electronics Ltd). This flight simulator is an exact replica of the Boeing 747-400 aircraft flight deck, and incorporates advanced effects to achieve a high level of realism. All pilots employed by the commercial airline had received training and license renewals inside the simulator. All simulator sessions were videotaped.

Scenario

Members of the scientific community, subject matter experts (SMEs), and representatives from the commercial airline assisted in the development of the scenario. The scenario was designed to systematically assess crews' ability to manage various operational distractions and to make a critical decision that had no single correct answer. Specifically, crews were required to fly from Sydney to Melbourne, Australia including pre-flight checks, take-off, cruise, descent, approach, and landing. This flight has a flying time of approximately 60–70 minutes. Results for crews' management of various operational threats are beyond the scope of this chapter and will be reported in future publications. The current chapter will focus explicitly on the critical decision event:

- At the outset, crews plan to head for Runway 16 in Melbourne. They have the Engine No. 3 Thrust Reverser locked out, meaning that the plane has reduced braking capability on the ground. Crews receive a valid weather forecast for Melbourne that indicates that landing in Melbourne is acceptable.
- At approximately 10 minutes before the top of descent, Air Traffic Control (ATC) issues a change of weather conditions for Melbourne.
- The change of weather conditions indicates that winds in Melbourne are above crosswind limitations for Runway 16 such that landing on this runway is no longer legal, and
- Furthermore, for the alternative runway in Melbourne, Runway 27, figures from the aircraft performance manual indicate that the landing distance required is very close to the landing distance available.

Essentially, the change in weather conditions for Melbourne presents the crew with three critical issues that they must resolve:

- Can they land on the original runway in Melbourne (16) given that there is now a strong crosswind and the runway is wet?
- Can they land on the perpendicular runway in Melbourne (27) which has no cross-wind, but it is a short runway, it is wet, and the plane has reduced braking capability due to the Engine No. 3 Thrust Reverser being locked out?
- Can they land at a different airport considering the weather at other airports and fuel constraints?

Decision variables

With the assistance of SMEs, important tasks that crews could or should undertake to resolve the critical decision event were identified. Subsequently, an observer who was blind to crews' sleep and duty history viewed and coded each of the simulator sessions from the video data. The list includes over 50 variables, but only 10 of the key variables are reported here. The first eight variables are categorical variables and reflect the decision-making process, while the final two variables in the list are timed measures:

- Landing distance: Whether crews correctly calculated the landing distance required on Runway 27
- Alternate airport: Whether crews established an alternate airport if they decided to land in Melbourne, and the time taken to determine this
- Weather updates: Whether crews requested weathers including forecasts and trends for Melbourne
- Options: The number of options considered by crews
- Contingency plans: Whether crews discussed contingency plans after finalizing their decision but before taking positive action towards that decision
- Subjective evaluation: Whether crews viewed the initial change in weather conditions in a positive or negative light. Notably, the importance of optimism-bias in crew decision-making has been highlighted by (Wilson and Fallshore, 2001)
- Diversion: Whether crews decided to continue to land in Melbourne or divert to another airport
- Diversion fuel: Correct calculation of diversion fuel
- Decision finalisation: The time taken from the end of the change of weather conditions (issued by ATC) to the time crews verbally decided their plan of action
- Positive action: The time from the end of the change of weather conditions to the time that crews took positive action towards implementing the decision (*e.g.*, the execution of the flight plan in the Flight Management Computer (FMC) or the request for diversion).

Procedure

Pilots attended the simulator session in Sydney, Australia. Simulator sessions consisted of three phases: i) pre-simulator (40 minutes), (ii) simulator (1.5 hours), and (iii) post-simulator (40 minutes).

During the pre-simulator phase, crews were briefed about the session, completed the Samn Perelli scale, and were required to complete a five-minute PVT. In addition, crews attended a pre-flight briefing where they were informed of the flight details. Crew received information that the sector was a daytime flight from Sydney to Melbourne and would take approximately 1.5 hours. They were also informed of other flight details such as passengers, NOTAMs, and fuel loads. During the simulator phase crews flew a full flight based on the scenario described above. During the post-simulator phase, pilots were required to complete subjective questionnaires and the five-minute PVT. Additionally, to determine the decision strategies used by crews, they were interviewed for approximately 30 minutes using Klein's (Klein, Calderwood and Macgregor, 1989) Critical Decision Method. Results from the post-simulator session will not be reported in the current chapter.

Rested pilots arrived at the simulator session at approximately 14:00h following four consecutive local nights rest. Non-rested pilots arrived at the simulator session immediately after returning from an international trip. Sleep/wake data were collected from all crews using the activity monitors and sleep and duty diaries approximately five days before their simulator session. Practice and baseline measures were also obtained from each pilot approximately 10 days before their simulator session. Non-rested crews' sleep/wake recovery data were collected for four consecutive nights following the simulator session.

All pilots were asked to abstain from consuming coffee 24 hours before their simulator session. To ensure the Captain and First Officers were familiar with each other at the time of the simulator session, pilots had flown an international pattern at least 12 days before their simulator session.

Statistical analyses

For the analyses, the data were split in two different ways (i) on the basis of recent work/rest history, and (ii) on the basis of recent sleep/wake history. In the first analyses, the independent variable was duty with two levels, rested (18 crews) and non-rested (18 crews). Rested crews completed their simulator session after having at least four consecutive local nights rest. Non-rested crews completed their simulator session immediately after returning to Sydney from a US trip, or a Europe trip.

For the second set of analyses, the independent variable was sleep with two levels – high sleep (24 crews) and low sleep (12 crews). High sleep' crews were defined as those in which both the Captain and First Officer had obtained at least five hours sleep in the 24 hours before the start of the simulator session. Low sleep' crews were defined as those in which at least one of the crew had obtained less than five hours sleep in the 24 hours before the start of the simulator session. Less-than five hours sleep was chosen as the cut-off point for the low-sleep level as previous research indicates that performance is typically impaired following this amount of restricted sleep (Caldwell and Caldwell, 2003).

The dependent variables in all analyses were (1) subjective fatigue at the start of the simulator session, (2) PVT performance (i.e., mean reciprocal response time: speed) at the start of the simulator session, and (3) the decision variables. One-way factorial ANOVAs were conducted on subjective fatigue, PVT performance, and the timed decision variables (*i.e.*, decision finalization, positive action). Cross-tabulation χ^2 analyses were conducted on the categorical decision variables (*i.e.*, landing distance, alternate airport, weather updates, options, contingency plans, subjective evaluation, diversion, and diversion fuel).

Results

Analysis impact of fatigue associated with work/rest factors

Subjective fatigue One-way factorial ANOVA found a significant effect of duty on the subjective fatigue of Captains ($F_{1,35} = 40.4$, $p < 0.001$) and First Officers ($F_{1,35} = 72.0$, $p < 0.001$). Specifically, Captains in the non-rested group (M = 5.0, SD = 0.5) rated themselves as being significantly more fatigued than Captains in the rested group (M = 2.9, SD = 1.3). Similarly, First Officers in the non-rested group (M = 4.7, SD = 0.8) rated themselves as being significantly more fatigued than First Officers in the rested group (M = 2.4, SD = 0.9).

Psychomotor vigilance performance One-way factorial ANOVA found no significant effect of duty on Captains' mean reciprocal response time. However, analyses revealed a significant effect of duty on First Officers' mean reciprocal response time ($F_{1,34} = 5.9$, $p < 0.05$). Specifically, First Officers in the non-rested group (M = 4.5 m, SD = 0.63) responded significantly slower than First Officers in the rested group (M = 4.1 m, SD = 0.43).

Decision Variables Analyses revealed no significant effect of duty on any of the non-timed or timed decision variables (see Table 23.1).

Table 23.1 Statistical analyses of the decision variables with duty as the independent variable

Decision variable	Statistical analysis	Statistic (d.f.)	p
Correct calculation of landing distance	χ^2	1.41 (1)	0.19
Alternative airport established	χ^2	1.63 (1)	0.28
Request of Melbourne weather updates	χ^2	3.60 (1)	1.0
Number of options considered	ANOVA	0.71 (1.35)	0.10
Discussion of back-up plans	χ^2	0.51 (1)	0.74
Subjective evaluation of weather change	χ^2	0.13 (1)	0.80
Diversion to another airport	χ^2	3.47 (1)	0.53
Correct calculation of diversion fuel	χ^2	0.87 (1)	0.78
Time to finalize decision	ANOVA	5.30 (1.35)	0.09
Time to positive action towards decision	ANOVA	9.17 (1.35)	0.17

Analysis[2] impact of fatigue associated with sleep/wake factors

Subjective fatigue One-way factorial ANOVA found a significant effect of sleep on subjective fatigue for both Captains ($F_{1,35} = 6.4$, $p < 0.05$) and First Officers ($F_{1,35} = 7.6$, $p < 0.05$). Specifically, Captains in the low-sleep group rated themselves as being significantly more fatigued (M = 4.8, SD = 0.8) than Captains in the high-sleep group (M = 3, SD = 1.6). Similarly, First Officers in the low-sleep group rated themselves as being significantly more fatigued (M = 4.1, SD = 0.9) than First Officers in the high-sleep group (M = 3.2, SD = 1.4).

Psychomotor vigilance performance One-way factorial ANOVA found no significant effect of sleep on Captains' mean reciprocal response time. However, analyses revealed a significant effect of sleep on First Officers' mean reciprocal response time ($F_{1,34} = 5.0$, $p < 0.05$). Specifically, First Officers in the low-sleep group responded significantly slower (M = 4.3 m, SD = 0.49) than First Officers in the high-sleep group (M = 3.9 m, SD = 0.56).

Decision Variables Analyses revealed no significant effect of sleep on any of the non-timed decision variables (see Table 23.2). Interestingly though, there was a significant effect of sleep on the timed decision variables, decision finalization ($F_{1,35} = 5.3$, $p < 0.05$) and positive action towards decision ($F_{1,35} = 9.2$, $p < 0.01$). Specifically, low sleep crews took significantly longer (M = 12.6 minutes, SD = 9.5) to finalize their decision than high sleep crews (M = 7.3 minutes, SD = 4.2). Similarly, low sleep crews took significantly longer (M = 16.1 minutes, SD = 7.9) to take positive action to enact their decision than high sleep crews (M = 9.6 minutes, SD = 5.0).

Table 23.2 Statistical analyses of the decision variables with sleep as the independent variable

Decision variable	Statistical analysis	Statistic (d.f.)	p
Correct calculation of landing distance	χ^2	1.41 (1)	0.17
Alternative airport established	χ^2	1.63 (1)	0.20
Request of Melbourne weather updates	χ^2	3.60 (1)	0.06
Number of options considered	ANOVA	0.71 (1.35)	0.41
Discussion of back-up plans	χ^2	0.51 (1)	0.47
Subjective evaluation of weather change	χ^2	0.13 (1)	0.72
Diversion to another airport	χ^2	3.47 (1)	0.18
Correct calculation of diversion fuel	χ^2	0.87 (1)	0.32

Decision variable	Statistical analysis	Statistic (d.f.)	p
Time to finalize decision	ANOVA	5.30(1,35)	0.03*
Time to positive action towards decision	ANOVA	9.17 (1.35)	0.01*

* Indicates significance at p <0.05 level discussion

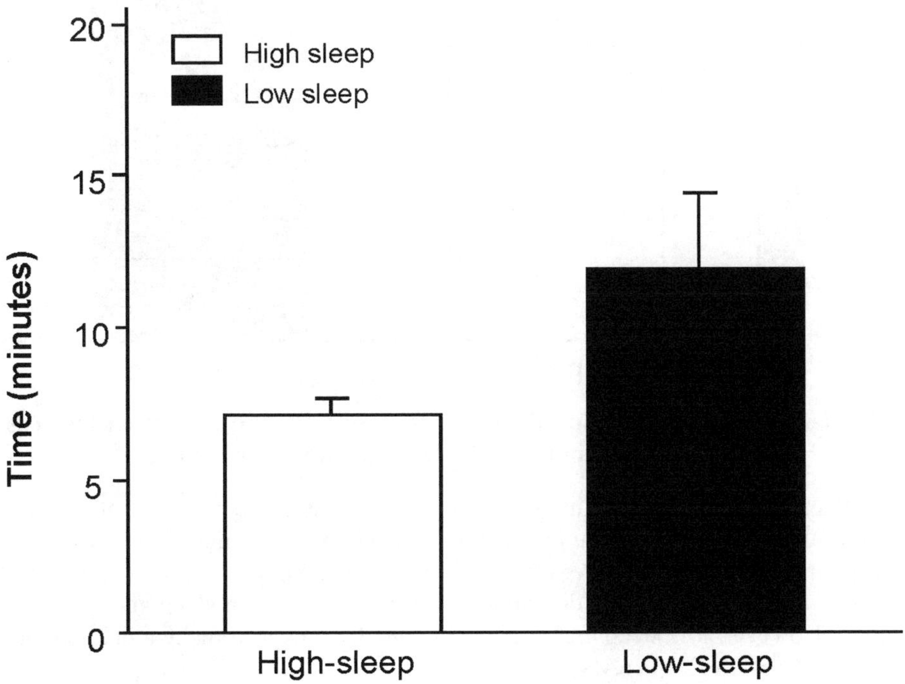

Figure 23.1 Mean times (in mins) for crews to finalize their decision in the high-sleep and low-sleep groups

Discussion

In summary, when crews are divided into two groups on the basis of their recent work/rest history, non-rested crews have higher self-rated fatigue than rested crews, and non-rested First Officers have poorer PVT performance than rested First Officers, and there is little difference between the decision-making of rested and non-rested crews.

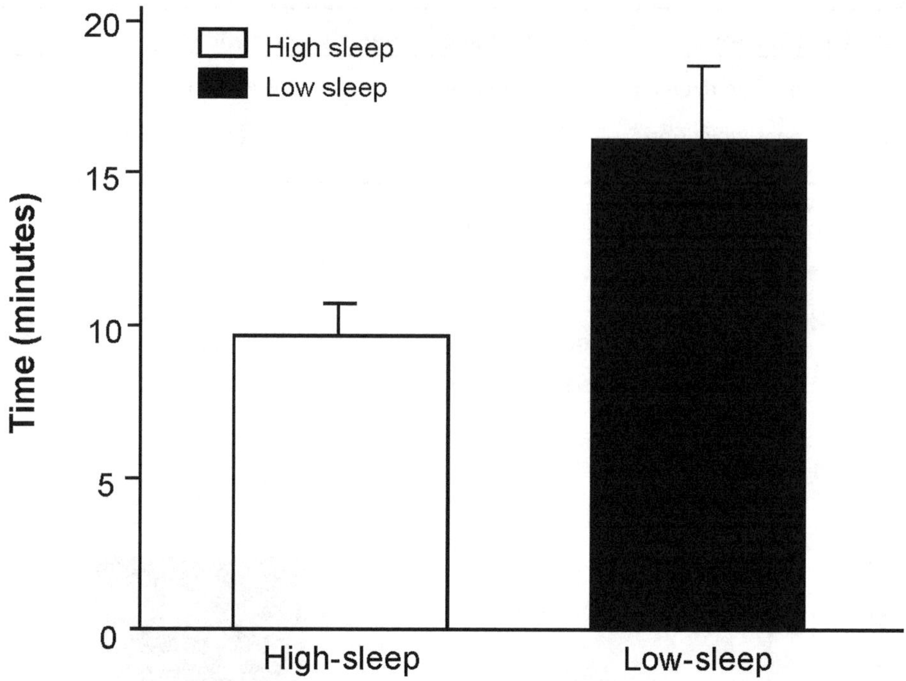

Figure 23.2 Mean times (in mins) for crews to take positive action towards their decision in the high-sleep and low-sleep groups

When crews are divided into two groups on the basis of their recent sleep/wake history, low sleep crews have higher self-rated fatigue than high sleep crews, low sleep First Officers have poorer PVT performance than high sleep First Officers, and there is no difference in the decision-making process of high sleep and low sleep crews. However, low sleep crews take substantially longer to make a decision and enact it than high sleep crews.

These preliminary analyses indicate that fatigued crews may be choosing to sacrifice speed for accuracy. That is, fatigued crews use a similar decision-making process to the non-fatigued crews, but this process takes considerably longer, which appears to be a speed-versus-accuracy trade-off. This may be the best fatigue-coping strategy in an aviation environment as it is more important to make a good decision than to make a quick decision as the consequences of making a poor decision can be catastrophic.

Importantly, these results highlight a dynamic regulatory system underlying human adaptive responses to task demands where behaviour is modified so that internal goals can be maintained. Hockey (1997) presents a theory for the management of effort and performance under stress which proposes that people control the effectiveness of task performance by the management of effort. This allows them to adopt 'performance protection strategies' to maintain priority task goals within acceptable limits through incurring costs. Findings from this study indicate that when at least one pilot is fatigued, the crew may engage in this type of performance protection strategy. Performance is maintained (such as operating the aircraft safely) at the cost of the time taken to make the decision. Importantly though, this raises the question of whether slowed decision-making incurs time-pressure costs at later stages of flight, and in turn, does this increase the risk to the operation? This is an area of research our laboratory is currently pursuing.

Other fundamental research considerations are: the identification of countermeasures to improve pilots' decision-making when they are fatigued; possible improvements to the amount/quality of sleep that pilots obtain before and during duty periods, whether regulating for sleep outcomes to complement (or replace) prescriptive duty and rest limits is required; whether threat management and error rates are influenced by work/rest and sleep/wake factors; the incorporation of crew characteristics into fatigue models, and potential insights from the post-flight interviews.

Conclusions

Effective flight crew decision-making is an important factor in the safety of airline operations. Preliminary results from the current study indicate that fatigue associated with sleep/wake factors impacts crew decision-making. This has important implications for current flight and duty limitations as these are generally based on work/rest factors and not sleep/wake factors. It is anticipated that the results of the current study will be integrated into a fatigue risk safety management system that may be more defensible for managing fatigue than traditional prescriptive systems.

Acknowledgements

We would like to thank our research partners, the Civil Aviation Safety Authority, Qantas, and the Australian and International Pilots Association, and the Australian Research Council for supporting this research.

References

Arendt, J., Stone, B. and Skene, D. (2000), Jet lag and Sleep Disruption, in *Principles and Practice of Sleep Medicine (Vol. 3)* edited by: Kryger, M. H., Philadelphia: W.B. Saunders.

Australian Transport Safety Bureau (ATSB) (2004), Final ATSB Report into the 24 July 2004 Boeing 737 Ground Proximity Caution near Canberra, Aviation Safety Investigation Report 200402747.

Belenky, G., Wesensten, N.J., Thorne, D.R., Thomas, M.L., Sing, H.C. and Redmond, D.P. *et al.* (2003), Patterns of Performance Degradation and Restoration during Sleep Restriction and Subsequent Recovery: A Sleep Dose–Response Study, *Journal of Sleep Research*, **12**, 1–12. [PubMed 12603781] [DOI: 10.1046/j.1365-2869.2003.00337.x]

Caldwell, J.A. (2005), Fatigue in Aviation, *Travel Medicine and Infectious Disease*,

3, 2, 85–96. [DOI: 10.1016/j.tmaid.2004.07.008]

Caldwell, J.A. and Caldwell, J.L. (2003), *Fatigue in Aviation*, Aldershot: Ashgate Publishing.

Dawson, D. and McCulloch, K. (2005), Managing Fatigue: It's About Sleep, *Sleep Medicine Reviews*, **9**, 5, 365–380. [PubMed 16099184] [DOI: 10.1016/j.smrv.2005.03.002]

Graeber, R.C. (1988), Fatigue and Circadian Rhythmicity, in *Aircrew Fatigue and Circadian Rhythmicity* eds Weiner, E. L. and Nagel, D. C., New York: Academic Press.

Hockey, G.R.J. (1997), Compensatory Control in the Regulation of Human Performance under Stress and High Workload: A Cognitive-Energetical Framework, *Biological Psychology*, **45**, 73–93. [PubMed 9083645] [DOI: 10.1016/S0301-0511%2896%2905223-4]

Klein, G., Calderwood, R.A. and Macgregor, D. (1989), Critical Decision Method for Eliciting Knowledge, *IEEE Transactions on Systems, Man, and Cybernetics*, **19**, 3, 462–472.

Lamond, N., Dawson, D. and Roach, G.D. (2005), Fatigue Assessment in the Field: Validation of a Hand-Held Electronic Psychomotor Vigilance Task, *Aviation Space and Environmental Medicine*, **4**, 486–489.

National Transportation Safety Board, (2002), 'Aircraft accident report: Collision with Trees On Final Approach Federal Express Flight 1478 Boeing 727-232, N497FE' Tallahassee. Florida.

Samel, A., Wegmann, H. and Vejvoda, M. (1997), Aircraft Fatigue in Long-Haul Operations, *Accident, Analysis and Prevention*, **29**, 4, 439–452. [PubMed 9248502] [DOI: 10.1016/S0001-4575%2897%2900023-7]

Swiss Aircraft Accident Investigation Bureau, (2001), Status of the Accident Investigation of Crossair Flight CRX 3597 of 24 November 2001 near Bassersdorf/zh. Press Release of 30 November 2001.

Wilson, D.R. and Fallshore, M. (2001), *Optimistic and Ability in Biases in Pilots' Decisions and Perceptions of Risk Regarding VFR Flight into IMC, Paper Presented at the 11th International Symposium on Aviation Psychology,* Columbus, OH.

Chapter 24

Medical Team Resource Management and Error Prevention Training

David G. Newman
Flight Medicine Systems Pty Ltd

The parallels between the critical performance levels required of airline teams and medical teams are becoming increasingly recognized. This chapter will discuss a new Medical Team Resource Management and Error Prevention Training program that is now available. This program is designed to improve medical team performance and prevent medical errors, by using techniques derived from the aviation industry. The program involves learning about team dynamics and performance limitations, building essential team skills for handling critical events, improving individual and team performance, and developing leadership skills. The program combines rigorous case study analysis, group discussions, and practical experience through real-time simulations.

Introduction

The medical and aviation environments share some important similarities. The consequences of error in either of these environments can be catastrophic, resulting in injury or death. The aviation industry has for many years been attempting to reduce the adverse impact of human error on flight safety. The medical environment can learn much from how the aviation industry has sought to address the issue of human error. This chapter discusses a training program for medical team members that is designed to improve their team performance and reduce the rate of medical error through adaptation of the aviation safety model to the medical environment.

Human error

Human error is a complex issue. We all know that humans make errors, the consequences of which can range from the trivial and un-noticed through to disasters of enormous proportions. The nature of human error has occupied the attention of researchers for many years, and continues to do so. The emphasis is on understanding the nature of human errors, why such errors are made, and as a result how to design strategies and techniques to minimize errors and/or their consequences.

At this point it is useful to define what is meant by human error. A complex and generic term, human error can be defined as an occurrence in which a planned sequence of mental or physical activities fails to achieve its intended outcome.

Why do errors occur? The answer to this fundamental question is multifactorial. A host of factors lead to human error. Space does not permit a thorough exposition of the various factors known to contribute to human error, but a number of key factors can be mentioned here. Firstly, humans demonstrate considerable individual variation, in terms of concentration, intelligence, memory, attention, information processing capability and so on. This individual variation in cognitive abilities between different people can also be seen in the same individual at different times. This intra-individual variation in cognitive ability can result from several factors, such as fatigue and illness.

Secondly, there may be issues relating to the human–machine interface that sets the human operator up to make an error. Design traps may lead an individual to commit an error in a given set of circumstances. It is possible that the actual outcome of this event was never foreseen by the designer and the engineering department, and may in fact have been present for several years before a given set of circumstances conspired on the day of the event to lead to an error. Latent conditions such as this can lie dormant for considerable time before combining with other factors to result in an error (Reason, 2000). Clearly it is imperative for designers of equipment to avoid such conditions in the design and construction of the equipment, but it most be remembered that designers and operators all suffer from the same individual variation mentioned above.

Thirdly, the environment in which the individual is operating may contribute to error formation. Issues such as training, operating procedures, supervision and organizational culture may all be significant contributors to an error being committed by an operator of equipment. Indeed, latent conditions may also be hidden within procedures and training systems.

Fourthly, poor communication can make a significant contribution to the risk of an error being committed. Assumptions, misunderstandings, confusion and misinterpretations can clearly lead to errors.

Types of errors

Three types of errors have been defined:

- Automatic or skill-based errors
- Rule-based errors
- Knowledge-based errors

Skill-based errors occur when the intention of the operator was correct but the execution of the task was not. Examples include switch settings being incorrect or a reversion to past learning (*e.g.*, a pilot in a new aircraft type experiencing an emergency may automatically act according to past learning in a different aircraft type, with such actions being inappropriate for the current aircraft being flown).

Slips and lapses are examples of skill-based errors. A slip can be defined as a failure of attention, in which an action may not be completed as intended. A lapse can be defined as a failure of memory, where an intended action is forgotten and not carried out. This may be due to the operator being interrupted at their task.

Rule-based errors are a consequence of 'if this, then that' action rules. Errors of this type may occur when there is insufficient training or currency with the operating environment. The rule may be inappropriately applied to the given situation.

Knowledge-based errors occur in novel, unplanned for situations where the knowledge of the operator is called upon to solve a problem which may not have been experienced before. In attempting to solve the problem, an error may occur. A classic aviation example of such an approach (with a favourable outcome in this instance) is United Airlines Flight 232, a DC-10 that crashed on 19 July 1989 at Sioux City, Iowa. At an altitude of 37,000 ft, an explosive failure of the tail-mounted engine resulted in the loss of hydraulic lines serving the flight control surfaces. The aircrew, unprepared for such a rare event, managed to continue flying using the novel approach of using differential thrust of the two remaining wing-mounted engines. Such a knowledge-based approach to a dire situation allowed 185 people to survive a catastrophic in-flight failure. Clearly in this example the application of knowledge to a novel situation produced a largely successful outcome. Sometimes, however, the outcome can be an error with significant consequences.

Rule and knowledge-based errors are also collectively termed mistakes. They represent intentional actions being incorrectly carried out. Sometimes the error and its evidence are ignored and dismissed due to the operator's belief that they have in fact acted correctly.

Medical error

Medical practice is a complex field of endeavour. In these days, medical science is increasingly sophisticated, with a myriad of high-technology equipment, procedures, investigations, tests, and surgical interventions. Intensive care medicine and surgery are becoming increasingly complicated, as medical science increases our understanding of the human body and technical innovations permit more complex surgery to be undertaken to treat various illnesses and conditions.

Yet it is well known that despite the advances in medical science and the ever-increasing degree of surgical sophistication, errors occur. Such errors can increase the morbidity and mortality of patients receiving medical care. In some cases, such errors may increase a patient's stay in hospital, at the other end of the spectrum can result in the death of a patient (Leape, 1994; Altman, 1999).

How big is the problem of medical error? The US Institute of Medicine has estimated that up to 98,000 people die annually as a result of medical error in the USA (Kohn, Corrigan and Donaldson, 2000). This figure has been put at 180,000/year in other studies. Medical error has been cited as the eighth most common cause of death. According to Zhan and Miller (2003), medical error is one of the 10 most common causes of death in the USA. The Harvard Medical Practice Study in 1991 found that 4 per cent of hospital patients had additional injuries as a result of medical treatment (Leape *et al.*, 1991), two-thirds of these injuries were due to medical errors, and almost 14 per cent resulted in the patient's death. Adverse drug events have been shown to contribute 2.2 days to the length of a hospital admission (Bates, 1997)

Learning from aviation

Medical practice can learn a lot from how the aviation industry has approached the issue of errors. Seventy per cent of aviation accidents have some human error contribution. Aviation has sought to address error, in an effort to make the entire aviation industry safer. Crew training has been identified as an important factor that can lead to a safer flight environment. This training is different from basic technical training, which is more to do with how to operate the aircraft. Airline crews tend to receive standardized technical training, with regular simulator training sessions to demonstrate compliance with technical objectives. In the safety management area, crew training involves how different individuals can work together to create a safe, effective and efficient flight deck team. The training to achieve this result is known as crew resource management (CRM), and has evolved into other types of related training such as multi-crew coordination training (MCC) and line-orientated flight training (LOFT) 3 (Helmreich, Merritt and Wilhelm, 1999; Helmreich, 2000; Helmreich and Merritt, 2001; Campbell and Bagshaw, 2002). All of this training is designed to equip the crews with the knowledge, skills and abilities to operate as an effective team that functions in predictable ways.

CRM training involves developing an understanding of various issues such as personality, communication, leadership and team dynamics. Pilots are taught how to effectively manage the flight deck team, as well as how to listen effectively to other team members and make use of their sometimes different insights into a given problem. The importance of standardized procedures, appropriate use of checklists, and the best ways to interact as a flight deck team are all emphasized. Today, much of this training is run in-house by airlines, and often in combination with other training done in flight simulators. The use of flight simulators allows various scenarios to be created which give the crew the opportunity to put their CRM training to use in dealing with an in-flight emergency.

So, what can the medical world learn from aviation? How can the principles of CRM, MCC and LOFT be used in a medical setting? In order to answer these questions, it is worth considering for a moment the inherent similarities between the medical and aviation environments. Both of these environments are sophisticated, complex, safety-critical environments, with a high degree of technology and automation. The individuals who operate in these environments are all highly trained, with a high level of technical knowledge. Errors in both environments can lead to significant injury and/or death, although the potential numbers of victims with any single occurrence are obviously larger with an aircraft accident. In both environments, resolving an emergency requires quick-thinking, rapid and appropriate decision-making, and an overall high level of performance in a time-critical situation.

As in aviation, medical personnel often work as teams, in operating theatres, emergency rooms and trauma units. In this respect, a team of doctors and nurses working in an operating theatre are no different from the crew operating a regular public transport airline flight. The team members may or may not have worked together before, and they will begin their task with a particular end-point in mind (either a successful arrival at the destination or a successful surgical outcome). How the team members work together to achieve these objectives, and how they can do so safely with minimal errors, is fundamentally important. Aviation has sought to address these issues with CRM training. It stands to reason that adapting CRM training to the medical environment should bring similar advantages in terms of iatrogenic injury reduction, medical error reduction and patient safety enhancement (Helmreich, 2000; Hugh, 2002; Grogan *et al.*, 2004).

Medical team resource management

Flight Medicine Systems (FMS) has developed an innovative approach to adapting the aviation flight safety training model to the aviation environment. The principles of CRM so regularly and widely used in the aviation setting have been adapted to the medical environment, and the program is known as medical team resource management (TRM). Combined with error prevention strategies, this training has been designed to improve medical team performance and prevent medical errors. The program is available to all medical team members, including doctors, nurses, and technicians. Surgical teams, intensive care teams, emergency room teams and trauma unit teams are ideally placed to extract enormous benefits from this training program.

FMS's 2-day intensive medical TRM program introduces team members to the basic concepts of human performance, as used by the aviation industry in its attempts to understand and prevent crew errors, incidents and accidents. Using a variety of actual airline accident and medical incident case studies, program participants will learn about crew resource management principles, and how CRM training has achieved positive safety results within the international airline industry.

The key objectives of the program are for participants to learn about human performance limitations and team dynamics, to build essential team and communication skills for handling critical events (using the aviation model), to develop leadership skills, and to improve individual and team performance.

The parallels between the critical performance levels required of airline teams and medical teams are highlighted. Techniques for improving medical team performance and preventing medical errors are considered. The program combines rigorous aviation and medical case study analysis and group discussions, with an emphasis on examining all of the relevant causative and contributory factors, such as organizational errors, human performance limitations, and authority gradients. This examination leads to the course participants developing error prevention techniques and strategies, as well as critical decision-making skills and improved communication abilities.

The unique aspect of this program is the dedicated use of a full-motion airline flight simulator. Participants will have the rare opportunity to put their newly developed team management skills into practice during realistic operational flight in a full-motion airline training flight simulator. The program uses several scenarios in the simulator, which is crewed by pilots and program participants. Medical team members can witness all of the principles of good CRM by observing how professional airline crews handle critical in-flight emergency situations. Furthermore, in some of the more novel scenarios, medical team members have the opportunity to fly the simulator and assist the crew as valuable flight deck crew members, thus putting their TRM training to good effect.

Conclusion

The underlying emphasis of the program is the enhancement of team performance in the medical environment. All members of a medical team need to work together as a well-structured, well-functioning team that has as its main aim patient safety. Failure to act as a well-coordinated team that understands all the issues present and communicates freely and accurately can lead to detrimental consequences for the patient. Medical TRM and error prevention training is designed to prevent such occurrences.

References

Altman, L.K. (1999), Chemotherapy Overdoses Lead to Review of Nurses, *New York Times*, 6 January, 1999.

Bates, D. *et al.* (1997), The Cost of Adverse Drug Events in Hospitalized Patients, *Journal of the American Medical Association*, Vol. 277, No. 4, 307–311.

Campbell, R.D. and Bagshaw, M. (2002), *Human Performance and Limitations in Aviation*, third edn, Oxford: Blackwell Science.

Grogan, E.L., Stiles, R.A., France, D.J., Speroff, T. and Morris, J.A. Jr *et al.* (1996), The Impact of Aviation-Based Teamwork Training on the Attitudes of Health Professionals, *Journal of the Amercian College of Surgeons*, **199,** 6, 843–48.

Helmreich, R.L. (2000), On Error Management: Lessons from Aviation, *British Medical Journal*, 2000, No. 320, 781–785. [DOI: 10.1136/bmj.320.7237.781]

Helmreich, R.L. and Merritt, A.C. (2001), What Surgeons Can Learn from Pilots, *Flight Safety Australia*, **5**, 58.

Helmreich, R.L., Merritt, A.C. and Wilhelm, J.A. (1999), The Evolution of Crew Resource Management Training in Commercial Aviation, *Journal of Aviation Psychology*, **9,** 1:19–32.

Hugh, T.B. (2002), New Strategies to Prevent Laparoscopic Bile Duct Injury – Surgeons Can Learn from Pilots, *Surgery 2002*, No. 132, 826–835.

Kohn, L.T., Corrigan, J.M. and Donaldson, M.S. (2000), *To Err is Human: Building a Safer Health System*, Washington, DC: National Academies Publishing.

Leape, L.L. (1994), Error in Medicine, *Journal of the American Medical Association*, 1994, Vol. 272, No. 23, 1851–1857. [DOI: 10.1001/jama.272.23.1851]

Leape, L.L., Brennan, T.A., Laird, N.M., Lawthers, A.G. and Localio, A.R. *et al.* (1991), Incidence of Adverse Events and Negligence in Hospitalized Patients: Results of the Harvard Medical Practice Study II, *NEJM*, 1991, No. 324, 377–

384.

Reason, J. (2000), Human Error: Models and Management, *British Medical Journal*, 2000, No. 320, 768–770.

Sexton, J.B., Thomas, E.J. and Helmreich, R.L. (2000), Error, Stress, and Teamwork in Medicine and Aviation: Cross Sectional Surveys, *British Medical Journal*, 2000, No. 320, 745–749. [DOI: 10.1136/bmj.320.7237.745]

Zhan, C. and Miller, M.R. (2003), Excess length of stay, Charges, and Mortality Attributable to Medical Injuries during Hospitalization, *Journal of the American Medical Association*, 2003, No. 290, 1868–1874.

Chapter 25

Driver Distraction: Review of the Literature and Suggestions for Countermeasure Development

Kristie L. Young and Michael A. Regan
Monash University Accident Research Centre

There is converging evidence that certain activities, objects and events inside and outside the motor vehicle can, and do, distract drivers, leading to degraded driving performance, increased crash risk and crashes. Findings from a recent US study suggest that up to 80 per cent of road crashes are due to driver inattention, of which distraction is a major factor. This chapter discusses what is known about driver distraction, what it being done to manage it in Australia, and what should be done to limit its detrimental impact on driver performance and safety.

Introduction

Despite the complexity of driving, drivers commonly engage in a range of non driving-related activities. Some of these, such as talking to passengers, have long been considered part of everyday driving. Other secondary activities, such as using mobile phones, have recently found their way into the vehicle cockpit. Any activity that competes for the driver's attention while driving has the potential to degrade driving performance and compromise road safety.

Driver distraction occurs when a driver engages, willingly or unwillingly, in a secondary activity which interferes with performance of the primary driving task. There are three main mechanisms of distraction: visual distraction, whereby the secondary activity can take the driver's eyes off the road; attentional distraction, whereby the driver's attention is taken away from the driving task; and physical interference, whereby secondary activities physically interfere with control of the vehicle (Regan, 2005).

There is converging evidence that driver distraction is a significant road safety issue and as more communication, entertainment and driver assistance systems proliferate the vehicle market, the incidence of distraction-related crashes is expected to escalate. In North America, Europe and Japan, driver distraction is a priority issue in road safety. However, the significance of driver distraction as road safety issue has only recently been recognized in Australia.

This chapter outlines recent and emerging developments in vehicle technologies and their potential impact on driving performance. It also discusses what is currently being done to manage driver distraction in Australia and describes various countermeasures that could be adopted to limit its impact on driver safety.

Sources of driver distraction and their impact on driving

There are many devices inside the vehicle that have potential to distract drivers (Young, Regan and Hammer, 2003; Regan, 2005). These include mobile phones, route guidance systems, radios, CD players, email and DVD players. In addition, there are many everyday activities undertaken inside the vehicle that can distract the driver, including eating and drinking and talking to passengers. The impact of each of these sources of distraction on driving performance and safety is reviewed in the following sections.

Technology-based distractions

Many new in-vehicle technologies exist or are being developed to enhance safety, increase mobility and provide drivers with information and entertainment. However, these technologies also have the potential to distract drivers and, in doing so, degrade driving performance and safety. As more in-vehicle devices proliferate the market, there has been growing concern regarding the safety implications of using such devices while driving. In response, a rapidly growing body of research has examined the impact of devices, particularly mobile phones, on driving performance (Young and Regan, 2005).

Mobile phones

A large body of knowledge exists regarding the impact of mobile phone use on driver performance and safety. This knowledge is limited, however, to that relating to use of phones to converse and, to a lesser extent, to send text messages.

A wide range of driving performance measures is adversely affected while using a mobile phone. As a broad overview, research has collectively shown that using a mobile phone while driving:

- Impairs drivers' ability to maintain the correct lane position
- Impairs drivers' ability to maintain an appropriate and predictable speed
- Results in longer reaction times to detect and respond to unexpected events
- Results in drivers missing traffic signals
- Reduces the functional visual field of view, which has been shown to be correlated with increased crash involvement
- Results in shorter following distances to vehicles in front
- Results in people accepting gaps in traffic streams that are not large enough
- Increases mental workload, resulting in higher levels of stress and frustration
- Encourages drivers to look straight ahead rather than scanning around the

road ahead, and
- Reduces drivers' awareness of what is happening around them in time and space.

Similar impairments have been demonstrated for hand-held and hands-free phones, refuting the widespread belief that conversing on hands-free phones is safer than hand-held phones. However, the use of hands-free phones has been shown to be less distracting than hand-held phones for some tasks. Hands-free phones, particularly voice-activated phones, have been shown to be slightly less distracting than hand-held phones when dialling (Schreiner, Blanco and Hankey, 2004). Using a hand-held phone to dial has been shown to be more distracting than conversing on a hand-held phone, but dialling a hands-free phone has a similar impact on driving as conversing on this phone type (Lamble *et al.*, 1999; Tornros and Bolling, 2005).

Only two published studies known to the authors have investigated the impact of text messaging on driving. A small-scale simulator study conducted in Sweden by Kircher and colleagues (2004) found that retrieving text messages increased braking reaction times to a motorcycle hazard, but found no other effects. Most recently, MUARC conducted a simulator study to examine the effects of both sending and retrieving text messages on the driving performance of young novice drivers (Hosking, Young and Regan, 2005). The study found that retrieving and, in particular, sending text messages adversely affects driving performance. When sending and retrieving text messages, the amount of time drivers spent with their eyes off the road increased by 400 per cent. Lane position variability also increased when text messaging and drivers made a greater number of lane excursions and incorrect lane changes when retrieving and sending text messages. The results did, however, reveal that some drivers attempted to compensate for these impairments by increasing their following distance.

Route guidance

In-vehicle route guidance systems are designed to guide drivers along the most direct route to a particular destination. Research on route guidance systems suggests that the extent to which they distract drivers depends on the mode of destination information entry and presentation of navigation instructions they employ. Generally, entering destination information is believed to be the most distracting task associated with the use of a route guidance system, however use of voice input technology can reduce the distraction associated with this task (Tijerina, Parmer and Goodman, 1998; Tsimhoni, Smith and Green, 2004). Similarly, route guidance systems that present navigation instructions using voice output are less distracting and more usable than systems that only present the information visually on a display. Turn-by-turn route guidance instructions are also deemed more acceptable and usable by drivers than more complex holistic maps, particularly when coupled with voice-activation. Finally, well-designed route guidance systems have been found to be less distracting than using a paper map (Dingus *et al.*, 1995; Srinivasan and Jovanis, 1997).

Email

Only two known studies have examined the effects on driving performance of retrieving, reading and responding to email messages (Lee *et al.*, 2001; Jamson *et al.*, 2004). These studies revealed that interacting with a speech-based email system increases drivers' reaction time to a braking lead vehicle by up to 30 per cent, reduces the number of corrective steering wheel movements made and also increases drivers' subjective workload. Drivers do, however, appear to compensate for the increased workload associated with using an email system by adopting longer following distances to vehicles ahead.

Radios and CD players

Despite being equipped to almost every car on the road, surprisingly little research has directly examined the distracting effects of interacting with, or listening to, a car radio. The little research that does exist, however, suggests that tuning a radio while driving appears to have a detrimental effect on driving performance, particularly for inexperienced drivers. Wilkman, Nieminen and Summala (1998) found that drivers spent more time looking away from the road when tuning a radio than when dialling a mobile phone, which adversely affected lane control. Another study by Horberry *et al.* (2003) found that tuning the radio resulted in increased subjective estimates of workload, degraded speed control and delayed responses to unexpected hazards. Interestingly, research also suggests that simply listening to radio broadcasts while driving can impair driving, by degrading lane-keeping performance (Janke *et al.*, 1994).

The distracting effects of interacting with CD players while driving have also been examined. Generally, research has found that the tasks involved in using a CD player (e.g., selecting and inserting CDs, changing tracks) results in poor lane keeping ability, more glances away from the road and greater variation in speed control than dialling a mobile phone (Jenness *et al.*, 2002). However, there is some evidence that the use of voice-control may minimize the distraction associated with using CD players (Gartner, Konig and Wittig, 2002).

Non technology-based distractions

Drivers often engage in a number of 'everyday' activities, which have the potential to distract them from the driving task. The research examining the distracting nature of non technology-based activities has focused on eating, smoking and passengers.

Eating

Despite the frequency with which drivers eat and drink in the vehicle, surprisingly little research has examined the impact of these activities on driving performance. The one study available suggests that, compared to driving with no distractions, eating is associated with a greater number of lane deviations and 'minimum speed violations', whereby drivers reduce their speed 10 mph below the limit (Jenness *et al.*, 2002).

Smoking

No performance studies have directly examined the impact on driving of smoking. Rather, research has focused on examining the increased crash risk associated with smoking. Generally, research has found that smokers have a 1.5 times greater risk of being involved in a crash than non-smokers and smoking while driving further increases this risk (Christie, 1991; Violanti and Marshall, 1996). A range of explanations for the association between smoking and increased crash risk has been advanced, ranging from smoking being a physical distraction to decrements in driving performance being due to high levels of carbon-monoxide intake. The causal mechanisms underlying this increased risk are unclear.

Passengers

There is evidence that passengers can distract drivers. It appears that younger drivers are more vulnerable to the effects of distraction from passengers than experienced drivers, and that crash risk increases significantly for young drivers when they carry one or more young passengers (Regan and Mitsopoulos, 2001; Williams, 2003). Very little is known about the actual interactions between drivers and passengers that cause distraction, how this affects driving performance, and why more experienced drivers are better able to deal with this distraction. There is evidence, however, that talking to a passenger is distracting, although less distracting than talking on a mobile phone (Consiglio *et al.*, 2003).

Impact of distraction on crash risk

A large body of evidence demonstrates that interacting with in-vehicle devices while driving can impair driving performance on a number of safety critical measures. But does this degradation in driving performance translate into an increase in crashes and crash risk? A number of studies have been conducted to examine this question.

Research by the National Highway Traffic Safety Administration (NHTSA) estimates that driver inattention, in its various forms, contributes to approximately 25 per cent of police-reported crashes. Driver distraction is one form of driver inattention and is claimed to be a contributing factor in over half of inattention crashes (Stutts *et al.*, 2001).

In relation to mobile phones, epidemiological studies have found that using a mobile phone while driving increases crash risk by anywhere between 4 and 9 times, regardless of whether the phone is hand-held or hands-free, and that the risk is greater for inexperienced drivers (Redelmeier and Tibshirani, 1997; Violanti, 1998; McEvoy *et al.*, 2005). This increase in crash risk is equivalent to driving with a Blood Alcohol Concentration (BAC) of about 0.08. Crash-based studies have also found that, of all crashes in which in-vehicle distraction is a contributing factor, 1.5 to 12 per cent can be attributed to mobile phones (Stutts *et al.*, 2001; Gordon, 2005).

There is, however, very little research examining the impact of interacting with devices other than mobile phones or performing everyday activities (*e.g.*, talking to passengers) on crash risk. For instance, despite the amount of research that has been undertaken to determine the impact of use of route navigation systems on driving performance, there exists no data that links use of these devices to crashes or increased/decreased crash risk. There is also no data on the impact on crash risk of accessing email or using Personal Digital Assistants (PDAs).

There is very little evidence linking radio or CD players to crashes, and no known research evidence linking use of these devices to increased crash risk. Stutts *et al.* (2001) found that, of the crashes examined, 8.3 per cent were the result of the driver being distracted by some event, object or activity inside or outside the vehicle, and 11.4 per cent of these crashes occurred when the driver was adjusting the radio, audiocassette or CD player. No individual data was reported for radio use.

There is some evidence from epidemiological research that smokers and young novice drivers who carry their peers as passengers, are also at increased risk of crashing. Young novice drivers appear to be up to 5 times more likely to crash if they carry two or more friends as passengers (Regan and Mitsopoulos, 2001; Williams, 2001). Smoking has been found to increase crash risk by up to 1.5 times (Brison, 1990). In both cases, distraction has been cited as a contributory factor. Research also suggests that eating and drinking account for anywhere between 1.7 and 4.2 per cent of distraction-related crashes.

Factors that mediate driver's vulnerability to distraction

The potential for a non-driving task to distract a driver is determined by the complex interaction of a number of factors including task complexity, current driving demands, driver experience and skill and the willingness of the driver to engage in the task. A non-driving task that distracts drivers and degrades driving in one situation may not do so in another situation and, similarly, non-driving tasks may differentially affect drivers from different driving populations.

Research has shown that the design of a device, the complexity and/or emotionality of the secondary task being performed, the complexity of the driving environment and driver characteristics, such as age and driving experience level, can all influence the potential for non-driving tasks to distract drivers. Generally, this research has found that as the difficulty of the secondary and/or driving tasks increases, the potential for the task to degrade driving performance also increases. Phone conversations which are complex or emotional have a greater detrimental impact of driving than less complex conversations (Harbluk, Noy, Eizenman, 2002). Similarly, performing a secondary task while driving in adverse weather conditions or heavy traffic has been shown to impair driving to a greater extent than when driving in good weather or less traffic (Cooper and Zheng, 2002; Strayer, Drews, Johnston, 2003).

Older drivers and young novice drivers have also been shown to be more susceptible to the distracting effects of engaging in secondary tasks while driving than experienced or middle-aged drivers. Young novice drivers are more vulnerable because they have not yet automated many driving activities, and hence have less spare attentional capacity to devote to secondary tasks. They are also probably less effective in self-regulating their driving performances across tasks. Older drivers, on the other hand, require more glances at mobile phones and other devices to read information, require more time to complete tasks, require more time to move their eyes between the road and displays inside the vehicle, and have less attention to distribute between competing tasks.

Driver distraction in Australia: the status quo

In Australia, very little is being done at present, relative to other developed countries, to address the issue of driver distraction. Only two of the Australian Road Rules relate to distraction, and even these relate only to a limited number of in-vehicle devices.

Australian Road Rule 300 states that the driver of a vehicle (except an emergency vehicle or police vehicle) must not use a hand-held mobile phone while the vehicle is moving, or is stationary but not parked. The Rule does not include a CB radio or any other two-way radio.

Australian Road Rule 299 prohibits TV screens and video display units from being seen by drivers while the vehicle is in motion, or stationary but not parked – and the device must not distract other drivers who are nearby. Australian Design Rule 42 (Section 18) states that all visual display units must not obscure the driver's vision, or impede driver or passenger movement in the vehicle and must not increase the risk of occupant injury. This rule also states that, unless a driver's aid, no part of the image on the display should be visible to the driver from the normal driving position.

Victorian Police officers also have discretion under the Road Safety Act (1986) to reprimand drivers who they think are driving dangerously (under Section 64) or carelessly (under Section 65). Section 65 is sometimes used to charge drivers in circumstances where a driver has been distracted and a crash occurs. Occasionally, it is used to prosecute drivers even where a crash does not occur if it can be seen that the driver is distracted. However, Section 65 of the Victorian Road Safety Act appears to be used rarely for this purpose.

Countermeasures to mitigate the effects of driver distraction

There is a lot that can be done to prevent driver distraction from becoming a major road safety problem in Australia.

Data

The current lack of crash data in Australia is preventing an accurate assessment of the number of people being killed and injured in distraction-related crashes. Police report forms, therefore, need to be amended to record data about distracting activities. Many new vehicles are equipped with event data recorders that could also be used to automatically record information about the use of telematics systems (*e.g.*, what controls were being operated just before a crash). This would help to clarify the role of these devices in crashes. In the meantime, regular exposure surveys need to be developed, administered and analysed to determine what, when, where, why, and how drivers engage in distracting activities.

Education

Governments, police, motoring clubs and other relevant agencies should conduct education and publicity campaigns to raise public awareness of the relative dangers associated with engaging in distracting activities, how to minimize the effects of distraction, and the penalties associated with engaging in distracting activities where these exist. Several Australian road authorities and telecommunication companies provide advice on the safe use of mobile phones when driving, however no advice is provided for other potentially distracting devices or activities.

As a priority, the Australian public should be made aware that text messaging is potentially more dangerous than using a hand-held phone, and that hands-free phones are just as risky as hand-held phones. The immediate focus should be on those groups most vulnerable to distraction (*e.g.*, inexperienced drivers).

Training

Learner drivers should be trained in how to safely manage distraction. It needs to be determined at what stage in their training it is most appropriate to expose Learner drivers to distracting activities, such as talking to passengers. Learner drivers should also be trained in how to optimally self-regulate their driving to reduce the effects of distraction, and the optimal modes in which to program and interact with systems. Young drivers also need to be made self-aware and calibrated, through training, of the effects of distraction on their driving performance.

Legislation

There is currently very little regulation in Australia governing the design and use of vehicle technologies that have potential to distract the driver. There is a need to review existing legislation and, where necessary, create new legislation to limit driver exposure to distracting activities. Regulatory measures currently being considered by Transport Canada (2003) that could also be considered in Australia, include regulations that:

- Require manufacturers to follow a specified driver-system integration process when designing and testing new technologies to ensure that ergonomic design issues are properly addressed during the design process

- Require all devices known to be highly distracting – for example manual destination entry for route guidance systems – to be automatically disabled when a vehicle is in motion or travelling above a certain speed
- Require manufacturers to adhere to some or all of the performance requirements specified in North American, Japanese and European human factors and ergonomics guidelines for the design of telematics systems
- Prohibit or limit open system architectures, re-configurable interfaces and the design and number of functions available through multifunction devices, and
- Could be introduced to specifically prohibit the installation of devices known to adversely compromise safety or, alternatively, ban drivers from using such devices (as is the case with hand-held mobile phones).

Enforcement

Intelligent transport system technologies now exist that could significantly enhance the ability of Police to enforce traffic laws. For example, it should be possible to configure mobile phones so that they can only be used if the phone is travelling at less than a particular low speed, or when stationary. These technologies should be developed and used where possible to support enforcement activities.

Vehicle design

One of the most effective courses of action to limit the distraction associated with emerging technologies is to optimize their ergonomic design. There exists one Australian Design Rule that attempts to limit distraction through regulation. Apart from that, Australia is reliant on the automotive industry and its suppliers to develop and apply voluntary safety standards for the ergonomic design of vehicle cockpit systems and interfaces. There should be a requirement for manufacturers to establish that they have met a duty of care obligation before a product is allowed to be used by drivers in a moving vehicle. In particular, relevant Federal and State government authorities should enter into a Memorandum of Understanding with industry which requires the manufacturers of vehicles to implement a driver-system integration design process to minimize the potentially adverse safety effects of new and existing technologies entering the vehicle. Relevant Federal and State government authorities should also support the development by local industry of procedures and standards for testing the level of distraction imposed by new technologies, and work with stakeholders to develop tools and techniques for measuring driver distraction and defining criteria and limits on distraction from new devices.

Research

There are a number of priority areas for research on driver distraction. Areas in which our knowledge is particular limited are: driver exposure to distraction; the self-regulatory strategies that drivers use to cope with distraction; ergonomic design of the human-machine interface to limit distraction; the quantification of crash risk; the definition and measurement of distraction; identifying levels of performance degradation due to distraction that constitute safety impairment; and estimating the costs and benefits of regulatory approaches to managing the issue.

Conclusions

There is a large and converging body of evidence to suggest that driver distractions of various kinds can and do degrade driving performance and cause crashes. New and emerging in-vehicle technologies have tremendous potential to enhance the safety, mobility and enjoyment of driving. It is important, however, that these systems are ergonomically designed to accommodate driver limitations and capabilities and that any negative effects on driving performance that they might induce, such as distraction, are minimized before system deployment. This chapter outlined a range of measures that can be taken to limit the potentially adverse effects of distraction. In managing driver distraction, however, we need to be practical. It is impractical, for example, to ban drivers from engaging in everyday activities, such as eating and drinking while driving. Researchers and road authorities also need to recognize the positive benefits to users that derive from various in-vehicle technologies, such as the alerting benefits of mobile phones to tired drivers. In Australia, we are fortunate to be at an early enough stage in the evolution of the vehicle cockpit to prevent distraction from becoming a greater problem than it already is.

References

Christie, R. (1991), Smoking and Traffic Accident Involvement: A Review of the Literature, *GR/91-3. VicRoads*, Victoria: Australia.

Consiglio, W., Driscoll, P., Witte, M. and Berg, W.P. (2003), Effect of Cellular Telephone Conversations and other Potential Interference on Reaction Time in a Braking Response, *Accident, Analysis and Prevention*, **35**, 495–500. [PubMed 12729813] [DOI: 10.1016/S0001-4575%2802%2900027-1]

Cooper, P.J. and Zheng, Y. (2002), Turning Gap Acceptance Decision-Making: Impact of Driver Distraction, *Journal of Safety Research*, **33**, 321–335. [PubMed 12404996] [DOI: 10.1016/S0022-4375%2802%2900029-4]

Dingus, T., McGehee, D., Hulse, M., Jahns, S. and Manakkal, N. (1995), Travtrek Evaluation Task C3 – Camera Car Study, Report No. FHWA-RD-94-076. McLean, VA: Office of Safety and Traffic Operations.

Gartner, U., Konig, W. and Wittig, T. (2002), Evaluation of Manual vs. Speech Input when Using a Driver Information System in Real Traffic, On-line Paper, http://ppc.uiowa.edu/driving-asse…nt%20Papers/02_Gartner_Wittig.htm.

Gordon, C. (2005), A Preliminary Examination of Driver Distraction Related Crashes in New Zealand, in eds Faulkes, I. J., Regan, M. A., Brown, J., Stevenson, M. R. and Porter, A., *Driver Distraction: Proceedings of an International Conference on Distracted Driving, Sydney, Australia*:, 2–3 June. Canberra, ACT: Australasian

College of Road Safety.

Harbluk, J.L., Noy, Y.I. and Eizenman, M. (2002), *The Impact of Cognitive Distraction on Driver Visual Behaviour and Vehicle Control (TP No. 13889 E)*, Ottawa, Canada: Transport Canada.

Horberry, T., Anderson, J., Regan, M.A., Triggs, T.J. and Brown, J. (forthcoming), Driver Distraction: the Effects of Concurrent In-Vehicle Tasks, Road Environment Complexity and Age on Driving Performance, *Accident; Analysis and Prevention*, Jan 2006.

Hosking, S., Young, K. and Regan, M. (2005), The Effects of Text Messaging on Young Novice Driver Performance, in *The Proceedings of the International Driver Distraction Conference, 2-3 June*, Australia: Sydney.

Jamson, A., Westerman, S., Hockey, G. and Carsten, O. (2004), Speech-based E-Mail and Driver Behaviour: Effects of an In-Vehicle Message System Interface, *Human Factors*, **46**, No. 4, 625-39.

Jancke, L., Musial, F., Vogt, J. and Kalveram, K.T. (1994), Monitoring Radio Programs and Time of Day Affect Simulated Car Driving Performance, *Perceptual and Motor Skills*, **79**, 484–486.

Jenness, J.W., Lattanzio, R.J., O'Toole, M. and Taylor, N. (2002), Voice-activated Dialling or Eating a Cheeseburger: Which is more Distracting during Simulated Driving?, *Proceedings of the Human Factors and Ergonomics Society 46th Annual Meeting, Pittsburgh, PA*.

Kircher, A., Vogel, K., Tornros, J., Bolling, A., Nilsson, L., Patten, C., Malmstrom, T. and Ceci, R. (2004), *Mobile Telephone Simulator Study*, Linköping, Sweden: Swedish National Road and Transport Research Institute.

Lamble, D., Rajalin, S. and Summala, H. (2002), Mobile Phone Use while Driving: Public Opinions on Restrictions, *Transportation*, **29**, 223–236.

Lee, J.D., Caven, B., Haake, S. and Brown, T.L. (2001), Speech-based Interaction with In-Vehicle Computers: The Effects of Speech-Based E-Mail on Drivers' Attention to the Roadway, *Human Factors*, **45**, 631–639.

McEvoy, S.P., Stevenson, M.R., McCartt, A.T., Woodward, M., Haworth, C., Palamara, P. and Cercarelli, R. (2005), Role of Mobile Phones in Motor Vehicle Crashes Resulting in Hospital Attendance: a Case-Crossover Study, *British Medical Journal*, BMJ, doi:10.1136/bmj.38537.397512.55.

Redelmeier, D.A. and Tibshirani, R.J. (1997), Association between Cellular Telephone Calls and Motor Vehicle Collisions, *The New England Journal of Medicine*, **336**, 453–458.

Regan, M.A. (2005), Driver Distraction: Reflections on the Past, Present and Future, in *Driver Distraction: Proceedings of an International Conference on Distracted Driving, 2–3 June*. Canberra, ACT eds Faulkes, I. J., Regan, M. A., Brown, J., Stevenson, M. R. and Porter, A., Sydney, Australia: Australasian College of Road Safety.

Regan, M.A. and Mitsopoulos, E. (2001), Understanding Passenger Influences on Driver Behaviour: Implications for Road Safety and Recommendations for Countermeasure Development, Report No. 180, Monash University Accident Research Centre, Victoria, Australia.

Regan, M.A., Oxley, J.A., Godley, S.T. and Tingvall, C. (2001), Intelligent Transport

System: Safety and human Factors Issues, Report No. 01/01. Report prepared by the Monash University Accident Research Centre for the Royal Automobile Club of Victoria (RACV).

Schreiner, C., Blanco, M. and Hankey, J.M. (2004), Investigating the Effect of Performing Voice Recognition Tasks on the Detection of forward and Peripheral Events, *Proceedings of the Human Factors and Ergonomics Society 48th Annual Meeting*, Louisiana: New Orleans.

Srinivasan, R. and Jovanis, P.P. (1997), Effect of In-Vehicle Route Guidance Systems on Driver Workload and Choice of Vehicle Speed: Findings from a Driving Simulator Experiment, in *Ergonomics and Safety of Intelligent Driver Interfaces* edited by: Noy, Y. I., New Jersey: Lawrence Erlbaum Associates.

Strayer, D.L., Drews, F.A. and Johnston, W.A. (2003), Cell-phone Induced Failures of Visual Attention during Simulated Driving, *Journal of Experimental Psychology: Applied*, **9**, 23–32.

Stutts, J.C., Reinfurt, D.W., Staplin, L. and Rodgman, E.A. (2001), *The Role of Driver Distraction in Traffic Crashes*, *Report Prepared for* Washington, DC: AAA Foundation for Traffic Safety.

Tijerina, L., Parmer, E. and Goodman, M.J. (1998), Driver Workload Assessment of Route Guidance System Destination Entry while Driving: A Test Track Study, *Proceedings of the 5th ITS World Congress, Seoul, Korea*.

Tornos, J.E.B. and Bolling, A.K. (2005), Mobile Phone Use – Effects of Hand-Held and Hands-Free Phones on Driving Performance, *Accident; Analysis and Prevention*, **37**, 902–909.

Transport Canada (2003), *Strategies for reducing driver distraction from in-vehicle telematics devices: A discussion document*. Transport Canada: Ottawa.

Tsimhoni, O., Smith, D. and Green, P. (2004), Address Entry while Driving: Speech Recognition versus a Touch-Screen Keyboard, *Human Factors*, **46**, 600–610. [PubMed 15709323] [DOI: 10.1518/hfes.46.4.600.56813]

Violanti, J. (1998), Cellular Phones and Fatal Traffic Collisions, *Accident, Analysis and Prevention*, **30**, 519–524.

Violanti, J.M. and Marshall, J.R. (1996), Cellular Phones and Traffic Accidents: an Epidemiological Approach, *Accident; Analysis and Prevention*, **28**.

Williams, A.F. (2001), *Teenage Passengers in Motor Vehicle Crashes: A Summary of Current Research*, Arlington, VA: Insurance Institute for Highway Safety.

Williams, A.F. (2003), Teenage Drivers: Patterns of Risk, *Journal of Safety Research*, Voll. **34**, No. 1, 5–15. [PubMed 12535901] [DOI: 10.1016/S0022-4375%2802%2900075-0]

Young, K. and Regan, M. (2005), Driver Distraction: A Review of the Literature, in eds Faulkes, I. J., Regan, M. A., Brown, J., Stevenson, M. R. and Porter, A., *Driver Distraction: Proceedings of an International Conference on Distracted Driving, 2–3 June.* Canberra, ACT Sydney, Australia: Australasian College of Road Safety.

Young, K., Regan, M. and Hammer, M. (2003), Driver Distraction: A Review of the Literature, Report No. 206, Monash University Accident Research Centre, Australia: Victoria.

Chapter 26

Assessing and Managing Human Factors Risks – Practical Examples in the Australian Rail Context

Julia Clancy
Lloyd's Register

This chapter presents three examples of recent human factors projects in the Australian Rail context. The first example shows how human factors can assist in the decision making process in the selection of a new train protection system by contributing to risk assessments and cost-benefit analyses. The second example illustrates how training needs analyses can be conducted and prioritized according to risk. Finally, the application of human factors in the management of organizational change is discussed, using an example from the rail sector.

Introduction

In the railway context, there is a strong reliance on humans in the system. Train drivers operate complex, heavy equipment to safely transport hundreds of commuters to their destination. Signallers and train controllers must manage the traffic on the network safely and efficiently. Station staff take actions to manage the crowds of passengers, to maintain safety on platforms.

Most of the time, the humans in the system act as a key defence against incidents and accidents due to their ability to make complex decisions and maintain flexibility in a changing situation. This is in contrast to comparatively rigid technical systems that lack the extent of real time flexibility available to humans. However, sometimes people make mistakes. Human error is inevitable. The discipline of human factors is aimed at understanding human behaviour and how we can optimize the performance of humans in the system to maximize safe operations.

Risk is defined as the chance of something happening that has undesirable consequences. Human factors is aimed at minimizing the chance of human failures occurring, by understanding the underlying causes and putting in place strategies to address these. Human failures are human actions that can lead to undesirable events. They include unintentional human errors as well as violations, which are non-compliances with procedures or rules. Although violations are usually intentional, the undesirable event is usually unintentional.

One of the key benefits of integrating human factors into risk assessment is proactive identification and prevention of human failure, whether it be an error or violation. As with all risk management, the level and depth of analysis and the time and resources should be commensurate with the level of risk involved.

In this chapter, three examples of assessing and managing human factors risks are presented. They are all examples in the Australian rail context.

Example 1: train protection system

A rail organization was considering a number of train protection options. These options varied in their design, for example, speed tracking automatic devices *versus* in-cab warning alerts. The basic principle of the system in all options was to initiate a brake application if a train is detected to be at risk of passing a signal at danger (SPAD) and / or exceeding the speed limit in a particular location. The design options varied in the degree of automation *vs.* driver involvement.

By including human factors in design decisions, the following benefits can be achieved:

- Improving the reliability of a socio-technical system
- Increased user acceptance
- Getting the design right the first time
- Designing or selecting a system that meets users' needs
- Reduced costs, and
- Reduced potential for human error / human violation.

A preliminary human factors assessment was conducted to identify and evaluate the key risks from the human factors perspective. The system design options were studied to:

- Examine the extent and nature of the human interaction with the various system design options. To what extent is human intervention possible and what is the nature of this?
- Identify the specific human factors that are applicable (for example, monitoring, alertness, reliance on automation and complacency) and the potential human failures that might arise, and
- Provide a semi-quantitative assessment of the level of risk that the human factors issues pose, including empirical evidence where available.

From a risk perspective, there are two issues that are important to assess when selecting an optimal design: system design and system availability.

System demand

Will the introduction of the system result in a change in behaviour that could result in an increased demand on the system to mitigate risk? For example, risk homeostasis theory (Wilde, 2001) suggests that individual has an inbuilt target level of acceptable risk which does not change. This level varies between individuals. When the level of acceptable risk in one part of the individual's life changes, there will be a corresponding rise/drop in acceptable risk elsewhere. The same is true of larger human systems (*e.g.*, a population of drivers). For example, in a Munich taxicab study (Aschenbrenner and Biehl, 1994), half of a fleet of cabs was equipped with antilock braking system (ABS) brakes, while the other half had older brake systems. The accident rate for both types of car (ABS and non-ABS) remained the same, because ABS-car drivers took more risks, assuming that ABS would 'take care of them'. They raised their risk taking, assuming the ABS would then lower the real risks, leaving their 'target level' of risk unchanged. The non-ABS drivers drove the same way, thinking that they had to be more careful, since ABS would not be there to help in case of a dangerous situation.

System availability – When the system is called upon, how likely is it to be effective; that is, operate as the design intended. While this will depend to some extent on the reliability of the system components, it will also vary depending on human behaviour. This will be particularly the case the more that the system is reliant on manual operation as opposed to being automated.

The human factors risks from the perspective of (1) system demand, and (2) system availability, were identified. This was achieved by reviewing the system design options in detail with appropriate subject-matter experts and reviewing existing human factors studies of similar systems where available. The causes and consequences of the human factors risks were also identified. The tools used to identify the human factors risks included the following:

1. Identification of the tasks involved in operation of each system design option:
   - Tasks and subtasks
   - Mental (cognitive) processes involved in the task, and
   - Physical activities associated with the task.
2. Identification of potential human failures:

Step 1 Identify potential failures of the system at each stage of the cognitive process of the user. That is, what could go wrong during:

- Information detection, *e.g.*, failure to pay attention to the appropriate information
- Information processing and interpretation, e.g. failure to interpret the information correctly
- Decision making, *e.g.*, failure to select the appropriate action to take, and
- Action execution, *e.g.*, failure to execute the action correctly.

Step 2 Identify how the failure could occur using a set of human failure guideword prompts. In this case, the Human HAZOP (Hazard and Operability) (Kirwan and Ainsworth, 1994) tool was applied.

Some of the human factors issues identified are listed here.

Decreased monitoring and complacency Decreased monitoring and vigilance due to a perception that the new train protection system will prevent an accident. Unintentional 'complacency creep' may occur where the drivers may become over reliant and overly 'trusting' of the 'back-up' system. This may result in less attention being paid to signalling or speed information. Evidence from the aviation industry suggests that 77 per cent of incidents in which over-reliance on automation were suspected involved a failure in monitoring (Mosier, Skitka and Korte, 1994).

Intentional risk taking Intentionally, pushing the boundaries of safe operation, e.g., driving faster or decreased vigilance towards signal sighting, because of the knowledge that the 'system will protect me'. There are several decision biases which contribute to this:

- Risk perception: I do not perceive there to be a significant risk because 'the train protection system will protect me', and
- Risk tolerance: There is a conscious change in behaviour, for example, an increase in speed, which is seen to be tolerable.

Non-acceptance / annoyance / habitual response Some of the design options involved varying degrees of in-cab warning alerts. The design of the warnings may not strike the appropriate balance of capturing attention *versus* the warnings being overly irritating. This can lead to drivers becoming excessively irritated with the system leading to attempted circumvention / dismantling of the system or deliberately ignoring the warning. Alternatively, it can lead to habitual conditioning where a driver habitually responds to the warning alert but does not register or process the information consciously and adapt behaviour if required (*e.g.*, by reducing speed).

Finally, once the issues had been identified, comparative assessment was undertaken for the various system design options. This involved estimating the likelihood and consequences of the various human factors risks for each system design option. This analysis was undertaken in workshops with a group of subject-matter experts including current train drivers.

The comparative assessment enabled each option to be rated according to the two key risk factors described above: (1) system demand; how likely is it that the system functionality (*e.g.*, train protection) will be required, and (2) system availability: how effective is the system in doing the job for which it was designed.

This information was used as a decision aid in determining the most appropriate train protection system from the human factors perspective. It enabled decision makers to understand which system would be most human error resilient and violation resistant. The information assisted in understanding which system would be most likely to be accepted by users (*e.g.*, drivers). Importantly, all design decisions are based on trade-offs amongst competing variables. That is, what you gain with one decision, you may lose with another. Therefore, this information made those trade-offs transparent and gave information to decision makers to make the most informed, risk-based judgement.

Example 2: risk-based training needs analysis

All rail organizations spend a large amount of resources and time in the training and competence assurance of their personnel. However, there are varying degrees of comfort in relation to whether personnel are actually trained adequately for the tasks that they are required to perform, particularly those personnel in safety critical operational roles, such as train drivers or signallers. There is a danger that organizations 'over-train' on skills that are of low risk or importance, or 'under-train' in areas that are of critical importance. Over-training can lead to inefficiencies in time and resources as well as taking away from other more important activities. Under-training can lead to an increase in an organizations' risk exposure, particularly where risk management controls are reliant on human behaviour.

Both the Glenbrook enquiry and Waterfall enquiry reports recognized the importance of training in a safety management system. Two key recommendations of the Waterfall report were recommendation 69 and 70. These related to the importance of conducting Task Analysis (TA) and Training Needs Analysis (TNA).

Further to this, there is increasing effort to better understand the risks that rail organizations face and the risk management controls that are, and / or, need to be, put in place to manage these risks. In many operations the human component is still a key defence against incidents and accidents occurring. We cannot eliminate the human from the system. Therefore, we need to be confident that our frontline personnel are equipped with the skills, knowledge and attributes required to be able to perform the required tasks safely. More specifically, we need to train our personnel such that we reduce the likelihood and consequences of human error to As Low As Reasonably Practicable (ALARP).

To achieve this, we need to take a risk-based approach to understanding and addressing training needs. A risk-based approach involves understanding the tasks that are performed in a given role, and understanding the risks associated with 'getting it wrong'. In essence, those tasks with the highest likelihood and / or consequence of not being performed correctly should be prioritized in the organizations' training approach. This risk-based prioritization should be applied not only to initial training but to competence assessment and assurance as well as ongoing refresher training.

The Risk-Based TNA is used to understand why there is a need for training, who needs to be trained, what needs to be trained, the priorities for the training (based on the risk of not training or 'getting it wrong'). Risk-Based TNA also identifies the training gaps and provides training options.

When to apply Risk-Based TNA:

- Creating new job roles
- Changes in roles or structural reorganization
- Changes in systems, processes or equipment
- Proactively for improving job performance, and
- Reactively for responding to an incident where human performance was implicated.

We have developed an approach to risk-based training needs analysis together with our client. It is currently in the process of being validated and refined; however, initial results have indicated that it is a useful and valuable approach to identifying and prioritizing training needs according to risk. It is based on a number of leading practice standards and guidelines, including RS/220 Railway Safety (UK) – Good Practice in Training October 2002 and BR 8,420 The Royal Naval Systems Approach to Training Quality Standard. The basic approach is outlined here.

Step 1 Scoping – Understand 'why' and identify 'who' needs to be trained. It is important to ascertain whether training is required because of a change or whether it is an improvement initiative. The context should be well understood to avoid training being implicated where it might not be the appropriate approach. For example, in many incident investigations, 'training' is listed as a recommended action; however, a redesign of the system might be the more appropriate approach for minimizing future risk. If it is a change, consider what roles are affected. Consider 'trainee' training as well as 'trainer' training that may be required.

Step 2 Task analysis – Understand 'what' needs to be trained. A task analysis should be performed to determine the tasks and activities that are performed in the role. It may be that you are analysing the entire role (for example, for a new role) or you may simply be analysing parts of the role that are affected by a change (for example, a new piece of equipment that the user is required to operate). The task analysis should specify the tasks and subtasks, the technical skills required to perform the task, the knowledge requirements and the performance standards expected. It is also important to identify the non-technical skills required for the task, *e.g.*, communication, vigilance, decision making, and so on. These are usually described as behavioural markers and are often just as critical (sometimes more critical) as technical skills to perform the task correctly, particularly under conditions of time pressure or high workload. Again, in the context of a change, only those tasks that have the potential to be affected by the change need to be explored as a priority.

Step 3 Establish the training priorities – Determine the difficulty, importance and frequency (DIF) for each task. A DIF analysis is a well-established approach for high risk industries. It enables tasks to be prioritized according to three important factors. The first factor – Difficulty – involves estimating how difficult the task is to perform using ratings from subject-matter experts (*e.g.*, personnel with a high level of experience in the role). Difficulty can simply be rated as Yes or No, or you may use a numeric scale if required. The 'importance' factor here should be considered as 'risk' – the likelihood and consequence of 'getting the task wrong'. Ideally, risk is determined from an organization's risk register. This contains information about the risks that the organization faces and the causes and consequences of these risks. It also contains quantitative information regarding the likelihood and consequences based on historical information and / or workshop ratings with knowledgeable personnel. From this, each task can be investigated to determine the ways in which it may go wrong and the associated outcomes. A quantitative risk estimate can then be assigned to each task, along with an understanding of the causes and the risk management controls.

If this information is not available, risk workshops should be held with a range of experienced personnel in order to estimate the risks associated with each task. The organizations' risk matrix should be used to estimate the risk. Once the risk information is obtained, it can be used as the Importance rating for the DIF, where high risk equates to high importance.

Finally, the Frequency component should be estimated in terms of how frequently the task is performed. Again, this is best achieved on a two- or three-point scale.

Once the Difficulty, Importance and Frequency ratings have been obtained, an algorithm can be used to determine the Priority scores. We have used the algorithm outlined in RS/220 Railway Safety (UK) – Good Practice in Training, October 2002. This calculates five priority levels, from 1 (highest priority) to 5 (lowest priority). For example, a high priority task would be one that is difficult to perform and high risk.

Step 4 Gap analysis – At this point, you should have a set of tasks that are required to be performed in the role and the priorities associated with the tasks. It may be the entire set of tasks for that role or a subset related to a change that is occurring. It is important at this point that a comparison is carried out to assess whether training currently exists and in what form for each of the training needs identified; this results in identification of further training curriculum design requirements.

Step 5 Training needs recommendations – Using the training priority results and the gap analysis results, recommendations can be made on a number of variables. These include:

- What needs to be trained?
- Who needs to be trained?
- How long to spend on each of the training components, depending upon the priority (difficulty, risk and frequency)?
- What training media to use? For example, high priority training should use media that are as closely matched as possible to the real setting in which the task is required to be performed. This is known as training fidelity
- The tasks that need to be assessed to be assured of competence, *e.g.*, higher priority elements should be part of the competence assessment and assurance, and
- The tasks that need regular refresher training, *e.g.*, those that are infrequently performed but high risk and difficult to perform.

Example 3: human factors in the management of risk during organizational changes

A change, whether it is an engineering design change or a change to organizational system or operational process, can potentially impact on safe operations. The impact on safety can often be an unpredicted side effect. Most changes have the potential to impact in some way on the people in the system. If the human component is not adequately considered in a change, the risks of the change may not be well understood. It would be difficult to demonstrate that you have reduced the risk to ALARP.

There are many benefits associated with integrating human factors into the change management process. Human factors analysis enables the proactive identification of the personnel affected by the change, particularly those in safety critical roles. It assists in the identification of user needs associated with the change, particularly during the lead-up and transition period following implementation. Human factors analysis of the change can enable the identification and prevention of human failure associated with the change. And finally, it can reduce resistance to change.

One example of the application of human factors in the management of organizational change was the introduction of a new, significantly re-designed standard working timetable at a large rail organization. The introduction of a new timetable has significant potential to introduce risk due to the pervasive nature of the change. It involves a change in the frequency and timing of services across the network and therefore has the ability to impact on virtually all operational roles.

In an organizational change such as this, it is important to identify, as early as possible, the potential human factors risks. In this project, this was achieved by seeking evidence to answer the following key questions.

- Who are the people that are affected by the change?
- How will they be affected by the change? How will their tasks change?
- What is the potential for human failure?
- How can these risks be managed?

Guidewords or prompts can be useful to promote lateral thinking. In this project, two approaches were used. The first approach involved using an organizational change model (Thomas, 2001), which comprises eight key elements of change. These elements included: Work processes, Organizational Structure, Communication, Interrelationships, Rewards, Leadership, Technology and Group Learning. These areas assisted in identifying the types of impacts that the change can have on the various roles within the organization. The second approach involved the creating of a matrix containing a summary list of the key changes occurring, *e.g.*, 'increased stabling at X location,' 'decreased frequency of services on X line,' and so on. This was mapped against a summary list of the key roles within the organization, enabling a change impact assessment to be conducted for each role.

This initial human factors screening approach identified a number of key areas where further human factors studies were required. This included studies such as workload and manning assessments, training needs analyses and assessments of human error potential due to issues such as changes to run numbers. The actions taken and status of these issues were tracked as part of the overall project hazard log.

The benefits of integrating human factors into the management of organizational change such as this example is that it identified a number of critical issues that if not addressed appropriately, could have resulted in delays in project implementation and / or safety exposures during the implementation and transition.

Conclusion

The management of railway system safety cannot ignore the human component. Human factors risks are not as straightforward as technical risks to predict and manage, given that humans are unpredictable. As the human factors discipline continues to mature, we are developing more sophisticated methods for accurately assessing and managing human factors risks. Given the degree of reliance on humans in the railway system, Human Factors is becoming more widely recognized as an integral part of railway safety management.

References

Aschenbrenner, M. and Biehl, B. (1994), Improved Safety through Improved Technical Measures?, Empirical Studies Regarding Risk Compensation Processes in Relation to Anti-Lock Braking Systems. In R. M. Trimpop and G.J.S. Wilde, *Challenges to Accident Prevention: The Issue of Risk Compensation Behaviour*, Groningen, the Netherlands: Styx Publishing.

BR 8,420. The Royal Naval Systems Approach to Training Quality Standard.

Engineering and Safety Management Yellow Book. Application Note 3. Human Error Causes, Consequences and Mitigations. Issue 1.0.

Kirwan, B. and Ainsworth, L. (1994), *A Guide to Practical Human Reliability Assessment*, London: Taylor & Francis.

Mosier, K.L., Skitka, L.J. and Korte, K.J. (1994), Cognitive and Psychological Issues in Flight Crew / Automation Interaction, in *Human Performance in Automated Systems: Current Research and Trends* eds Mouloua, M. and Parasuraman, R., Hillsdale, NJ: Lawrence Erlbaum Associates.

Rail and Safety Standards Board (RSSB), Tracer for the Rail Industry: Rail Specific Human Reliability Assessment Technique for Driving Tasks.

RS/220 (2002) Railway Safety UK (2002), Good Practice in Training, October.

Special Commission of Inquiry into the Waterfall Accident, New South Wales, Final Report, January, 2005.

Thomas, S.J. (2001), *Successfully Managing Change in Organizations: a User's Guide*, New York: Industrial Press.

Wilde, G.J.S. (2001), *Target Risk 2: A New Psychology of Safety and Health*. Toronto, ON: PDE Publications.

Chapter 27

Anaesthetic Registrars' Stress Mediators in a Simulated Critical Incident

Kate Fraser, Matthew J.W. Thomas and Renée Petrilli
University of South Australia

Anaesthetic registrars often work shift work hours, administering drugs while fatigued and under stress. This paper presents the results of a simulator-based study designed to investigate mediators of anaesthetic registrars' subjective and objective stress responses during critical incidents. Results indicate that despite fatigue-related impairment on simple reaction time tasks following three consecutive night shifts, subjective and objective stress do not seem to be impacted during the management of a critical incident. Further, other factors such as personality traits and attitudes towards stress also show no significant effect as mediators of stress. The study points to the importance of fatigue management in anaesthesia.

Introduction

Anaesthetists' work involves both stress and fatigue. Stress occurs for several reasons, including the frequent administration of life-affecting drugs, the necessity for rapid action during operations, and the high levels of concentration required during a shift. Variable shifts, long working hours and extended on-call periods are customary in the medical field, causing disrupted circadian rhythm, sleep loss, and fatigue (Howard *et al.*, 2002a; Stoller *et al.*, 2005). Around the clock health care means many anaesthetists must work shift work hours at odds with their normal biological rhythms (Howard *et al.*, 2002a). Research has shown that shift workers are more likely to experience sleep disturbances, reduced alertness, and cognitive impairment compared daytime workers (Rajaratnam and Arendt, 2001). These consequences of shift work can contribute to preventable errors (Åkerstedt, 2003; Folkard and Tucker, 2003).

Up to 80 per cent of incidents in anaesthesia are thought to be caused by human factors (Fletcher *et al.*, 2002). Furthermore, fatigue related human error is increasingly being recognized as contributing to errors, incidents and accidents in anaesthesia (Howard and Gaba, 1997; Howard *et al.*, 2002b; Murray and Dodds, 2003). Complex drug administration often occurs for anaesthetists 'under conditions of stress' (Abeysekera *et al.*, 2005). Their stress is compounded by the haste in which procedures must be carried out in the operating room (OR). Stress can also result from fatigue caused by the high levels of concentration required in multiple consecutive operations in the daily OR timetable (Abeysekera *et al.*, 2005; Wheeler and Wheeler, 2005). For many anaesthetists, compounded stress from these aspects of their work has become unmanaged and excessive, compromising patient care and safety (Jackson, 1999). However, despite ample evidence that fatigue and stress have a negative effect on performance, research has shown significant numbers of anaesthetists feel they are unaffected by it, and that tiredness does not impact on their performance (Flin *et al.*, 2003). Studies looking at attitudes towards stress, however, have shown that attitude is a good predictor of performance; it indicates the extent to which people will place themselves in circumstances likely to generate accidents or incidents (Sexton, Thomas and Helmreich, 2000). Therefore, investigations into these attitudes may provide insights into better training for those at risk.

In addition to the significance of attitudes towards stress, personality may also be a factor, since research has shown various personality traits are associated with different perceptions of stress and different coping styles (Costa, Somerfield and McCrae, 1996; Arnstein, 1997). Therefore, it may play a role in mediating stress. Personality traits may be important because some tendencies, *e.g.* neuroticism, may inhibit ability to cope well with stress, while others, *e.g.*, extraversion, may assist coping (Ferguson, 2001; McManus, Keeling and Paide, 2004). Although personality traits of anaesthetists have been investigated in several studies (Kluger *et al.*, 2002), a literature search has not uncovered any such study of Australian anaesthetists. Therefore, an investigation into personality and stress responses was included in this study.

The principal aim of the present study was to examine the effects of fatigue on anaesthetists' responses to stress. The secondary aim of the study was to examine the effects of other mediating factors – attitudes towards stress, and personality – on anaesthetists' perceived stress. The results of the study may assist in the possible design and re-structuring of anaesthetists' work rosters, as well as provide insights to assist anaesthetists in their management of stress and fatigue.

Method

Sample

Ten volunteer anaesthetic registrars from the Royal Adelaide Hospital (7 male, 3 female), aged between 26 and 37 years ($M = 33.89$, SD = 2.09), participated in the study, who had been working irregular hours for an average of 7.5 years (SD = 2.69). Years of anaesthetic training varied between 0.5 and 7 years ($M = 3.94$, SD = 1.61).

Materials and measures

Samn-Perelli fatigue scale – A seven-point Samn-Perelli fatigue scale (1982) was used to obtain subjective alertness ratings. Answers ranged from 1 – 'Fully alert, wide awake,' to 7 – 'Completely exhausted, unable to function effectively'. Participants circled the number that corresponded to their level of fatigue before the simulator phase.

Psychomotor vigilance task Objective fatigue level was measured using a psychomotor vigilance task (PVT, Walter Reed version), a 10-minute computerized vigilance test on a hand-held palm pilot which measures response time upon presentation of a visual stimulus (target). Stimuli were presented randomly, in two- to 10-second intervals. Participants pressed the response button as quickly as possible after the visual stimulus was shown. Response speed was then displayed for a 500ms epoch. Registrars were under instruction to avoid anticipating the presentation of the stimulus.

Attitudes towards stress and fatigue Attitudes towards stress and fatigue were measured using the Safety Attitudes Questionnaire: operating room Version (SAQOR), containing 10 statements pertaining specifically to stress and fatigue, four of which have been validated (Sexton *et al.*, undated). Participants read each statement and rated it on a five-point Likert scale.

Personality measures Measures of Neuroticism and Extraversion were obtained from the NEO Five Factor Inventory, a 60-item personality questionnaire. Participants indicated on a five-point Likert scale their degree of agreement with statements. Answers correlated to a number between zero and four, with 12 statements pertaining to each personality trait. Raw scores were converted into T scores using the Profile Form S (Adult), provided with the questionnaire, to determine which trait the participants scored higher on.

Ambulatory measures Measures of heart rate and RMSSD (heart rate variability) were measured by the Vrÿe Universiteit Ambulatory Monitoring System (Ams) (version 4.6, TD-FPP, Vrÿe Universiteit, Amsterdam). Alcohol swabs were used to clean the skin, and six electrodes were positioned to measure ECG and impedance cardiogram (ICG) activity.

Interviews A semi-structured interview based on the Critical Decision Method (CDM; Klein, Calderwood and Macgregor, 1989) followed each simulator session, to gain information about participants' physiological and psychological stress responses during the clinical scenario.

Visual analogue scale and stress management techniques

Concurrently, with the interview, subjective ratings of stress and stress management were obtained using a bipolar linear visual analogue scale (VAS). One comprised the question: 'How stressed did you feel during [the critical] event?' . Participants marked a 100 mm line, with anchors 'not at all stressed' at the left end and 'very stressed' at the right. The second asked: 'How well do you feel you managed your stress during this event?' Participants again marked a 100 mm line, this time with anchors 'not at all well' at the left, and 'very well' at the right.

Following these, registrars indicated which of a list of 10 stress-management techniques they employ in everyday life, which included the option 'No stress management strategy'.

Simulator scenarios Using the METI patient simulator at the RAH in collaboration with a consultant anaesthetist, two scenarios were designed to be sufficiently difficult to potentially elicit some sort of stress response in participants. Scenarios took place in a real OR, and real monitoring equipment and drug ampoules were used. For further realism, an anaesthetic nurse was present, with the researchers as scrub nurse and surgeon.

Each scenario comprised a technical sabotage and a critical incident of a clinical nature. The timing of these varied slightly depending on how long the registrar took to begin the rapid sequence induction (securing the patient's airway), how long it took to detect the sabotage, and measures were taken if the sabotage was not detected (*e.g.*, the anaesthetic nurse would 'notice' the problem and fix it). Protocols for the rested and non-rested conditions were the same, with scenarios counterbalanced to avoid learning effects.

A trained observer assessed whether participants had detected the sabotage and remedied the leak, and whether the critical incident was correctly diagnosed and treated. Times were also recorded as to how long registrars took to treat the problem after they had made a diagnosis.

Protocol

The study utilized a repeated measures crossover design with two experimental conditions: (i) rested condition, and (ii) non-rested condition. Participants were sent a Palm Pilot seven days before participating in their first simulator session to obtain baseline scores of PVT performance. They were required to complete a PVT at 10:00h, 13:30h and 17:30h on a rested day to obtain baseline scores of sustained alertness.

At 08:00h, participants arrived at the simulator unit following either three consecutive night shifts or three consecutive days off (see Figure 27.1). Participants did not consume coffee before the session, which was broken up into the pre-simulator phase, simulator phase, and post-simulator phase. For both conditions, participants indicated their subjective alertness levels and completed a PVT before the simulator phase. Following this, a 10-minute baseline measure of heart rate and RMSSD was recorded for comparison to measures of these during the simulator scenario.

The simulator phase itself consisted of either of the straightforward clinical scenarios of approximately 30 minutes duration. The participant was briefed by the consultant anaesthetist, and informed to verbalize any observations and any drugs that were administered during the scenario. As intubation of the patient was performed, the anaesthetic nurse sabotaged the equipment. Once this was detected and resolved, the onset of the critical incident was programmed into the patient simulator via the control computer. After the participant had treated/attempted to treat the critical incident, the scenario ended.

The post-simulator phase followed, comprising subjective alertness ratings, a PVT, and a semi-structured interview regarding the registrar's decision making, stress and fatigue during the scenario. Two VAS were completed, and any general stress management techniques used by the registrar were recorded. The two conditions were separated by at least a 10-day interval.

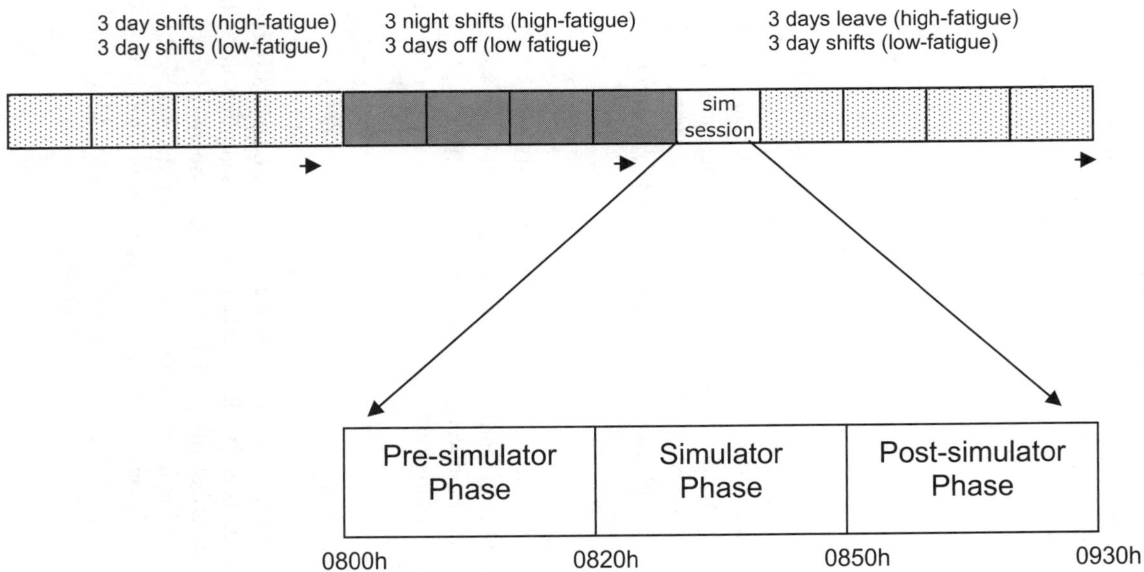

Figure 27.1 Experimental protocol with three night shifts or three days' leave prior to simulator session

Statistical analyses

T-tests were conducted to compare the rested and non-rested sleep and subjective alertness data, and to look at differences in PVT performance between the two conditions, relative to baseline performance. Repeated measures ANOVA were used to determine the effects of fatigue on RMSSD, compare the effects of fatigue on RMSSD and on heart rate between the rested and non-rested conditions, and determine whether attitudes towards stress or personality mediated the effects of fatigue on HRV. Correlation analyses were used to investigate the relationship between personality traits and perceived stress.

Results

Sleep

Paired samples T-tests showed that neither total sleep time for the 24-hour period leading up to the simulator session (M_{diff} = 0.17 hour, t_9 = 0.08, p = 0.94) nor for the 48-hour period before the simulator session (M_{diff} = 2.76 hour, t_9 = 1.62, p = 0.14) differed significantly between rested and non-rested participants. Results of a further paired samples T-test showed that prior wake data for the rested condition and non-rested condition were not significantly different (M_{diff} = –1.71 hour, t_9 = –1.05, p = 0.32).

Subjective alertness scores

Paired samples T-test results indicated that subjective alertness scores between the rested and non-rested conditions were significantly different (M_{diff} = –2.4, t_9 = –7.9, p <0.00), demonstrating that registrars felt more fatigued after working three consecutive night shifts.

Psychomotor vigilance task

Repeated measures ANOVA carried out on the data revealed that differences between the conditions were unlikely to have arisen due to sampling error ($F_{1,18}$ = 8.39, p = 0.004). An overall effect size of 0.48 (Partial Eta squared) showed that almost 50 per cent of the variation in error scores can be accounted for by differing fatigue levels. Post-hoc comparisons revealed that mean reaction time was significantly slower before the non-rested simulation than at baseline (M_{diff} = 0.44, t_9 = 2.53, p = 0.032).

Objective stress measures

Separate repeated measures ANOVAS examined RMSSD during the simulation as compared with baseline RMSSD in two conditions. Time had a significant effect on RMSSD in both the rested ($F_{1,18}$ = 4.65, p = 0.003) and non-rested ($F_{1,18}$ = 6.96, p = 0.004) conditions. Post-hoc comparisons revealed RMSSD was significantly lower at every other point during the simulation than at baseline (all $p < 0.05$) for both conditions.

An additional repeated measure ANOVA compared RMSSD between the two conditions. Time had a significant effect on RMSSD ($F_{1,18} = 9.69$, $p < 0.000$), however, condition did not. A visual inspection of the data shows HRV decreases sharply after the baseline, and decreases further at the time of the critical incident in the rested condition, but increases slightly at time of critical incident in the non-rested condition (see Figure 27.2). Post-hoc analyses revealed that RMSSD was significantly different at baseline, as compared to RMSSD at every other point during the simulation.

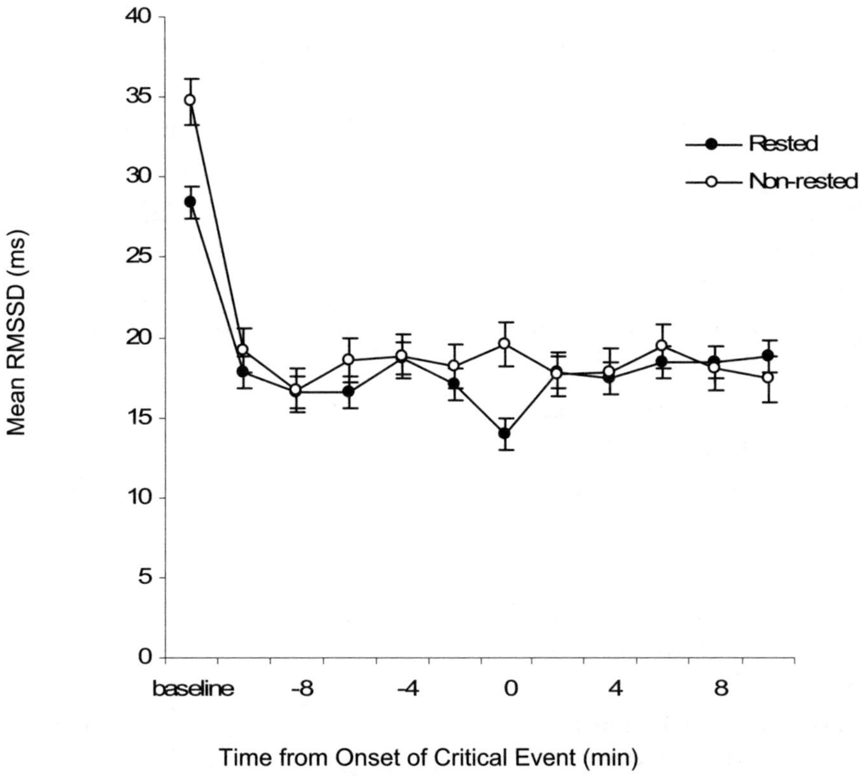

Figure 27.2 Mean RMSSD (±SEM) for baseline and 10 minutes prior to and post onset of the critical event for the rested condition and non-rested condition

Paired samples T-tests showed neither baseline RMSSD (M_{diff} *f* = –6.26, t_9 = –1.12, p = 0.29), nor RMSSD at time of the critical incident (M_{diff} = –5.7, t_9 = –1.26, p = 0.24) were significantly different for the rested and non-rested conditions.

Repeated measures ANOVA results comparing heart rate across the two conditions also revealed time had a significant effect ($F_{1,18} = 20.83$, p = 0.000), and similarly, condition did not. However, visual inspection of the data indicates mean heart rate appears to have been consistently lower in the non-rested condition either side of the critical event.

Personality measures and attitudes towards stress and fatigue

Table 27.1 summarizes registrars' responses to the stress- and fatigue-related questions from the SAQOR. A repeated measures ANOVA comparing RMSSD data across the two conditions revealed no significant difference between the registrars who scored more highly on the SAQOR and those who scored lower ($F_{1,18} = 0.128$, $p = 0.73$). A separate repeated measures ANOVA was conducted to determine whether a significant difference would be found on including the remaining six statements. However, results again showed no significant difference in RMSSD between registrars with higher scores and those with lower scores ($F_{1,18} = 1.46$, $p = 0.262$).

Table 27.1 Summary of attitudes to stress and fatigue (% response)

Items	Disagree	Neutral	Agree
When my workload becomes excessive, my performance is impaired.*	10	0	90
I am more likely to make errors in tense or hostile situations.*	0	0	100
Fatigue impairs my performance during emergency situations.*	10	40	50
I am less effective at work when fatigued.*	0	0	100
Stress from personal problems adversely affects my performance.	30	10	60
Truly professional personnel can leave personal problems behind When working.	20	10	70
I feel fatigued when I get up in the morning and have to face another Day on the job.	20	10	70
I feel burned out from my work.	50	40	10
I feel I am working too hard on my job.	60	30	10
During emergency situations (*e.g.* emergency resuscitations), my performance is not affected by working with inexperienced or less capable personnel.	50	30	20

*Asterisked items have been statistically validated (Sexton *et al.*, n.d.).

Regarding personality traits, a correlation matrix revealed a strong, positive correlation between extraversion and perceived stress in the rested condition ($r = 0.695$, $p < 0.05$), but no correlation in the non-rested condition. A separate repeated measures ANOVA showed no significant effect of higher neuroticism or higher extraversion on registrars' perceived stress between the rested and non-rested conditions ($F_{1,18} = 1.18$, $p = 0.31$).

Interviews and subjective stress

A paired samples T-test revealed no significant difference in perceived stress VAS scores between the conditions (rested $M = 51.3$, SD = 26.98; non-rested $M = 49$, SD = 21.47). Moreover, participants indicated their stress management during the simulator session was more than sufficient for the level of perceived stress, as shown by mean stress management VAS scores (rested $M = 60.3$, SD = 19.13; non-rested $M = 63.3$, SD = 9.53).

A qualitative analysis of post-simulator interviews revealed anaesthetic registrars experience a range of physiological and psychological symptoms when stressed, with increased heart rate the most reported physiological symptom. For psychological symptoms, a number of participants mentioned increased anxiety, need for concentration, and becoming less tolerant when in a stressful situation. Of the 20 simulator sessions, only six were identified as having elicited some sort of stress.

Several themes regarding stress emerged from the registrars' comments – working with consultants, the simulator, and self-monitoring as a form of stress management in the OR. Several participants highlighted that their stress levels increase when working with certain consultants. One noted that this stress can have a negative impact on performance.

Participants also indicated the simulated scenario was not very stressful because the patient was not real and there were no real-life repercussions from their actions:

> Maybe it's that I know it's not a real patient, maybe it would affect me differently with a real patient.

Self-reassurance, keeping focused on the task at hand, and constantly checking the management of the patient were mentioned by several registrars as ways to combat stress during operations. However one participant noted that, despite experiencing a little psychological stress during the scenario, he did not try to overcome it because 'it seemed appropriate'.

Discussion

Effects of fatigue on performance and stress

Although participants obtained similar amounts of sleep in both conditions, results clearly show that working night shifts impacts upon fatigue levels. Registrars acknowledge they feel significantly more fatigued in the non-rested condition, and the effects of fatigue are also evident from the non-rested PVT results. Furthermore, mean reciprocal reaction time for non-rested participants was 3.52 (standard deviation 0.57), indicating a fatigue-related impairment similar to a blood alcohol level of 0.05 (Dawson and Reid, 1997). This suggests that fatigue contributes to impaired performance. The current results indicate, however, that fatigue does not seem to have a similar effect on stress, whether it is a measure of physiological stress (RMSSD) or perceived stress. Fatigue is known to affect both low-level cognitive skills employed when testing reaction time and high-level cognitive and decision-making skills used during the simulation (Howard *et al.*, 2002a). However, it may be that the PVT is markedly more sensitive to the effects of fatigue. Alternatively, the registrars were well-enough equipped with coping strategies for the incident not to have been a problem for them, and for their stress not to have worsened when they were fatigued. For example, the majority mentioned the technique of heightening concentration to avoid stress.

Indeed, it was noted during the post-simulator interviews that, in general, the scenarios were not stressful, and potential reasons for this were extrapolated from the data. These reasons included the absence of a consultant and the use of a simulator. It also came to light that the participants use a combination of strategies to help them deal with their stress. This suggests a possibility that there were no significant differences in registrars' stress responses between the rested and non-rested conditions because having these coping mechanisms in place reduces their stress levels generally.

Effects of attitude and personality on physiological stress

Registrars from this sample tended to have a realistic view of stress and fatigue in the workplace. In other words, a majority of the participants agreed with statements in the Safety Attitudes Questionnaire, such as 'When my workload becomes excessive, my performance is impaired,' and 'I am more likely to make errors in tense or hostile situations,' and disagreed with the statement 'During emergency situations, my performance is not affected by working with inexperienced or less capable personnel'. However, most of the registrars also agreed that 'Truly professional personnel can leave personal problems behind when working,' suggesting they believe they are to some extent immune to the effects of stress. This did not have an effect on HRV, however, with no significant difference in RMSSD for the rested and non-rested conditions.

Scoring higher on neuroticism as compared to extraversion also did not impact on perceived. However, extraversion did correlate positively with perceived stress in the rested condition, but not in the non-rested condition. This result could be due to effects of fatigue dampening effects of stress during the scenario. That is, the registrars who scored higher on the trait of extraversion than on neuroticism could perceive less stress in the non-rested scenario because higher levels of fatigue may contribute to complacency. Some researchers believe that the issue of consistency of coping with regard to personality is not resolved, and that several questions need further exploration. One such question is whether consistency reflects the influence of personality traits. Thus, another possibility is that participants scoring higher on the trait of openness to experience may be less consistent in their selection of coping strategies (Suls, David and Harvey, 1996). Evidence for variation in coping consistency has been found in studies (Atkinson and Violato, 1994), as well as results indicating that people scoring highly on neuroticism were found to be significantly less consistent in which coping strategies they used. Other possible explanations are that the scenario was insufficiently stressful to evoke any differences or that the registrars had sufficiently effective coping skills to deal with the stress present. However, the effects of stress may be more subtle due to the sample size, in which case a larger sample would have provided a better chance for obtaining significant results.

Future research

It is anticipated that the results from the current study will be built up as part of an ongoing study of fatigue in the medical industry to be used eventually as a comparison for fatigue-related studies in the aviation industry. Given the PVT results, it seems that fatigue did have some impact on performance. Therefore, it might be useful to educate anaesthetists about fatigue-related error. One way to decrease possible impairment from fatigue is to encourage use of self-assessment of competency. The aviation industry has adopted a competency self-assessment ('IMSAFE') that pilots are encouraged to use before flying (Arnstein, 1997; Flin *et al.*, 2003). This method checks for Illness, Medication, Stress, Alcohol, Fatigue, and Eating, and is a routine that could easily be transferred to anaesthesia (Arnstein, 1997), to encourage more vigilance, making self-monitoring a regular part of anaesthetists' daily routines.

Conclusions

The results of this simulator-based study reveal that while low-level cognitive skills, such as those needed for the PVT are vulnerable to the effects of fatigue, the fatigue-related effects of shift work do not impact on the perceived stress, or indeed the heart rate variability of anaesthetic registrars. That is, fatigue seemed to affect some aspects of performance, but did not increase stress in the participants. Since it is possible that the fatigue may have off-set some of the potential stress, it can be concluded that there is still a need for formal education on fatigue-related error in anaesthesia.

Acknowledgements

Thanks to my fellow authors and the Centre for Sleep Research. Many thanks also to Dr Simon Jenkins, Dr Gerald Toh, and all of the registrars at the Royal Adelaide Hospital without whom this study would not have been possible.

References

Abeysekera, A., Bergman, I.J., Kluger, M.T. and Short, T.G. (2005), Drug Error in Anaesthetic Practice: a Review of 896 Reports from the Australian Incident Monitoring Study Database, *Anaesthesia*, **60**, 3, 220–227.

Åkerstedt, T. (2003), Shiftwork and Disturbed Sleep/Wakefulness, *Occupational Medicine*, **53**, 2, 89–94.

Arnstein, F. (1997), Catalogue of Human Error, *British Journal of Anaesthesia*, **79**, 645–656. [PubMed 9422906]

Atkinson, M.J. and Violato, C. (1994), Neuroticism and Coping with Anger: The Trans-Situational Consistency of Coping Responses, *Personality and Individual Differences*, **17**, 6, 769–782.

Costa, P.T., Somerfield, M.R. and McCrae, R.R. (1996), Personality and Coping: A Reconceptualization, in *Handbook of Coping. Theory, Research, Applications* eds Zeidner, M. and Endlers, N. S., New York: John Wiley & Sons.

Dawson, D. and Reid, K. (1997), Fatigue, Alcohol and Performance Impairment, *Nature*, **388**, 6639, 235.

Ferguson, E. (2001), Personality and Coping Traits: A Joint Factor Analysis, *British Journal of Health Psychology*, **6** (4), 311–325. [DOI: 10.1348/135910701169232]

Fletcher, G.C.L., McGeorge, P., Flin, R.H., Glavin, R.J. and Maran, N.J. (2002), The Role of Non-Technical Skills in anaesthesia: a Review of Current Literature, *British Journal of Anaesthesia*, **88** (3), 418–429. [PubMed 11990277] [DOI: 10.1093/bja%2F88.3.418]

Flin, R., Fletcher, G., McGeorge, P., Sutherland, A. and Patey, R. (2003), Anaesthetists' Attitudes to Teamwork and Safety, *Anaesthesia*, **58**, 3, 233–242.

Folkard, S. and Tucker, P. (2003), Shiftwork, Safety and Productivity, *Occupational Medicine*, **53**, 5, 95–101.

Gander, P.H., Merry, A.F., Millar, M.M. and Weller, J. (2000), Hours of work and fatigue-related error: A survey of New Zealand anaesthetists, *Anaesthesia and Intensive Care*, **28**, 178–183.

Howard, S.K. and Gaba, D.M. (1997), Human Performance and Patient Safety, in *Patient Safety in Anaesthetic Practice* eds Morell, R. C. and Eichorn, J. H., New York: Churchill Livingstone.

Howard, S.K., Gaba, D.M., Rosekind, M.R. and Zarcone, V.P. (2002a), The Risks and Implications of Excessive Daytime Sleepiness in Resident physicians, *Academic Medicine*, **77**, 10, 1019–1025. [PubMed 12377678] [DOI: 10.1097/00001888-200210000-00015]

Howard, S.K., Rosekind, M.R., Katz, J.D. and Berry, A.J. (2002b), Fatigue in Anaesthesia: Implications and Strategies for Patient and Provider Safety, *Anesthesiology*, **97**, 5, 1281–1294.

Jackson, S.H. (1999), The Role of Stress in Anaesthetists' Health and Well-Being, *Acta Anaesthesiol Scandinavica*, **43**, 583–602.

Klein, G., Calderwood, R. and Macgregor, D. (1989), Critical Decision Method for Eliciting Knowledge, *IEEE Transactions on Systems, Man, and Cybernetics*, **19**

(3), 462–472. [DOI: 10.1109/21.31053]

Kluger, M.T., Watson, D., Laidlaw, T.M. and Fletcher, T. (2002), Personality Testing and Profiling for Anaesthetic Job Recruitment: Attitudes of Anaesthetic Specialists/ Consultants in New Zealand and Scotland, *Anaesthesia*, **57**, 2, 116–122.

McManus, I.C., Keeling, A. and Paide, E. (2004), Stress, Burnout and Doctors' Attitudes to Work Are Determined by Personality and Learning Style: A twelve year Longitudinal Study of UK Medical Graduates, BMC, *Medicine*, **2**, 29.

Murray, D. and Dodds, C. (2003), The Effect of Sleep Disruption on Performance of Anaesthetists – A Pilot Study, *Anaesthesia*, **58**, 6, 520–525.

Rajaratnam, S.M.W. and Arendt, J. (2001), Health in a 24-h Society, *The Lancet*, **358**, 999–1005. [DOI: 10.1016/S0140-6736%2801%2906108-6]

Samn, S.W. and Perelli, L.P., (1982), 'Estimating aircrew fatigue: a technique with application to airlift operations', Brooks AFB, USAF School of Aerospace Medicine.

Sexton, J.B., Thomas, E.J. and Helmreich, R.L. (2000), Error, Stress, and Teamwork in Medicine and Aviation: Cross Sectional Surveys, *British Medical Journal*, **320**, 7237, 745–749.

Sexton, J.B., Thomas, E.J., Helmreich, R.L., Neilands, T.B., Rowan, K., Vella, K., Boyden, J. and Roberts, P.R. (undated), Frontline Assessments of Healthcare Culture: Safety Attitudes Questionnaire Norms and Psychometric Properties', *Technical Report 04-01*, The University of Texas Center of Excellence for Patient Safety Research and Practice.

Stoller, E.P., Papp, K.K., Aikens, J.E., Erokwu, B. and Strohl, K.P. (2005), Strategies Resident-physicians Use to Manage Sleep Loss and Fatigue, Medical Education, *Online*, **10**, 7.

Suls, J., David, J.P. and Harvey, J.H. (1996), Personality and Coping: Three Generations of Research, *Journal of Personality*, **64** (4), 711–735. [PubMed 8956512] [DOI: 10.1111/j.1467-6494.1996.tb00942.x]

Wheeler, S.J. and Wheeler, D.W. (2005), Medication Errors in Anaesthesia and Critical Care, *Anaesthesia*, **60**, 3, 257–273.

Chapter 28

Trying Before Buying: Human Factors Evaluations of New Medical Technology

J.M. Davies
Department of Anesthesia
University of Calgary, Canada

J.K. Caird
S. Chisholm
Cognitive Ergonomics Research Laboratory
University of Calgary, Canada

Human factors (HF) research has begun to determine how medical technology can be designed to minimize human error and increase safety. Relatively few studies have addressed common usability issues and design solutions of different types of intravenous (IV) infusion and patient controlled analgesia (PCA) pumps. Problems with both types of pumps have contributed to patients receiving the wrong fluid, drug, or dose, sometimes with fatal results. Human factors methodology that can be employed to evaluate both IV and PCA pumps include Heuristic Analysis and Usability Testing. Results of our evaluations showed that all pumps studied had unique strengths and weaknesses affecting their usability and safety. These findings illustrate the importance of pre-purchase HF analysis in the selection of medical devices. However, evaluation of medical devices does not end with purchasing. Rather, there is need for on-going and systematic surveillance, through implementation, use and replacement phases, as well as 'intelligent wariness' whenever medical devices are used in efforts to improve patient care.

Introduction

Medical devices are essential tools of health care. According to Health Canada, the term 'medical device' covers a 'wide range of health or medical instruments used in the treatment, mitigation, diagnosis or prevention of a disease or abnormal physical condition' (Medical Devices, Health Canada 2006). In Canada, 'medical devices' are only for human use (Medical Devices Regulations, Government of Canada 2006). This is in contrast with the definition of 'medical devices' given by the Food and Drug Administration of the United States of America, whereby such devices are 'intended to affect the structure or any function of the body of man or other animals' (US Food and Drug Administration 2006).

No matter whether or not the 'patient' is human or animal, such devices can contribute to injuries if used incorrectly. However, the 'incorrect' use of medical devices is often related to the design of the device (Lin *et al.,* 1998).

When a healthcare institution decides to purchase a medical device, the institution has the opportunity to make this decision based on several factors. These include cost, previous experience with the vendor (for such things as in-service and on-going product support) and potential for contribution to patient harm. Theoretically, purchasing decisions that consider whether or not the design of a medical device meets some minimum levels of safety can be balanced against the more traditional considerations of product cost and support. If a device is purchased that facilitates usability – the making of few or no errors when used – then patient safety is advanced. Ordinarily, product cost and support are the only dimensions considered when medical device are purchased.

This is in contrast to the purchasing requirements for a number of military establishments around the world, which must determine whether or not equipment meets human factors guidelines – or 'designing for human use' (Chapanis, 1995). For example, MIL-STD-1472F (1999) is probably the 'best known human engineering design standard' in the United States and contains the human factors guidelines for the U.S. military. The aim of these guidelines is to:

1. achieve required performance by operator, control, and maintenance personnel
2. minimize skill and personnel requirements and training time
3. achieve required reliability of personnel-equipment/software combinations, and
4. foster design standardization within and among systems (MIL-STD-1472F 2004).

If a piece of equipment does not meet human factors selection criteria, then it cannot be purchased. In contrast, purchasing decisions by healthcare providers, hospitals or regions are made at the level of the healthcare provider, hospital or region and do not ordinarily consider human factors criteria. A fundamental means to improve patient safety is to change purchasing so that decisions are made that consider human factors. In addition, the sharing of these human factors-based decisions among healthcare providers, hospitals and regions will avoid time intensive analyses being repeated ad infinitum. This paper presents a number of examples of how human factors methods have been used to support purchasing decisions related to specific types of medical devices in a large Canadian health region.

Materials

One important type of medical device used in healthcare is that of infusion pumps – both for intravenous (IV) delivery of fluids and drugs and for controlled delivery of pain-relieving drugs through patient controlled analgesia (PCA). Although these pumps have revolutionized some aspects of patient care, they are not without problems. In July 2003, the Institute for Safe Medication Practices Canada (ISMP

Canada) issued a Safety Bulletin that described several problems with infusion pumps, including the 'lack of free-flow protection mechanisms' on the pumps (ISMP Canada 2003). When 'free flow' of IV fluids occurs, the patient does not receive the usual precisely controlled infusion of fluid and/or drugs but may receive a bolus of the fluid and/or medication. The Bulletin also described 'decimal point errors', where errors were made with programming the pumps with respect to 'order of magnitude'. For example, instead of 1.0 ml being programmed, the pump would be programmed to deliver 10 ml or even 100 ml. In 2004, Health Canada issued a 'Notice to Hospitals' describing 'important safety information on infusion pumps'. The Notice stated that Health Canada had received reports of 425 separate incidents involving infusion pumps, a few of which had not occurred in Canada. Of these 425 incidents, 23 resulted in death, 135 in injury, and 127 had the potential to lead to injury or death. All of the injuries and all but three of the deaths were directly related to infusion pumps.

Common problems have also been described with PCA pumps. For example, Vicente and colleagues (2003) reported their analysis of events leading to the death of a patient. In that case, the patient received a five-fold dose of morphine over what had been ordered. In 2004, ISMP Canada described a case in which a patient received a ten-fold increase in the dose of morphine ordered because the nurse had entered '2 ml/hour' instead of '2 mg/hour' on the pump key pad (ISMP Canada 2004). In 2005, Santell described his classification of 5110 pump-related 'errors'. The most common included 'improper dose or quantity of the drug' (38.9%), 'unauthorized drug' (18.4%), and 'omission error' (17.6%).

Methods

As part of our Human Factors (HF) studies, we chose to study both IV and PCA pumps and our HF methodology included both Heuristic Evaluation and Usability Testing. With Heuristic Evaluation (HE), HF experts identify problems in an analytical fashion using design guidelines. With Usability Testing (UT), users of a device try to achieve specific goals to highlight potential usability issues based on their performance and difficulties.

Heuristic Evaluation methods

Heuristic Evaluation was used as a first step to evaluate the usability of three IV pumps for the Calgary Health Region. According to Nielsen (2003), three to five evaluators provide the best cost-benefit tradeoff. We chose to have five evaluators inspect each pump independently. Evaluators judged the human factors compliance of each pump with a set of human factors design guidelines. These guidelines included:

- Language comprehensibility
- Navigation cues
- System status and feedback

- Alerts and warnings, and
- Avoidance of error-producing defaults.

Positive features of the IV pumps included information about the pump's system status and feedback, through the provision of status lights. These lights showed:

- green (for when the pump was infusing)
- yellow (for when the pump was on stand-by), and
- red (for when the pump was in a state of alarm).

Another positive feature related to alerts and warnings. In that feature, a warning alerted the user to the fact that the free flow clamp was open when loading the tube.

Negative features related to navigation cues. On one multi-channel pump, the desired channel could not be selected by pressing the 'Channel Select' button on the corresponding channel. In fact, to select a channel, the user had to make a selection from the main screen (Gagnon *et al*., 2004).

Two PCA pumps were also evaluated using Heuristic Evaluation. One of the most important results of this comparison was that of language comprehensibility. Differences in terminology were found both between pumps (*e.g*., loading dose vs. set bolus) and also within one pump (*e.g*., 'basal' was interchanged with 'continuous'). There was a lack of consistency in navigation cues between the two pumps, potentially provoking confusion related to previous experience with one of the pumps. There were also problems with system status and feedback, with both visual and auditory displays. Visual displays were small and the display often timed-out too quickly (in an effort designed to reduce battery drain). Audible displays also differed, with one set of beeps being distracting, rather than alerting. Alerts and warnings also differed between the two pumps, with differences in sensitivity for both occlusion of the tubing and the presence of air in the line (Chisholm *et al.,* 2005).

Usability testing methods and results

For the usability testing, 13 volunteer expert nurses, all from the Foothills Medical Centre, were employed. These nurses had a mean age of 34.3 years; an average of 11.4 years of nursing experience, and at the time of the study worked an average of 30.8 hours each week. The study was carried out in the Shock Trauma Air Rescue Simulator (STARS) training facility, Calgary, where the Baxter Colleague CX, 3 channel infusion pump was studied using 'Stan', the Emergency Care Simulator. The nurses followed three use cases, employing a 'think aloud' verbal protocol. Digital video recordings were made of nurse interactions with the pump and 'Stan'.

Probably the most important finding was that of the nurses' failure to engage the 'Select New Patient' function. Almost three-quarters (72%) missed the prompt to select this function. The importance of this omission is related to the fact that if not selected, then the pump retains all the settings for the previous patient. As a result, defaults for doses and concentrations of previously administered medications are not changed (Lamsdale *et al*., 2005).

Discussion

There is no doubt that medical device design engineers do design good equipment, including that of IV and PCA pumps. However, no pump is perfect. Engineers are not users and therefore HF evaluations are critical. These evaluations can provide salient information in two critical areas.

First, with respect to the improvement of medical devices design and use, there is also no doubt that problems with equipment, including pumps, can be corrected if identified. Although it is vital that 'operators be thoroughly familiar with the device' (Weinger 2000), the most commonly espoused correction in healthcare is that of 'more training' (White, 1987; Smythe 1992; Cohen 1993; Lin, 1998) for the apparently recalcitrant healthcare workforce. However, another commonly espoused human factors aphorism is that 'Training is the last bastion of poor design'. Making changes to the software of a pump is relatively easy and inexpensive, especially in contrast to making changes to the hardware of a pump, which is not only more difficult and more expensive, but also takes longer.

Second, HF evaluations can also provide information for better decision-making with respect to purchasing. However, both the timing and acceptance of such evaluations is critical, with the goal being the integration of HF evaluations into the purchasing processes, rather than being seen as something separate and optional.

However, HF evaluation of medical devices does not end with their purchase. Every institution needs some method of on-going surveillance, for example, computer-based review of electronic medical records (Samore *et al.* 2004) and audit of clinical engineering logs (Small 2004). Ideally surveillance should be systematic and should encompass the device implementation phase, through on-going use, through to the replacement phase. When expert users interact with a new device, they often revert to previously learned habits, which may lead to errors with the new device (Besnard and Cacitti 2005). Information gathered about problems with devices and any harm to patients should be shared with other users, through national medical device problem reporting systems, such as those hosted by Health Canada (MED Effect, Health Canada), the United States Food and Drug Administration (Medical Device Safety, FDA) and the Therapeutic Goods Administration branch of the Australian Department of Health & Ageing (Recalls & Alerts, Australia). Another excellent source of shared information is that of the Institute for Safe Medical Practice (ISMP) for both Canada (ISMP Canada) and the United States, which issue electronic safety alerts. For example, on January 12, 2006, a Medication Safety Alert from ISMP highlighted the problem of 'double key bounce and double keying errors' with a particular IV infusion pump keypad (ISMP 2006). Safety alerts are also issued by ECRI, a non-profit agency in the United States, the focus of which is 'healthcare technology, patient safety, healthcare risk and quality management, and healthcare environmental management' (ECRI 2006). Communication is important for reducing safety issues, as often times the same errors are made repeatedly, due to the lack of knowledge.

An example of problems arising after purchase was that of the Baxter Colleague IV Pump. This pump had been marketed for some years but on March 15, 2005 was the subject of a company-issued Urgent Device Correction letter. This letter described how users inadvertently powered off the pump when they pressed the ON/OFF key instead of the START key. In the letter, Baxter stated that the company was 'modifying product design'. The same letter also described 'Pump Failure Codes'. These 'electronic failures' had 'occurred at an infrequent rate'. Because the pump was often used to control the intravenous infusion of drugs intended to control patients' blood pressure and heart rates, the company stated that users should 'have a contingency plan to mitigate any disruptions of infusions of life-sustaining drugs' (Baxter 2005/03/15). Four months later, Baxter issued a Recall of the Colleague Infusion Pump. The recall described a 'design issue … which may have been associated with a patient death' (Baxter 2005/07/21).

Our HF evaluations of the IV pumps predated these public announcements and were provided to the purchasing decision makers in the form of strengths and weaknesses of the pumps being considered for purchase. Initially the HF information was not used in the purchasing decision-making process. Since then, a closer working relationship has developed between the HF team and the purchasing decision-makers. Although our HF evaluations did not identify the same problems as described in the Urgent Device Correction letter and the Recall, the validity and usefulness of the HF evaluations is now accepted. Subsequent requests for human factors assistance with other medical devices have increased and the health region has hired a human factors specialist. Ideally, all medical devices would be evaluated or analyzed accordingly, but this is not realistic because of insufficient resources. Medical devices with the greatest potential for lethal patient harm might represent those that are primarily selected for additional evaluation.

Not all HF evaluations of medical devices will result in a clear decision for a certain device. In some instances, the process will produce a list of devices that have strengths and weaknesses and no particular product will stand out as the obvious choice. The selection of one device over another may or may not be beneficial from the viewpoint of patient safety. More time-intensive evaluations, such as patient simulation, may be required to produce the results necessary to support conclusively a particular decision. Unfortunately, most institutions do not have the luxury of time to perform these additional analyses, suggesting the further necessity for a ready source of such information.

Finally, HF evaluations of medical devices allow healthcare providers and institutions to be proactive about patient safety, rather than reacting to patient harm. However, HF evaluation is just the start and 'intelligent wariness' (Reason, 2000) is still required so as to ensure ever safer patient care.

Conclusion

Try before you buy! Whether or not a medical device is potentially easier to use and invokes fewer errors requires that an evaluation be conducted or referenced before purchase so that alternate devices can be considered. Once a purchase has been made, vendors are unlikely to accommodate requests for changes to their products. The corollary to 'try before you buy' is 'what you buy is what you get'. The healthcare provider or institution is, in effect, 'stuck' with that device for its lifecycle. If the medical device is badly designed, then it will contribute to the increased potential for patient harm.

Medical device manufacturers are sensitized to patient safety. Product marketing is likely to tout how safe a particular device is to improve sales. Whether a device achieves a higher level of usability is not dependent on the rhetoric of marketing. The necessary changes to the processes by which devices are developed and released into the market requires the hiring of human factors and biomedical engineers who are able to perform these critical development tasks. However, manufacturers are not likely to change their processes to improve patient safety unless their potential customers require that their products must meet minimum human factors criteria. Thus, healthcare providers may eventually get what they demand and their patients deserve.

Acknowledgements

We would like to thank a large number of individuals who conducted the studies, which are summarized above. These individuals are Allison Lamsdale, Roger Gagnon, Jonathan Histon, Carl Hudson, Jason LaBerge, and Stefanie Kramer. Mike Lamacchia and JN Armstrong facilitated the usability study in the Shock Trauma Air Rescue Simulation (STARS) Facility. Numerous individuals championed this work within the Calgary Health Region, including Steve Long, Bruce MacKenzie, Andrea Robertson, Glenn McCrae and Carmella Duchscherer.

References

Baxter (2005/03/15). RE: Colleague Volumetric Infusion Pump, Product Codes 2M8151, 2M8151R, 2M8161, 2M8161R, 2M8153, 2M8153R, 2M8163, 2M8163R. Urgent Device Correction. Baxter Healthcare Corporation. www.baxter.com (Accessed 2006/01/27).

Baxter (2005/07/21). RE: Colleague Volumetric Infusion Pump, Product Codes 2M8151, 2M8151R, 2M8161, 2M8161R, 2M8153, 2M8153R, 2M8163, 2M8163R, DNM 8151, DNM8151R, DNM8153, DNM8153R. Notice to Hospitals. Important Safety Information on the Recall of Certain Baxter Colleague Volumetric Infusion Pumps. http://www.hc-sc.gc.ca/dhp-mps/medeff/advisories-avis/prof/infusion-perfusion_pump (Accessed 2006/01/27).

Besnard D, Cacitti L. Interface changes causing accidents. An empirical study of negative transfer. *International Journal of Human-Computer Studies* 2005; **62**:105–25.

Chapanis, A. (1995), *Human Factors in Systems Engineering*. Toronto: John Wiley.

Chisholm S., Kramer S., and Lockhart, J. Heuristic comparison of two PCA pumps. Cognitive Ergonomics Research Laboratory, University of Calgary, 2005.

Cohen MR, Preventing errors associated with P.C.A. pumps. *Nursing* 1993, 23:17

ECRI (2006). http://www.ecri.org (Accessed 2006/01/27).

Gagnon, R., Laberge, J., Lamsdale, A., Histon, J., Hudson, C., Davies, J., & Caird, JK (2004), A user-centered evaluation of three intravenous infusion pumps.

Proceedings of the 48th Annual Human Factors and Ergonomics Society Meeting (pp. 1773–1777), Santa Monica, CA: Human Factors and Ergonomics Society.

Health Canada (2004). Health risks associated with use of infusion pumps. Notice to Hospitals. Health Canada endorsed important safety information on infusion pumps. April 16, 2004. http://www.hc-sc.gc.ca/dhp-mps/medeff/advisories-avis/prof/2004/infusion_pumps_nth-ah_e.html (Accessed 2006/01/27)

Health Canada. MED Effect. Advisories, Warnings & Recalls for Health Professionals. Medical Devices. http://www.hc-sc.gc.ca/dhp-mps/medeff/advisories-avis/prof/index_e.html (Accessed 2006/01/27)

Health Canada. Medical Devices, http://www.hc-sc.gc.ca/dhp-mps/md-im/index_e.html (Accessed 2006/01/27)

ISMP Canada (2003). Infusion pumps – opportunities for improvement. ISMP Canada Safety Bulletin, July 2003, Volume 3, Issue 7. ISMP Canada and HIROC. http://www.ismp-canada.org/download/ISMPCSB2003-07InfusionPumps.pdf (Accessed 2006/01/27)

ISMP Canada (2004). High alert drugs and infusion pumps: extra precautions required. ISMP Canada Safety Bulletin. April 2004. ISMP Canada and HIROC. http://www.ismp-canada.org/download/ISMPCSB2004-04.pdf (Accessed 2006/01/27)

ISMP (2006). Double key bounce and double keying errors. ISMP Medication Safety Alert! Acute Care. 2006/01/12. www.ismp.org (Accessed 2006/01/27)

Lamsdale, A., Chisholm, S., Gagnon, R., Davies, J., & Caird, JK (2005). A usability evaluation of an infusion pump by nurses using a patient simulator. Proceedings of the 49th Annual Meeting of the Human Factors and Ergonomics Society (pp. 1024–1028), Santa Monica, CA: Human Factors and Ergonomics Society.

Lin L, Isla R, Doniz K, Harkness H, Vicente KJ, Doyle DJ. (1998). Applying human factors to the design of medical equipment: patient-controlled analgesia. *Journal of Clinical Monitoring and Computing* 1998;14:253-63

Medical Devices Regulations, Food and Drug Act, Government of Canada. http://lois.justice.gc.ca/en/F-27/SOR-98-282/index.html (Accessed 2006/01/27).

MIL-STD-1472F (1999) Department of Defense Design Criteria Standard. HUMAN ENGINEERING. AMSC N/A. AREA HFAC. Army Aviation and Missile Command. United States of America. http://www.hf.faa.gov/docs/508/docs/milstd14.pdf (Accessed 2006/01/27)

MIL-STD-1472F (2004) Department of Defense Design Criteria Standard: Human Engineering SPONSOR: Army Aviation and Missile Command. http://www0.dtic.mil/matris/ddsm/srch/ddsm0135.html (Accessed 2006/01/27)

Nielsen J. (2003). Usability heuristics. In: *Usability Engineering*. San Diego: Academic press. Pp 115–63.

Reason J. (2005) Human error: Models and management. *BMJ* 2000;20:768–70

Recalls & Alerts. Therapeutic Goods Administration. Australian Government. Department of Health & Ageing. http://www.tga.gov.au/recalls/index.htm (Accessed 2006/01/27)

Samore MH, Evans RS, Lassen A, Gould P, Lloyd J, Gardner RM, Abouzelof R, Taylor C, Woodbury DA, Willy M, Bright RA. Surveillance of medical-device-related hazards and adverse events in hospitalised patients. *JAMA* 2004;291;325–

34.

Santell JP (2005). Preventing errors that occur with PCA pumps. US Pharmacist 2005:30:58–60.http://www.uspharmacist.com/index.asp?show-article&page=8_1414.htm (Accessed 2006/01/27)

Small S. Medical device-associated safety and risk. *JAMA* 2004;291:367-70

Smythe M. Patient-controlled analgesia: A review. *Pharmacotherapy* 1992;12:132-43

US Food and Drug Administration. http://fda/gov/cdrh/devadvice/312.html#link_2 (Accessed 2006/01/27)

US Food and Drug Administration. Medical Advice Safety. Center for Devices and Radiological Health. http://www.fda.gov/cdrh/patientsafety/ (Accessed 2006/01/27)

Vicente KJ, Kada-Bekhaled K, Hillel G, Cassano A, Orser BA. (2003). Programming error contribute to death from patient-controlled analgesia: Case report and estimate probability. *Canadian Journal of Anesthesia* 2003;50:328-32

Weinger MB. Designer's Notebook. User-centred design: A clinician's perspective. Medical Device & Diagnostic Industry Magazine. Medical Device Link. The Online Information Source for the Medical Device Industry. Originally published January 2000. http://www.devicelink.com/grabber.php3?URL=http://www.devicelink.com/mddi/archive/ (Accessed 2006/01/27)

White PF. Mishaps with patient-controlled analgesia. *Anesthesiology* 1987;66:81

Index